A-Level Chemistry

Exam Board: AQA

Revising for Chemistry exams is stressful, that's for sure — even just getting your notes sorted out can leave you needing a lie down. But help is at hand...

This brilliant CGP book explains **everything you'll need to learn** (and nothing you won't), all in a straightforward style that's easy to get your head around. We've also included **exam questions** to test how ready you are for the real thing.

There's even a free Online Edition you can read on your computer or tablet!

How to get your free Online Edition

Go to **cgpbooks.co.uk/extras** and enter this code...

1696 5285 8260 0211

This code only works for one person. If somebody else has used this book before you, they might have already claimed the Online Edition.

A-Level revision? It has to be CGP!

Published by CGP

Editors:
Katie Braid, Mary Falkner, Gordon Henderson, Emily Howe, Paul Jordin, Rachel Kordan, Sarah Pattison, Sophie Scott and Ben Train.

Contributors:
Mike Bossart, Robert Clarke, Vikki Cunningham, Ian H. Davis, John Duffy, Max Fishel, Emma Grimwood, Lucy Muncaster, Derek Swain, Paul Warren and Chris Workman.

ISBN: 978 1 78294 300 6

With thanks to Barrie Crowther and Karen Wells for the proofreading.
With thanks to Jan Greenway for the copyright research.

Cover Photo **Laguna Design**/Science Photo Library

With thanks to Science Photo Library for permission to reproduce the photograph used on page 98.

Clipart from Corel®
Printed by Elanders Ltd, Newcastle upon Tyne.

Based on the classic CGP style created by Richard Parsons.

Contents

If you're revising for the **AS exams**, you'll need Unit 1: Sections 1-5, Unit 2: Sections 1-2, Unit 3 Sections 1-4, and the Practical Skills section at the back.

If you're revising for the **A-Level exams**, you'll need the whole book.

Unit 1: Section 1 — Atomic Structure (AS)

Unit 1: Section 2 — Amount of Substance (AS)

Unit 1: Section 3 — Bonding (AS)

Unit 1: Section 4 — Energetics (AS)

Unit 1: Section 5 — Kinetics, Equilibria and Redox Reactions (AS)

Unit 1: Section 6 — Thermodynamics

Unit 1: Section 7 — Rate Equations and K_p

Unit 1: Section 8 — Electrode Potentials and Cells

Unit 3: Section 4 — Organic Analysis (AS)

Unit 3: Section 5 — Isomerism and Carbonyl Compounds

Unit 3: Section 6 — Aromatic Compounds and Amines

Unit 3: Section 7 — Polymers

Unit 3: Section 8 — Amino Acids, Proteins and DNA

Unit 3: Section 9 — Further Synthesis and Analysis

Practical Skills (AS)

Do Well In Your Exams

The Atom

This stuff about atoms and elements should be ingrained in your brain from GCSE. You do need to know it perfectly though, if you are to negotiate your way through the field of man-eating tigers that is Chemistry.

Atoms are made up of Protons, Neutrons and Electrons

All elements are made of **atoms**. Atoms are made up of 3 types of particle — **protons**, **neutrons** and **electrons**.

Electrons

1) Electrons have **–1** charge.
2) They whizz around the nucleus in **orbitals**. The orbitals take up most of the **volume** of the atom.

Nucleus

1) Most of the **mass** of the atom is concentrated in the nucleus.
2) The **diameter** of the nucleus is rather titchy compared to the whole atom.
3) The nucleus is where you find the **protons** and **neutrons**.

The mass and charge of these subatomic particles is **really small**, so **relative mass** and **relative charge** are usually used instead.

Subatomic particle	Relative mass	Relative charge
Proton	1	+1
Neutron	1	0
Electron, e^-	$\frac{1}{2000}$	–1

The mass of an electron is negligible compared to a proton or a neutron — this means you can usually ignore it.

Nuclear Symbols Show Numbers of Subatomic Particles

You can figure out the **number** of protons, neutrons and electrons in an atom from the **nuclear symbol**.

$${}^{A}_{Z}\mathrm{X}$$

Mass number — This tells you the **total** number of **protons** and **neutrons** in the nucleus.

Element symbol

Atomic (proton) number

1) This is the number of **protons** in the nucleus — it identifies the element.
2) **All** atoms of the same element have the **same** number of protons.

Sometimes the atomic number is left out of the nuclear symbol, e.g. ${}^{7}\mathrm{Li}$. You don't really need it because the element's symbol tells you its value.

1) For **neutral** atoms, which have no overall charge, the number of electrons is **the same as** the number of protons.
2) The number of neutrons is **mass number minus atomic number**, i.e. 'top minus bottom' in the nuclear symbol.

Nuclear symbol	Atomic number, Z	Mass number, A	Protons	Electrons	Neutrons
${}^{7}_{3}\mathrm{Li}$	3	7	3	3	7 – 3 = 4
${}^{79}_{35}\mathrm{Br}$	35	79	35	35	79 – 35 = 44
${}^{24}_{12}\mathrm{Mg}$	12	24	12	12	24 – 12 = 12

"Hello, I'm Newt Ron..."

Ions have Different Numbers of Protons and Electrons

Atoms form ions by losing or gaining electrons. **Negative** ions have **more electrons** than protons...

E.g. Br^- — The negative charge means that there's 1 more electron than there are protons. Br has 35 protons (see table above), so Br^- must have 36 electrons. The overall charge = +35 – 36 = –1.

...and **positive** ions have **fewer electrons** than protons. It kind of makes sense when you think about it.

E.g. Mg^{2+} — The 2+ charge means that there are 2 fewer electrons than there are protons. Mg has 12 protons (see table above), so Mg^{2+} must have 10 electrons. The overall charge = +12 – 10 = +2.

The Atom

Isotopes are Atoms of the *Same Element* with Different Numbers of *Neutrons*

Make sure you **learn** this definition and totally **understand** what it means:

Isotopes of an element are atoms with the same number of protons but different numbers of neutrons.

Chlorine-35 and chlorine-37 are examples of isotopes.

35 – 17 = 18 neutrons ← **Different** mass numbers mean different numbers of neutrons. → **37 – 17 = 20 neutrons**

$^{35}_{17}Cl$ The **atomic numbers** are the same. **Both** isotopes have 17 protons and 17 electrons. $^{37}_{17}Cl$

1) It's the **number** and **arrangement** of electrons that decides the **chemical properties** of an element. Isotopes have the **same configuration of electrons**, so they've got the **same** chemical properties.
2) Isotopes of an element do have slightly different **physical properties** though, such as different densities, rates of diffusion, etc. This is because **physical properties** tend to depend on the **mass** of the atom.

Here's another example — naturally occurring **magnesium** consists of 3 isotopes.

^{24}Mg (79%)	^{25}Mg (10%)	^{26}Mg (11%)
12 protons	12 protons	12 protons
<u>12</u> neutrons	<u>13</u> neutrons	<u>14</u> neutrons
12 electrons	12 electrons	12 electrons

The periodic table gives the atomic number of each element. The other number by an element's symbol in the periodic table isn't the mass number though — it's the relative atomic mass (see page 5).

Practice Questions

Q1 Draw a diagram showing the structure of an atom, labelling each part.

Q2 Where is the mass concentrated in an atom, and what makes up most of the volume of an atom?

Q3 Draw a table showing the relative charge and relative mass of the three subatomic particles found in atoms.

Q4 Using an example, explain the terms 'atomic number' and 'mass number'.

Q5 Define the term 'isotopes' and give examples.

Exam Questions

Q1 Hydrogen, deuterium and tritium are all isotopes of each other.

a) Identify one similarity and one difference between these isotopes. [2 marks]

b) Deuterium can be written as $^{2}_{1}H$. Determine the number of protons, neutrons and electrons in a deuterium atom. [1 mark]

c) Write the nuclear symbol for tritium, given that it has 2 neutrons. [1 mark]

Q2 This question relates to the atoms or ions A to D: A $^{32}_{16}S^{2-}$ B $^{40}_{18}Ar$ C $^{30}_{16}S$ D $^{42}_{20}Ca$

a) Identify the similarity for each of the following pairs, justifying your answer in each case.

i) A and B. [1 mark]

ii) A and C. [1 mark]

iii) B and D. [1 mark]

b) Which two of the atoms or ions are isotopes of each other? Explain your reasoning. [2 marks]

Got it learned yet? — Isotope so...

This is a nice page to ease you into things. Remember that positive ions have fewer electrons than protons, and negative ions have more electrons than protons. Get that straight in your mind or you'll end up in a right mess. There's nowt too hard about isotopes neither. They're just the same element with different numbers of neutrons.

Models of Atomic Structure

The model of the atom on the previous page is darn useful for understanding loads of ideas in chemistry. You can picture what's happening in your mind really well. But it is just a model. So it's not completely like that really.

The Accepted Model of the Atom Has Changed Throughout History

The model of the atom you're expected to know (the one on page 2) is one of the **currently accepted** ones. But in the past, **completely different** models were accepted, because they fitted the evidence available at the time. As scientists did more experiments, **new evidence** was found and the models were **modified** to fit it.

1) At the start of the 19th century **John Dalton** described atoms as **solid spheres**, and said that different spheres made up the different elements.

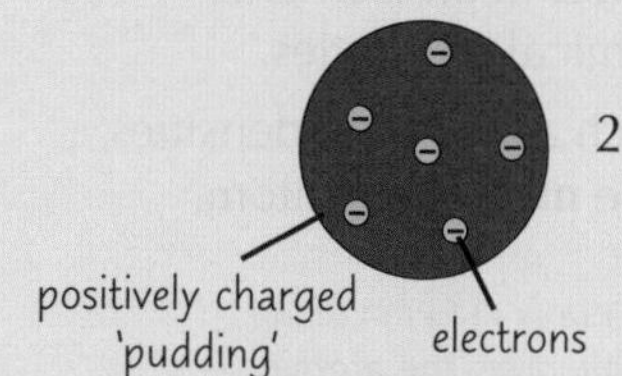

2) In 1897 **J. J. Thomson** discovered the electron. This showed that atoms **weren't** solid and indivisible. The 'solid sphere' idea of atomic structure had to be changed. The new model was known as the '**plum pudding model**'.

3) In 1909 **Ernest Rutherford** and his students **Hans Geiger** and **Ernest Marsden** conducted their famous **gold foil experiment**. They fired positively charged **alpha particles** at a very thin sheet of gold. The plum pudding model suggested that **most** alpha particles would be **slightly** deflected by the positive 'pudding' that made up most of the atom. In fact, most of the particles passed **straight through** the gold with only a small number being deflected **backwards**. The plum pudding model **couldn't be right**, so Rutherford developed the **nuclear model** of the atom. In this, a **tiny positively charged nucleus** is surrounded by a **'cloud'** of **negative electrons** — most of the atom is **empty space**.

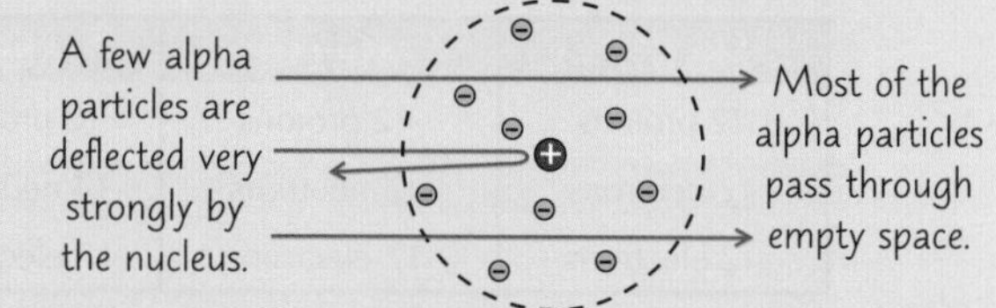

A few alpha particles are deflected very strongly by the nucleus.
Most of the alpha particles pass through empty space.

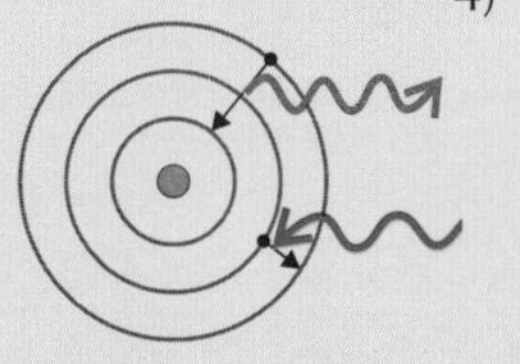

4) Scientists realised that electrons in a '**cloud**' around the nucleus of an atom, as Rutherford described, would quickly **spiral down** into the nucleus, causing the atom to **collapse**. **Niels Bohr** proposed a new model of the atom where electrons exist in **shells** or **orbits** of **fixed energy**. When electrons move between shells, **electromagnetic radiation** (with **fixed** energy or **frequency**) is **emitted** or **absorbed**. The Bohr model **fitted experimental observations** of the radiation emitted and absorbed by atoms.

5) Scientists later discovered that not all the electrons in a shell have the same energy. This meant that the Bohr model wasn't quite right. So, they **refined** it to include **sub-shells**. The refined Bohr model isn't perfect — more accurate models exist today, but it's a useful model because it's **simple** and explains many **experimental observations**, like bonding and ionisation energy trends.

Practice Questions

Q1 Why did Rutherford think that a new model of the atom was needed?

Q2 Who developed the 'nuclear' model of the atom?

Q3 Why is Bohr's model thought to be a truer description of the atom than Rutherford's?

Q4 Describe the model of the atom first proposed by Niels Bohr. How was it later refined?

Q5 More accurate models of the atom have been developed since the Bohr model.
Explain why the Bohr model is still used today.

These models are tiny — even smaller than size zero, I reckon...

The process of developing a model to fit the evidence available, looking for more evidence to show if it's correct or not, then revising the model if necessary is really important. It happens with all new scientific ideas. Even really long standing theories get changed or ditched sometimes in response to a new piece of evidence that someone has uncovered...

Relative Mass and the Mass Spectrometer

Relative mass...What? Eh?...Read on...

Relative Masses are Masses of Atoms Compared to Carbon-12

The actual mass of an atom is **very, very tiny**. Don't worry about exactly how tiny for now, but it's far **too small** to weigh. So, the mass of one atom is compared to the mass of a different atom. This is its **relative mass**. Here are some definitions to learn:

The **relative atomic mass**, A_r, is the **average mass** of an atom of an element on a scale where an atom of **carbon-12** is 12.

Relative atomic mass is an average, so it's not usually a whole number. Relative isotopic mass is usually a whole number. E.g. a natural sample of chlorine contains a mixture of ^{35}Cl (75%) and ^{37}Cl (25%), so the relative isotopic masses are 35 and 37. But its relative atomic mass is 35.5.

Relative isotopic mass is the mass of an atom of an **isotope** of an element on a scale where an atom of **carbon-12** is 12.

The **relative molecular mass**, M_r, is the average mass of a **molecule** on a scale where an atom of **carbon-12** is 12.

To find the relative molecular mass, just add up the relative atomic mass values of all the atoms in the molecule, e.g. $M_r(C_2H_6O) = (2 \times 12.0) + (6 \times 1.0) + 16.0 = 46.0$

Relative formula mass is used instead for compounds that are ionic (or giant covalent). To find the relative formula mass, just add up the relative atomic masses (A_r) of all the atoms in the formula unit. E.g. $M_r(CaF_2) = 40.1 + (2 \times 19.0) = 78.1$

"All those pies have made me relatively massive."

Relative Masses can be Measured Using a Mass Spectrometer

You can use a **mass spectrometer** to find out loads of stuff. It can tell you the **relative atomic mass**, **relative molecular mass**, **relative isotopic abundance** and your **horoscope** for the next fortnight.

There are **4** things that happen when a sample is squirted into a **time of flight (TOF) mass spectrometer**.

① **Ionisation** Two ways of ionising your sample are:
Electrospray ionisation — the sample is dissolved and pushed through a small nozzle at high pressure. A high voltage is applied to it, causing each particle to **gain an H^+ ion**. The sample is turned into a **gas** made up of **positive ions**.
Electron impact ionisation — the sample is vaporised and an 'electron gun' is used to fire high energy electrons at it. This **knocks one electron off** each particle, so they become **+1 ions**.

② **Acceleration** — the positively charged ions are accelerated by an **electric field** so that they all have the **same kinetic energy**. (This means that the lighter ions will end up moving faster than the heavier ions.)

③ **Ion Drift** — the ions enter a region with no electric field, so they just **drift** through it. Lighter ions will drift through faster than heavier ions.

④ **Detection** — because **lighter ions** travel at **higher speeds** in the drift region, they reach the detector in **less time** than **heavier ions**. The detector detects **charged** particles and a **mass spectrum** (see next page) is produced.

vacuum

lower mass/charge ion

higher mass/charge ion

ion detector

The detectors used in mass spectrometers detect charged particles. An electrical current is produced in the detector when a charged particle hits it.

Relative Mass and the Mass Spectrometer

A Mass Spectrum is Mass/Charge plotted against Abundance

The **y-axis** gives the **abundance of ions**, often as a percentage. For an element, the **height** of each peak gives the **relative isotopic abundance**, e.g. 75.5% of this sample is made up of the ^{35}Cl isotope.

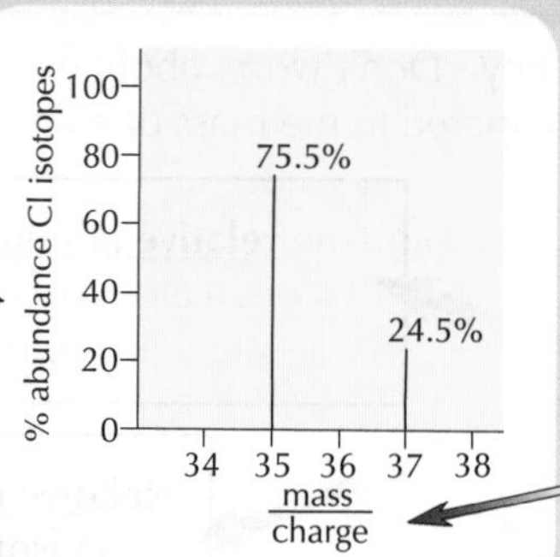

If the sample is an **element**, each line will represent a **different isotope** of the element.

The **x-axis** units are given as a **'mass/charge'** ratio. Since the charge on the ions is mostly **+1**, you can usually assume the x-axis is simply the **relative isotopic mass**.

You Can Work Out Relative Atomic Mass from a Mass Spectrum

You need to know how to calculate the **relative atomic mass** (A_r) of an element from the **mass spectrum**. Here's how to calculate A_r for magnesium, using the mass spectrum below:

1: For each peak, read the **% relative isotopic abundance** from the y-axis and the **relative isotopic mass** from the x-axis. **Multiply** them together to get the total mass for each isotope: 79 × 24 = 1896; 10 × 25 = 250; 11 × 26 = 286

2: Add up these totals. 1896 + 250 + 286 = 2432

3: Divide by 100 (as percentages were used). A_r(Mg) = 2432 ÷ 100 = 24.32 = 24.3 (3 s.f.)

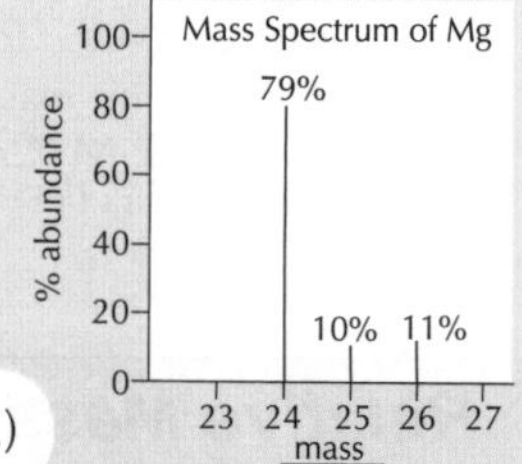

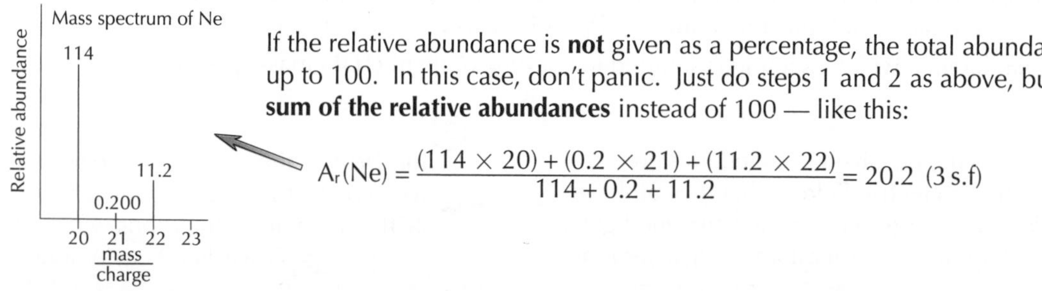

If the relative abundance is **not** given as a percentage, the total abundance may not add up to 100. In this case, don't panic. Just do steps 1 and 2 as above, but then divide by the **sum of the relative abundances** instead of 100 — like this:

$$A_r(\text{Ne}) = \frac{(114 \times 20) + (0.2 \times 21) + (11.2 \times 22)}{114 + 0.2 + 11.2} = 20.2 \text{ (3 s.f)}$$

Mass Spectrometry can be used to Identify Elements

Elements with different **isotopes** produce more than one line in a mass spectrum because the isotopes have **different masses**. This produces characteristic patterns which can be used as **'fingerprints'** to **identify** certain **elements**.

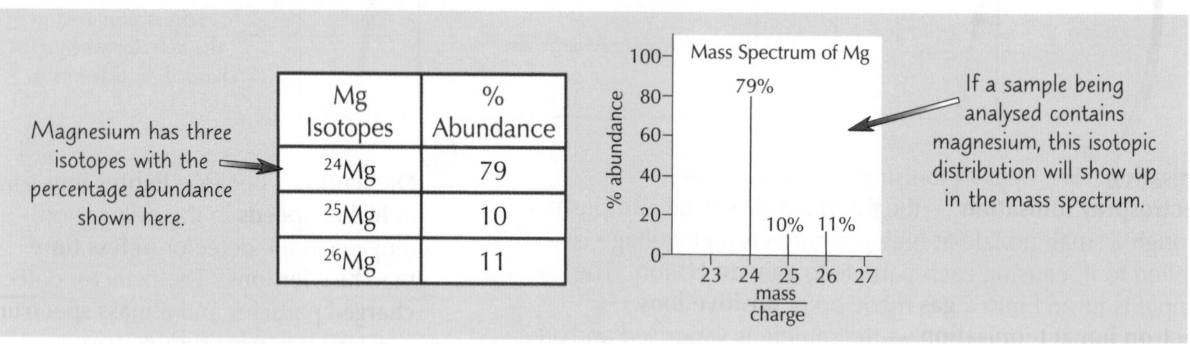

Mg Isotopes	% Abundance
^{24}Mg	79
^{25}Mg	10
^{26}Mg	11

Many elements only have one stable isotope. They can still be identified in a mass spectrum by looking for a line at their **relative atomic mass**.

Relative Mass and the Mass Spectrometer

Mass Spectrometry can be used to Identify Molecules

You can also get a mass spectrum for a **molecular sample**.

1) A **molecular ion**, M^+, is formed in the mass spectrometer when one electron is removed from the molecule.
2) This gives a peak in the spectrum with a mass/charge ratio equal to the **relative molecular mass** of the molecule.
3) This can be used to help **identify** an unknown compound.
   There's more about using mass spectrometry to identify compounds on page 157.

Example: A sample of a straight-chain alcohol is analysed in a mass spectrometer. The mass/charge ratio of its molecular ion is 46.0. Identify the alcohol.

The table on the right shows the M_r of the first three straight-chain alcohols. The mass/charge ratio of the molecular ion must **equal** the M_r of the alcohol in the sample. So the alcohol must be **ethanol, C_2H_5OH**.

Alcohol	M_r
methanol CH_3OH	32.0
ethanol C_2H_5OH	46.0
propanol C_3H_7OH	60.0

If you have a mixture of compounds with different M_r values, you'll get a peak for the molecular ion of each one.

Practice Questions

Q1 Explain what relative atomic mass (A_r) and relative isotopic mass mean.

Q2 Explain the difference between relative molecular mass and relative formula mass.

Q3 Describe how electrospray ionisation works.

Q4 Explain how a mass spectrum can be used to determine relative molecular mass.

Exam Questions

Q1 Copper, Cu, exists in two main isotopic forms, ^{63}Cu and ^{65}Cu.

a) Calculate the relative atomic mass of Cu using the information from the mass spectrum on the right. [2 marks]

b) Explain why the relative atomic mass of copper is not a whole number. [2 marks]

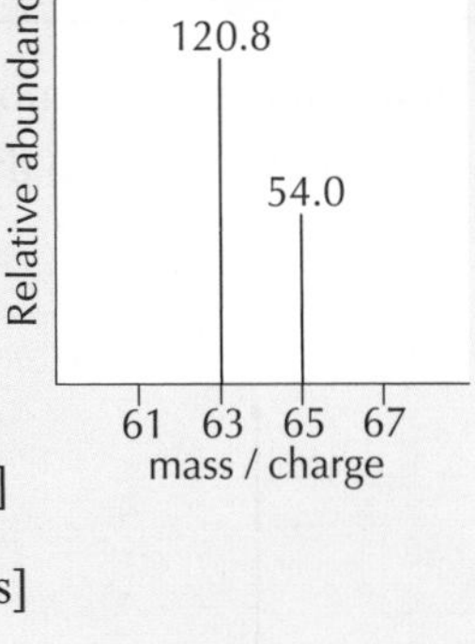

Q2 The percentage make-up of naturally occurring potassium is 93.11% ^{39}K, 0.12% ^{40}K and 6.77% ^{41}K.

a) What method is used to determine the mass and abundance of each isotope? [1 mark]

b) Use the information to determine the relative atomic mass of potassium. [2 marks]

Q3 A mixture containing chlorine (Cl), gallium (Ga), bromine (Br) and rubidium (Rb), was analysed in a time of flight mass spectrometer.

a) Why do samples need to be positively charged in time of flight mass spectrometry? [2 marks]

b) Explain how time of flight mass spectrometry distinguishes between ions with different masses. [4 marks]

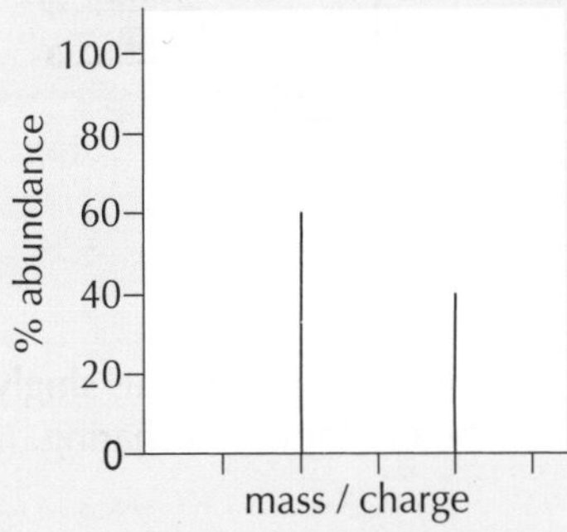

c) The abundance of the isotopes of the elements in the mixture are shown below. Which element is responsible for the part of the spectrum on the right?

A Chlorine: 75.8% ^{35}Cl, 24.2% ^{37}Cl B Gallium: 60.1% ^{69}Ga, 39.9% ^{71}Ga

C Bromine: 50.7% ^{79}Br, 49.3% ^{81}Br D Rubidium: 72.2% ^{85}Rb, 27.8% ^{87}Rb [1 mark]

You can't pick your relatives — you just have to calculate them...

All this mass spectrometry stuff looks a bit evil, but it really isn't that bad once you get your head round it. Make sure you've done the practice and exam questions, cos they cover all the stuff you need to be able to do, and if you can get them right, you've nailed it. Then you can go and do something much more fun, like cutting the lawn with nail scissors...

Electronic Structure

Those little electrons prancing about like mini bunnies decide what'll react with what — it's what chemistry's all about.

Electron Shells are Made Up of Sub-Shells and Orbitals

1) In the currently accepted model of the atom, electrons have **fixed energies**. They move around the nucleus in certain regions of the atom called **shells** or **energy levels**.
2) Each shell is given a number called the **principal quantum number**. The **further** a shell is from the nucleus, the **higher** its energy and the **larger** its principal quantum number.
3) **Experiments** show that not all the electrons in a shell have exactly the same energy. The **atomic model** explains this — shells are divided up into **sub-shells** that have slightly different energies. The sub-shells have different numbers of **orbitals** which can each hold up to **2 electrons**.

This table shows the number of electrons that fit in each type of sub-shell.

Sub-shell	Number of orbitals	Maximum electrons
s	1	$1 \times 2 = 2$
p	3	$3 \times 2 = 6$
d	5	$5 \times 2 = 10$
f	7	$7 \times 2 = 14$

And this one shows the sub-shells and electrons in the first four energy levels.

Shell	Sub-shell	Total number of electrons
1st	1*s*	2
2nd	2*s* 2*p*	$2 + (3 \times 2) = 8$
3rd	3*s* 3*p* 3*d*	$2 + (3 \times 2) + (5 \times 2) = 18$
4th	4*s* 4*p* 4*d* 4*f*	$2 + (3 \times 2) + (5 \times 2) + (7 \times 2) = 32$

5) The two electrons in each orbital spin in **opposite directions**.

Work Out Electron Configurations by Filling the Lowest Energy Levels First

You can figure out most electron configurations pretty easily, so long as you know a few simple rules —

1) Electrons fill up the **lowest** energy sub-shells first.

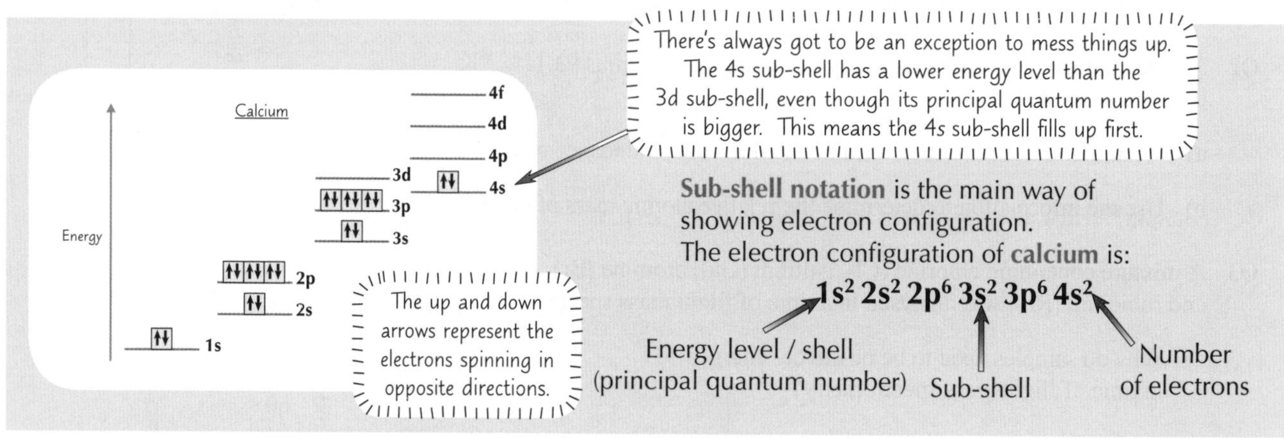

Calcium
Energy
4f
4d
4p
3d
4s
3p
3s
2p
2s
1s
There's always got to be an exception to mess things up. The 4s sub-shell has a lower energy level than the 3d sub-shell, even though its principal quantum number is bigger. This means the 4s sub-shell fills up first.
The up and down arrows represent the electrons spinning in opposite directions.
Sub-shell notation is the main way of showing electron configuration.
The electron configuration of calcium is:
1s² 2s² 2p⁶ 3s² 3p⁶ 4s²
Energy level / shell (principal quantum number)
Sub-shell
Number of electrons

2) Electrons fill orbitals **singly** before they start sharing.

3) For the configuration of **ions** from the **s** and **p** blocks of the periodic table, just **remove or add** the electrons to or from the highest energy occupied sub-shell.
E.g. Mg = $1s^2\ 2s^2\ 2p^6\ 3s^2$, so Mg^{2+} = $1s^2\ 2s^2\ 2p^6$. Cl = $1s^2\ 2s^2\ 2p^6\ 3s^2\ 3p^5$, so Cl^- = $1s^2\ 2s^2\ 2p^6\ 3s^2\ 3p^6$.

See the next page for more on the s and p blocks.

Watch out — **noble gas symbols**, like that of argon (Ar), are sometimes used in electron configurations. For example, calcium ($1s^2\ 2s^2\ 2p^6\ 3s^2\ 3p^6\ 4s^2$) can be written as [Ar]$4s^2$, where [Ar] = $1s^2\ 2s^2\ 2p^6\ 3s^2\ 3p^6$.

Electronic Structure

Transition Metals Behave Unusually

1) **Chromium** (Cr) and **copper** (Cu) are badly behaved. They donate one of their **4s** electrons to the **3d sub-shell.** It's because they're happier with a **more stable** full or half-full d sub-shell.
Cr atom (24 e^-): $1s^2\ 2s^2\ 2p^6\ 3s^2\ 3p^6\ 3d^5\ 4s^1$ Cu atom (29 e^-): $1s^2\ 2s^2\ 2p^6\ 3s^2\ 3p^6\ 3d^{10}\ 4s^1$
2) And here's another weird thing about transition metals — when they become **ions**, they lose their **4s** electrons **before** their 3d electrons.
Fe atom (26 e^-): $1s^2\ 2s^2\ 2p^6\ 3s^2\ 3p^6\ 3d^6\ 4s^2 \rightarrow$ Fe^{3+} ion (23 e^-): $1s^2\ 2s^2\ 2p^6\ 3s^2\ 3p^6\ 3d^5$

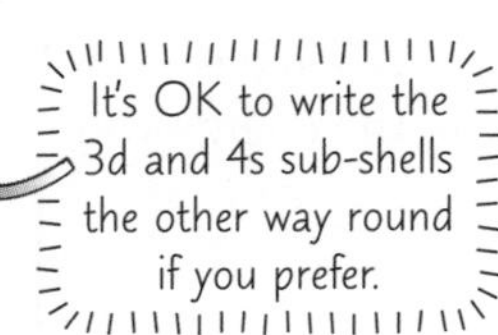

It's OK to write the 3d and 4s sub-shells the other way round if you prefer.

Electronic Structure Decides the **Chemical Properties** of an Element

The number of **outer shell electrons** decides the chemical properties of an element.

1) The **s block** elements (Groups 1 and 2) have 1 or 2 outer shell electrons. These are easily **lost** to form positive ions with an **inert gas configuration.**
E.g. Na: $1s^2\ 2s^2\ 2p^6\ 3s^1 \rightarrow Na^+$: $1s^2\ 2s^2\ 2p^6$
(the electronic configuration of neon).

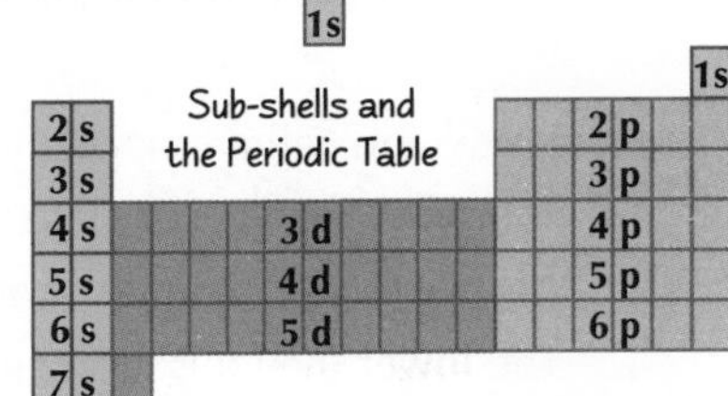

2) The elements in Groups 5, 6 and 7 (in the p block) can **gain** 1, 2 or 3 electrons to form negative ions with an **inert gas configuration.**
E.g. O: $1s^2\ 2s^2\ 2p^4 \rightarrow O^{2-}$: $1s^2\ 2s^2\ 2p^6$.
Groups 4 to 7 can also **share** electrons when they form covalent bonds.
3) Group 0 (the inert gases) have **completely filled** s and p sub-shells and don't need to bother gaining, losing or sharing electrons — their full sub-shells make them **inert.**
4) The **d block elements** (transition metals) tend to **lose** s and d electrons to form positive ions.

Practice Questions

Q1 Write down the sub-shells in order of increasing energy up to 4p.

Q2 How many electrons would each of these types of sub-shell contain when full: a) s, b) p, c) d?

Q3 Chromium and copper don't fill up their shells in the same way as other atoms. Explain the differences.

Q4 Which groups of the Periodic Table tend to gain electrons to form negative ions?

Exam Questions

Q1 Potassium reacts with oxygen to form potassium oxide, K_2O.

a) Give the full electron configurations of the K atom and K^+ ion. [2 marks]

b) Using arrow-in-box notation, give the electron configuration of the oxygen atom. [2 marks]

Q2 This question concerns the electron configurations in atoms and ions.

a) What is the full electron configuration of a manganese atom? [1 mark]

b) Using arrow-in-box notation, give the electron configuration of the Al^{3+} ion. [2 marks]

c) Identify the element with the electron configuration $1s^2 2s^2 2p^6 3s^2 3p^6 3d^{10} 4s^2 4p^2$. [1 mark]

d) Suggest the identity of an atom, a positive ion and a negative ion with the configuration $1s^2\ 2s^2\ 2p^6\ 3s^2\ 3p^6$. [3 marks]

She shells sub-shells on the sheshore...

The way electrons fill up the orbitals is like how strangers fill up seats on a bus. Everyone tends to sit in their own seat till they're forced to share. Except for the huge, scary, smelly man who comes and sits next to you. Make sure you learn the order the sub-shells are filled up, so you can write electron configurations for any atom or ion they throw at you.

Ionisation Energy

This page gets a trifle brain-boggling, so I hope you've got a few aspirin handy...

Ionisation *is the* ***Removal*** *of One or More* ***Electrons***

You might see 'ionisation energy' referred to as 'ionisation enthalpy' instead.

When electrons have been removed from an atom or molecule, it's been **ionised**. The energy you need to remove the first electron is called the **first ionisation energy**.

> The **first ionisation energy** is the energy needed to remove 1 electron from **each atom** in **1 mole** of **gaseous** atoms to form 1 mole of gaseous 1+ ions.

You have to put energy in to ionise an atom or molecule, so it's an **endothermic process** — there's more about endothermic processes on page 36.

You can write **equations** for this process — here's the equation for the **first ionisation of oxygen**:

$$O_{(g)} \rightarrow O^{+}_{(g)} + e^{-} \qquad \text{1st ionisation energy} = +1314 \text{ kJ mol}^{-1}$$

Here are a few rather important points about ionisation energies:

1) You **must** use the gas state symbol, **(g)**, because ionisation energies are measured for gaseous atoms.
2) Always refer to **1 mole** of atoms, as stated in the definition, rather than to a single atom.
3) The **lower** the ionisation energy, the **easier** it is to form an ion.

The ***Factors*** *Affecting Ionisation Energy are...*

Ways in which bears are like electrons #14 — the more warm apple pies there are cooling on a window sill, the tastier the window sill smells and the stronger the attraction for the bears.

Nuclear Charge — The **more protons** there are in the nucleus, the more positively charged the nucleus is and the **stronger the attraction** for the electrons.

Distance from Nucleus — Attraction falls off very **rapidly with distance**. An electron **close** to the nucleus will be **much more** strongly attracted than one further away.

Shielding — As the number of electrons **between** the outer electrons and the nucleus **increases**, the outer electrons feel less attraction towards the nuclear charge. This lessening of the pull of the nucleus by inner shells of electrons is called **shielding (or screening)**.

> A **high ionisation energy** means there's a **high attraction** between the **electron** and the **nucleus** and so **more energy** is needed to remove the electron.

Successive Ionisation Energies *Involve Removing* ***Additional*** *Electrons*

1) You can remove **all** the electrons from an atom, leaving only the nucleus. Each time you remove an electron, there's a **successive ionisation energy**.
2) The definition for the **second ionisation energy is** —

> The **second ionisation energy** is the energy needed to remove 1 electron from **each ion** in **1 mole** of **gaseous** 1+ ions to form 1 mole of gaseous 2+ ions.

And here's the equation for the **second ionisation of oxygen**:

$$O^{+}_{(g)} \rightarrow O^{2+}_{(g)} + e^{-} \qquad \text{2nd ionisation energy} = +3388 \text{ kJ mol}^{-1}$$

3) You need to be able to write equations for **any** successive ionisation energy. The equation for the ***nth*** **ionisation energy** is....

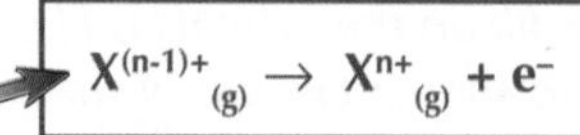

$$X^{(n-1)+}_{(g)} \rightarrow X^{n+}_{(g)} + e^{-}$$

Ionisation Energy

Successive Ionisation Energies Show **Shell Structure**

A **graph** of successive ionisation energies (like this one for sodium) provides evidence for the **shell structure** of atoms.

1) **Within each shell**, successive ionisation energies **increase**. This is because electrons are being removed from an **increasingly positive ion** — there's **less repulsion** amongst the remaining electrons, so they're **held more strongly** by the nucleus.
2) The **big jumps** in ionisation energy happen when a new shell is broken into — an electron is being removed from a shell **closer** to the nucleus.

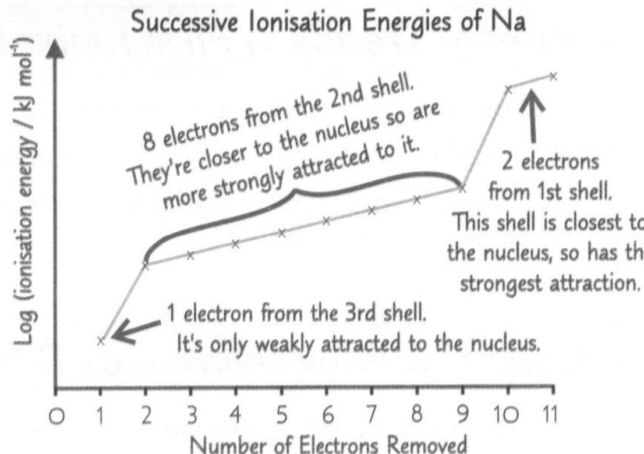

Successive Ionisation Energies of Na
Log (ionisation energy / kJ mol⁻¹)
8 electrons from the 2nd shell. They're closer to the nucleus so are more strongly attracted to it.
2 electrons from 1st shell. This shell is closest to the nucleus, so has the strongest attraction.
1 electron from the 3rd shell. It's only weakly attracted to the nucleus.
0 1 2 3 4 5 6 7 8 9 10 11
Number of Electrons Removed

1) Graphs like this can tell you which **group** of the periodic table an element belongs to. Just count **how many electrons are removed** before the first big jump to find the group number.

E.g. In the graph for sodium, **one electron** is removed before the first big jump — sodium is in **group 1**.

2) These graphs can be used to predict the **electronic structure** of elements. Working from **right to left**, count how many points there are before each big jump to find how many electrons are in each shell, starting with the first.

E.g. The graph for sodium has **2 points** on the right-hand side, then a jump, then **8 points**, a jump, and **1 final point**. Sodium has **2 electrons** in the first shell, **8** in the second and **1** in the third.

Practice Questions

Q1 Define first ionisation energy and give an equation as an example.

Q2 Describe the three main factors that affect ionisation energies.

Q3 How is ionisation energy related to the force of attraction between an electron and the nucleus of an atom?

Exam Questions

Q1 This table shows the nuclear charge and first ionisation energy for four elements.

Element	B	C	N	0
Charge of Nucleus	+5	+6	+7	+8
1st Ionisation Energy (kJ mol^{-1})	801	1087	1402	1314

a) Write an equation, including state symbols, to represent the first ionisation energy of carbon (C). [2 marks]

b) In these four elements, what is the relationship between nuclear charge and first ionisation energy? [1 mark]

c) Explain why nuclear charge has this effect on first ionisation energy. [2 marks]

Q2 This graph shows the successive ionisation energies of a certain element.

a) To which group of the periodic table does this element belong? [1 mark]

b) Why does it takes more energy to remove each successive electron? [2 marks]

c) What causes the sudden increases in ionisation energy? [1 mark]

d) What is the total number of electron shells in this element? [1 mark]

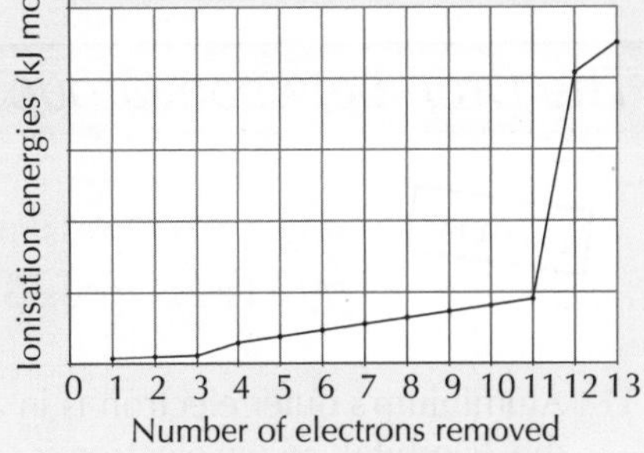

Ionisation energies (kJ mol⁻¹)
0 1 2 3 4 5 6 7 8 9 10 11 12 13
Number of electrons removed

Shirt crumpled — ionise it...

When you're talking about ionisation energies in exams, always use the three main factors — shielding, nuclear charge and distance from nucleus. Recite the definition of the first ionisation energies to yourself until you can't take any more.

Trends in First Ionisation Energy

Let joy be unconfined — it's another two pages about ionisation energy. This time though it's all about trends in first ionisation energies and what they can tell you about the structure of atoms.

There are **Trends** in **First Ionisation Energies**

1) The first ionisation energies of elements **down a group** of the periodic table **decrease**.
2) The first ionisation energies of elements **across a period generally increase**.
3) You need to know **how** and **why** ionisation energy **changes** as you go down **Group 2** and across **Period 3**. Read on to discover all...

Ionisation Energy **Decreases** Down Group 2

This graph shows the first ionisation energies of the elements in **Group 2**. It provides **evidence** that electron shells **REALLY DO EXIST** and that successive elements down the group have **extra**, **bigger**, **shells**...

1) If each element down Group 2 has an **extra electron shell** compared to the one above, the extra inner shells will **shield** the outer electrons from the attraction of the nucleus.
2) Also, the extra shell means that the outer electrons are **further away** from the nucleus, so the nucleus's attraction will be greatly reduced.

It makes sense that both of these factors will make it **easier** to remove outer electrons, resulting in a **lower ionisation energy**.

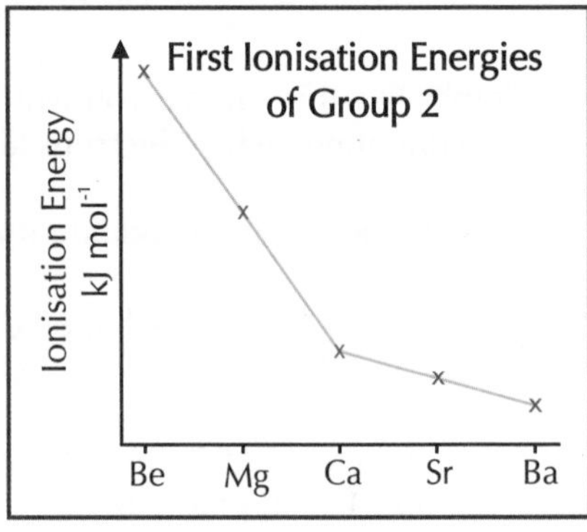

Ionisation Energy **Increases** Across a Period

The graph below shows the first ionisation energies of the elements in **Period 3**.

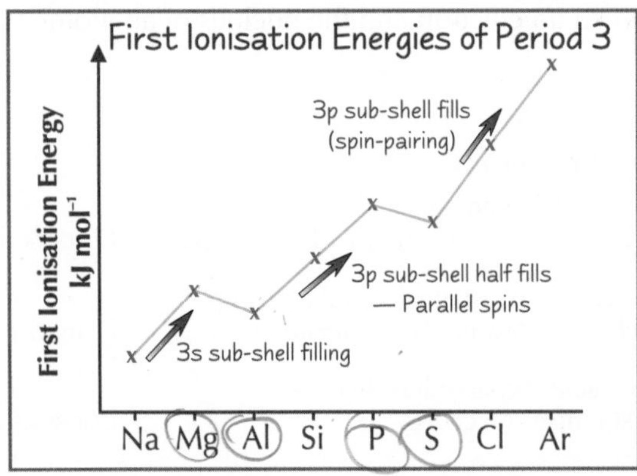

1) As you **move across** a period, the general trend is for the ionisation energies to **increase** — i.e. it gets harder to remove the outer electrons.
2) This can be explained because the number of protons is increasing, which means a stronger **nuclear attraction**.
3) All the extra electrons are at **roughly the same** energy level, even if the outer electrons are in different orbital types.
4) This means there's generally little **extra shielding** effect or **extra distance** to lessen the attraction from the nucleus.
5) But, there are **small drops** between Groups 2 and 3, and 5 and 6. Tell me more, I hear you cry. Well, alright then...

The Drop between Groups 2 and 3 Shows **Sub-Shell Structure**

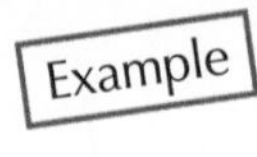

Mg	$1s^2 2s^2 2p^6 3s^2$	1st ionisation energy = 738 kJ mol^{-1}
Al	$1s^2 2s^2 2p^6 3s^2 3p^1$	1st ionisation energy = 578 kJ mol^{-1}

1) Aluminium's outer electron is in a **3p orbital** rather than a 3s. The 3p orbital has a **slightly higher** energy than the 3s orbital, so the electron is, on average, to be found **further** from the nucleus.
2) The 3p orbital has additional shielding provided by the **$3s^2$ electrons**.
3) Both these factors together are strong enough to **override** the effect of the increased nuclear charge, resulting in the ionisation energy **dropping** slightly.
4) This pattern in ionisation energies provides **evidence** for the theory of electron sub-shells.

Trends in First Ionisation Energy

The Drop between Groups 5 and 6 is due to Electron Repulsion

P	$1s^2\,2s^2\,2p^6\,3s^2\,3p^3$	1st ionisation energy = 1012 kJ mol^{-1}
S	$1s^2\,2s^2\,2p^6\,3s^2\,3p^4$	1st ionisation energy = 1000 kJ mol^{-1}

1) The **shielding is identical** in the phosphorus and sulfur atoms, and the electron is being removed from an identical orbital.
2) In phosphorus's case, the electron is being removed from a **singly-occupied** orbital. But in sulfur, the **electron** is being **removed** from an orbital containing two electrons.

   Phosphorus: [Ne] 3s [↑↓] 3p [↑][↑][↑]   Sulfur: [Ne] 3s [↑↓] 3p [↑↓][↑][↑]

   The **repulsion** between two electrons in an orbital means that electrons are **easier to remove** from shared orbitals.
3) Yup, yet more **evidence** for the electronic structure model.

Ways in which bears are like electrons #23 — the repulsion between two bears in a river means that bears are easier to remove from shared rivers.

Practice Questions

Q1 Describe the trend in ionisation energy as you go down Group 2.

Q2 Why do ionisation energies change down a group?

Q3 Which electron is easier to remove, a lone electron in an orbital or one in a pair?

Exam Questions

Q1 The first ionisation energies of the elements lithium to neon are given below in kJ mol^{-1}:

Li	Be	B	C	N	O	F	Ne
519	900	799	1090	1400	1310	1680	2080

a) Explain why the ionisation energies show an overall tendency to increase across the period. [3 marks]

b) Explain the irregularities in this trend for:

i) boron [2 marks]

ii) oxygen [2 marks]

Q2 First ionisation energy decreases down Group 2.

Explain how this trend provides evidence for the arrangement of electrons in levels. [3 marks]

Q3 a) Which of these elements has the highest first ionisation energy?

A Krypton B Lithium C Potassium D Neon [1 mark]

b) Explain your reasoning for the answer given in part a). [2 marks]

Q4 a) Which of these elements has the largest jump between its first and second ionisation energies?

A Sodium B Magnesium C Argon D Chlorine [1 mark]

b) Explain your reasoning for the answer given in part a). [3 marks]

First ionisation energies are so popular, they're trending...

If all these trends across and up and down the periodic table get confusing, just think back to basics — it's all about how strongly the outer electron is attracted to the nucleus. THAT depends on shielding, nuclear charge and distance from the nucleus and THEY depend on which shell, sub-shell and orbital the electron is in.

The Mole

It'd be handy to be able to count out atoms — but they're way too tiny. You can't even see them, never mind get hold of them with tweezers. Luckily, you can use the idea of moles to figure out how exactly much stuff you've got.

A Mole is Just a (Very Large) Number of Particles

In formulas, number of moles is often given the symbol n.

1) Amount of substance can be measured using a unit called the **mole** (**mol** for short).
2) One mole contains **6.02×10^{23} particles** — this number is known as **Avogadro's constant**.
3) It **doesn't matter** what the particles are. They can be atoms, molecules, electrons, ions, penguins — **anything**.

1 mole of **carbon** (C) contains **6.02×10^{23} atoms**. 1 mole of **methane** (CH_4) contains **6.02×10^{23} molecules**.
1 mole of **sodium ions** (Na^+) contains **6.02×10^{23} ions**. 1 mole of **electrons** contains **6.02×10^{23} electrons**.

4) You can use Avogadro's constant to convert between number of particles and number of moles.

Just remember this formula: **Number of particles = Number of moles × Avogadro's constant**

Example: How many atoms are in 0.450 moles of pure iron?

Number of atoms = $0.450 \times (6.02 \times 10^{23}) = \mathbf{2.71 \times 10^{23}}$

The Amount of a Substance in Moles can be Calculated from Mass and M_r

1) **1 mole** of any substance has a **mass** that's the same as its **relative molecular mass** (M_r) in **grams**.
For example, the M_r of water (H_2O) is $(1 \times 2) + 16 = 18$, so 1 mole of water has a mass of 18 g.
2) This means that you can work out how many **moles** of a substance you have from the **mass** of the substance and its **relative molecular mass** (M_r). Here's the formula you need:

$$\text{Number of moles} = \frac{\text{mass of substance}}{M_r}$$

Example: How many moles of aluminium oxide are present in 5.10 g of Al_2O_3?

M_r of $Al_2O_3 = (2 \times 27.0) + (3 \times 16.0) = 102.0$

Number of moles of $Al_2O_3 = \frac{5.10}{102.0} = \mathbf{0.0500\ moles}$

If you need to find M_r or mass instead, you can re-arrange the formula using this formula triangle:

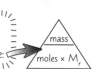

mass / moles × M_r

The Concentration of a Solution is Measured in mol dm^{-3}

1) The **concentration** of a solution is how many **moles** are dissolved per **1 dm³** (that's 1 litre) of solution. The units are **mol dm^{-3}**.
2) Here's the formula to find the **number of moles**:

Number of moles = Concentration (mol dm^{-3}) × Volume (dm³)

This one can go in a handy formula triangle too:

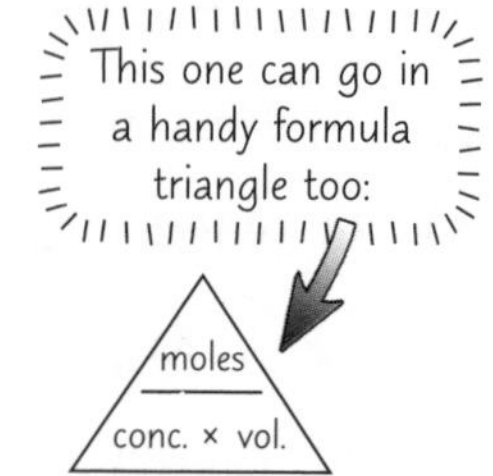

3) Watch out for the units — you might be given the volume in cm³ rather than dm³. If that's the case, you'll have to convert it to dm³ first.

Example: What mass of sodium hydroxide (NaOH) needs to be dissolved in water to give 50.0 cm³ of a solution with a concentration of 2.00 mol dm^{-3}?

Volume of solution in dm³ = $50 \div 1000 = 0.05$ dm³
Number of moles NaOH = 2.00 mol dm^{-3} × 0.0500 dm³ = 0.100
M_r of NaOH = $23.0 + 16.0 + 1.0 = 40.0$
Mass = number of moles × M_r = 0.100×40.0 = **4.00 g**

1 dm³ = 1000 cm³
So to convert from cm³ to dm³ you need to divide by 1000.

The Mole

Ideal Gas equation — pV = nRT

The **ideal gas equation** lets you find the **number of moles** in a certain volume of gas. You just need to know the temperature and pressure that the gas is at.

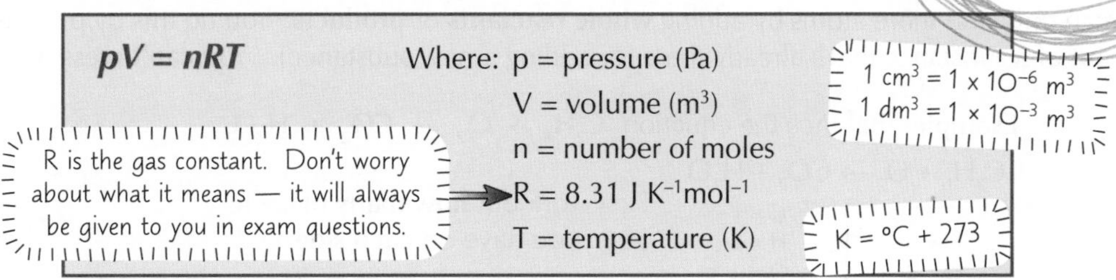

$pV = nRT$
Where: p = pressure (Pa)
V = volume (m^3)
n = number of moles
R = 8.31 J $K^{-1}mol^{-1}$
T = temperature (K)
R is the gas constant. Don't worry about what it means — it will always be given to you in exam questions.
1 cm^3 = 1 × 10^{-6} m^3
1 dm^3 = 1 × 10^{-3} m^3
K = °C + 273

Example:

At a temperature of 60.0 °C and a pressure of 250 kPa, a gas occupied a volume of 1100 cm^3 and had a mass of 1.60 g. Find its relative molecular mass. The gas constant is 8.31 J $K^{-1}mol^{-1}$.

Use the ideal gas equation to find the number of moles of gas present.
p = 250 × 1000 = 250 × 10^3 Pa, V = 1100 × (1 × 10^{-6}) = 1.10 × 10^{-3} m^3,
R = 8.31 J $K^{-1}mol^{-1}$, T = 60.0 + 273 = 333 K

Start off by putting everything into the right units.

$$n = \frac{pV}{RT} = \left(\frac{(250 \times 10^3) \times (1.10 \times 10^{-3})}{8.31 \times 333}\right) = 0.0994 \text{ moles}$$

M_r of gas = mass ÷ number of moles = 1.60 g ÷ 0.0994 = **16.1** (3 s.f.)

Practice Questions

Q1 How many molecules are there in one mole of ethane molecules?*

Q2 How many atoms are in there 0.500 moles of carbon?*

Q3 A student dissolves 0.100 moles of NaCl in 0.500 dm^3 of water. Find the concentration of the solution.*

Q4 Write down the ideal gas equation.

*Answers on page 213.

Exam Questions

Q1 How many moles of calcium sulfate are there in 34.05 g of $CaSO_4$? [1 mark]

Q2 Calculate the mass of 0.360 moles of ethanoic acid (CH_3COOH). [1 mark]

Q3 How many moles of nitric acid are there in 50.0 cm^3 of 0.250 mol dm^{-3} of HNO_3? [1 mark]

Q4 What mass of H_2SO_4 is needed to produce 60.0 cm^3 of 0.250 mol dm^{-3} solution? [2 marks]

Q5 Calculate the concentration, in mol dm^{-3}, of 3.65 g of HCl dissolved in 100 cm^3 of water. [2 marks]

Q6 What volume, in m^3, will be occupied by 88.0 g of propane gas (C_3H_8) at 25.0 °C and 100 kPa?
Give your answer to 3 significant figures. The gas constant is R = 8.31 J K^{-1} mol^{-1}. [3 marks]

Q7 At 301 K, a 35.2 g sample of CO_2 has a volume of 20.0 dm^3. What pressure is the gas at (to 3 significant figures)?

A 10 000 kPa B 100 kPa C 200 kPa D 10.0 kPa [1 mark]

Molasses — 6.02 × 10^{23} donkeys...

You need this stuff for loads of the calculation questions you might get, so learn it inside out. Before you start plugging numbers into formulas, make sure they're in the right units. If they're not, you need to know how to convert them or you'll be tossing marks out the window. Learn all the definitions and formulas, then have a bash at the questions.

Equations and Calculations

Balancing equations might cause you a few palpitations — as soon as you make one bit right, the rest goes pear-shaped.

Balanced Equations have **Equal Numbers** of each Atom on **Both Sides**

1) Balanced equations have the **same number** of each atom on **both** sides. They're... well... you know... balanced.
2) You can only add more atoms by adding **whole reactants** or **products**. You do this by putting a number **in front** of a substance that's already there (or adding a new substance). You **can't** mess with formulas — ever.

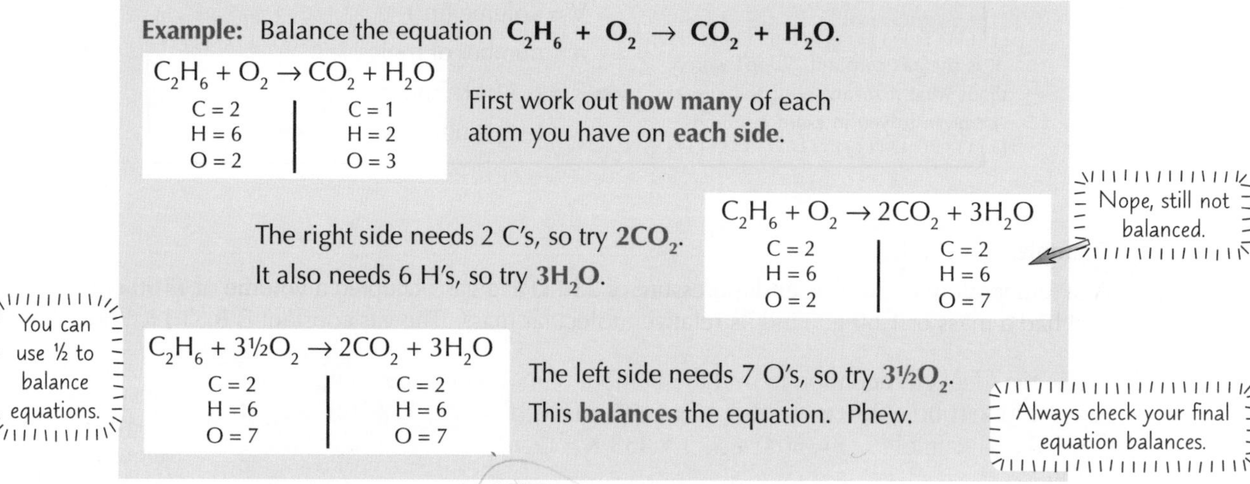

Example: Balance the equation $C_2H_6 + O_2 \rightarrow CO_2 + H_2O$.
$C_2H_6 + O_2 \rightarrow CO_2 + H_2O$
C = 2 | C = 1
H = 6 | H = 2
O = 2 | O = 3
First work out **how many** of each atom you have on **each side**.
The right side needs 2 C's, so try $2CO_2$. It also needs 6 H's, so try $3H_2O$.
$C_2H_6 + O_2 \rightarrow 2CO_2 + 3H_2O$
C = 2 | C = 2
H = 6 | H = 6
O = 2 | O = 7
Nope, still not balanced.
You can use ½ to balance equations.
$C_2H_6 + 3\frac{1}{2}O_2 \rightarrow 2CO_2 + 3H_2O$
C = 2 | C = 2
H = 6 | H = 6
O = 7 | O = 7
The left side needs 7 O's, so try $3\frac{1}{2}O_2$. This **balances** the equation. Phew.
Always check your final equation balances.

Ionic Equations Only Show the Reacting Particles

1) You can also write an **ionic equation** for any reaction involving **ions** that happens **in solution**.
2) In an ionic equation, only the **reacting particles** (and the **products** they form) are included.

Example: Here is the **full balanced equation** for the reaction of **nitric acid** with **sodium hydroxide**:

$$HNO_3 + NaOH \rightarrow NaNO_3 + H_2O$$

The **ionic** substances in this equation will **dissolve**, breaking up into ions in solution. You can rewrite the equation to show all the **ions** that are in the reaction mixture:

$$H^+ + NO_3^- + Na^+ + OH^- \rightarrow Na^+ + NO_3^- + H_2O$$

Leave anything that isn't an ion in solution (like the H_2O) as it is.

To get from this to the ionic equation, just cross out any ions that appear on **both sides** of the equation — in this case, that's the sodium ions (Na^+) and the nitrate ions (NO_3^-).

So the **ionic equation** for this reaction is:

$$H^+ + OH^- \rightarrow H_2O$$

An ion that's present in the reaction mixture, but doesn't get involved in the reaction is called a spectator ion.

3) When you've written an ionic equation, check that the **charges** are **balanced**, as well as the atoms — if the charges don't balance, the equation isn't right.

In the example above, the **net charge** on the left hand side is +1 + –1 = **0** and the net charge on the right hand side is **0** — so the charges balance.

State Symbols Give a bit More Information about the Substances

State symbols are put after each reactant or product in an equation. They tell you what **state of matter** things are in.

s = solid
l = liquid
g = gas
aq = aqueous (solution in water)

To show you what I mean, here's an example —

$$CaCO_{3\,(s)} + 2HCl_{(aq)} \rightarrow CaCl_{2\,(aq)} + H_2O_{(l)} + CO_{2\,(g)}$$

$CaCO_{3\,(s)}$	$2HCl_{(aq)}$	$CaCl_{2\,(aq)}$	$H_2O_{(l)}$	$CO_{2\,(g)}$
solid	aqueous	aqueous	liquid	gas

Equations and Calculations

Balanced Equations can be used to Work out Masses

This is handy for working out how much **reactant** you need to make a certain **mass of product** (or **vice versa**).

Example: Calculate the mass of iron oxide produced if 27.9 g of iron is burnt in air.

$$4Fe + 3O_2 \rightarrow 2Fe_2O_3$$

M_r of Fe = 55.8, so the number of moles in 27.9 g of Fe = $\frac{\text{mass}}{M_r} = \frac{27.9}{55.8}$ = 0.500 moles

From the equation: 4 moles of Fe gives 2 moles of Fe_2O_3, so 0.500 moles of Fe would give 0.250 moles of Fe_2O_3.

M_r of Fe_2O_3 = (2 × 55.8) + (3 × 16.0) = 159.6

Mass of Fe_2O_3 produced = moles × M_r = 0.250 × 159.6 = **39.9 g.**

Once you know the number of moles and the M_r of Fe_2O_3, it's easy to work out the mass.

That's not all... Balanced Equations can be used to Work Out Gas Volumes

It's pretty handy to be able to use the ideal gas equation to work out **how much gas** a reaction will produce, so that you can use **large enough apparatus**. Or else there might be a rather large bang.

Example: What volume of gas, in dm^3, is produced when 15.0 g of sodium reacts with excess water at a temperature of 25.0 °C and a pressure of 100 kPa? The gas constant is 8.31 J K^{-1} mol^{-1}.

$$2Na_{(s)} + 2H_2O_{(l)} \rightarrow 2NaOH_{(aq)} + H_{2\,(g)}$$

'Excess water' just means that you know all of the sodium will react.

M_r of Na = 23.0, so number of moles in 15.0 g of Na = $\frac{15.0}{23.0}$ = 0.652 moles

From the equation, 2 moles of Na produces 1 mole of H_2,
so 0.652 moles of Na must produce $\frac{0.652}{2}$ = 0.326 moles of H_2.

Volume of H_2 = $\frac{nRT}{p} = \frac{0.326 \times 8.31 \times 298}{100 \times 10^3}$ = 0.00807 m^3 = **8.07 dm^3** (3 s.f.)

Practice Questions

Q1 What is the difference between a full equation and an ionic equation?

Q2 Chlorine (Cl_2) reacts with potassium bromide (KBr) in solution to give potassium chloride (KCl) and bromine (Br_2).
a) Write a full balanced equation for this reaction.* b) Write an ionic equation for this reaction.*

Q3 What is the state symbol for a solution of sodium chloride dissolved in water?

*Answers on page 213.

Exam Questions

Q1 Calculate the mass of ethene required to produce 258 g of chloroethane, C_2H_5Cl.

$$C_2H_4 + HCl \rightarrow C_2H_5Cl$$

[3 marks]

Q2 15.0 g of calcium carbonate is heated strongly so that it fully decomposes. $CaCO_{3\,(s)} \rightarrow CaO_{(s)} + CO_{2\,(g)}$

a) Calculate the mass of calcium oxide produced. [3 marks]

b) Calculate the volume of gas produced in m^3 at 25.0 °C and 100 kPa.
The gas constant is 8.31 J K^{-1} mol^{-1}. [3 marks]

Q3 Balance this equation: $KI + Pb(NO_3)_2 \rightarrow PbI_2 + KNO_3$ [1 mark]

Don't get in a state about equations...

Balancing equations is a really, really important skill in Chemistry, so make sure you can do it. You will ONLY be able to calculate reacting masses and gas volumes if you've got a balanced equation to work from. I've said it once, and I'll say it again — practise, practise, practise... it's the only road to salvation. (By the way, exactly where is salvation anyway?)

Titrations

Titrations are used to find out the concentration of acid or alkali solutions. You'll probably get to do a titration or two (lucky you) and you'll definitely need to know all the ins and outs of how to do them for your exams. So read on...

*A **Standard Solution** Has a **Known** Concentration*

Standard solutions can also be called volumetric solutions.

Before you do a titration, you might have to make up a **standard solution** to use.
A **standard solution** is any solution that you **know** the **exact concentration** of. Making a standard solution involves dissolving a **known amount** of **solid** in a known amount of **water** to create a known concentration.

Example: Make 250 cm^3 of a 2.00 mol dm^{-3} solution of sodium hydroxide.

1) First work out how many **moles** of sodium hydroxide you need using the formula: **moles = concentration × volume**
$= 2.00 \text{ mol dm}^{-3} \times 0.250 \text{ dm}^3 = 0.500$ moles

Remember, the volume needs to be in dm^3 for this bit.

2) Now work out how many **grams** of sodium hydroxide you need using the formula: **mass = moles × M$_r$**
$= 0.500 \times 40.0 = 20.0$ g

M_r of NaOH: $23.0 + 16.0 + 1.0 = 40.0$

3) Place a weighing boat on a **digital balance** and weigh out this mass of solid. Tip it into a beaker. Now **re-weigh** the boat (which may still contain traces of the solid). Subtract the mass of the boat from the mass of the boat and solid together to find the **precise mass** of solid used.

4) Add **distilled water** to the beaker and **stir** until all the sodium hydroxide has **dissolved**.

5) Tip the solution into a **volumetric flask** — make sure it's the right size for the volume that you're making (250 cm^3 in this case). Use a **funnel** to make sure it all goes in.

6) **Rinse** the beaker and stirring rod with distilled water and add that to the **flask** too. This makes sure there's no solute clinging to the beaker or rod.

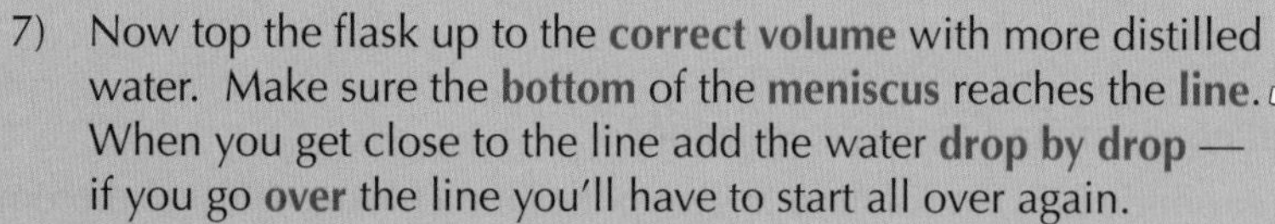

7) Now top the flask up to the **correct volume** with more distilled water. Make sure the **bottom** of the **meniscus** reaches the **line**. When you get close to the line add the water **drop by drop** — if you go **over** the line you'll have to start all over again.

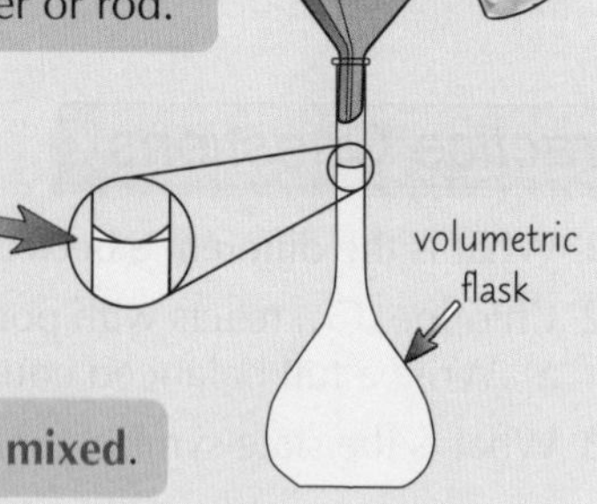

8) **Stopper** the flask and turn it upside down a few times to make sure it's **mixed**.

bottom of meniscus meets line

*Titrations let you work out the **Concentration** of an **Acid** or **Alkali***

1) A **titration** allows you to find out **exactly** how much acid is needed to **neutralise** a measured quantity of alkali (or the other way round).
2) You can use this data to work out the **concentration** of the alkali.
3) Start off by using a **pipette** to measure out a set volume of the solution that you want to know the concentration of. Put it in a flask.
4) Add a few drops of an appropriate **indicator** (see next page) to the flask.
5) Then fill a **burette** (see next page) with a **standard solution** of the acid — remember, that means you know its exact concentration.
6) Use a **funnel** to carefully pour the acid into the burette. Always do this **below eye level** to avoid any acid splashing on to your face or eyes. (You should wear **safety glasses** too.)
7) Now you're ready to **titrate**...

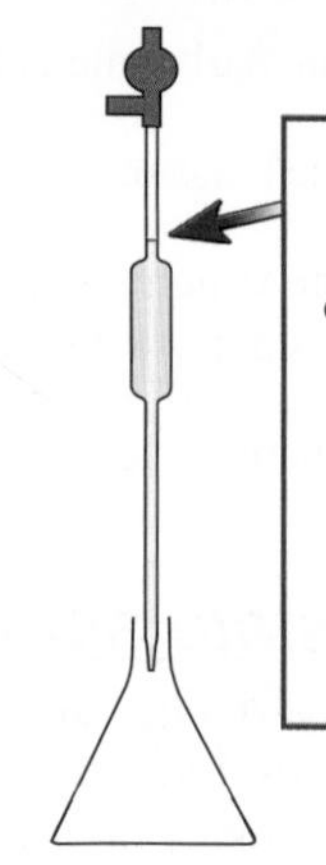

Pipette
Pipettes measure only one volume of solution. Fill the pipette to just above the line, then take the pipette out of the solution. Then drop the level down carefully to the line.

Titrations

Titrations need to be done **Really Accurately**

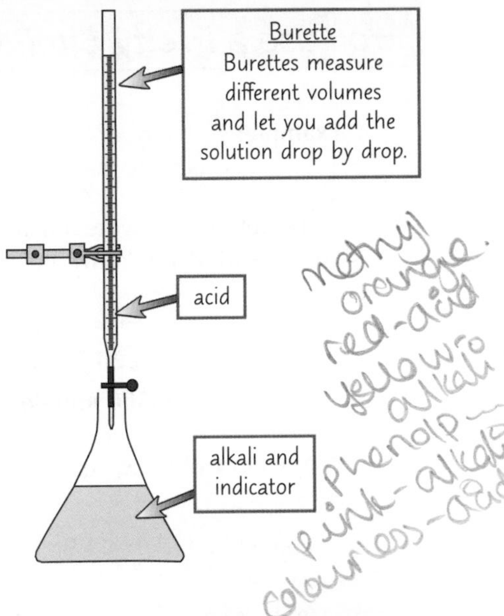

Burette
Burettes measure different volumes and let you add the solution drop by drop.
acid
alkali and indicator

1) First do a **rough titration** to get an idea where the **end point** (the exact point where the alkali is **neutralised** and the indicator changes colour) is. Add the **acid** to the alkali using a **burette**, giving the flask a regular **swirl**.
2) Now do an **accurate** titration. Take an initial reading to see exactly how much acid is in the burette. Then run the acid in to within 2 cm^3 of the end point. When you get to this stage, add it **dropwise** — if you don't notice exactly when the colour changes you'll **overshoot** and your result won't be accurate.
3) Work out the **amount** of acid used to **neutralise** the alkali. This is just the final reading minus the initial reading. This volume is known as the **titre**.
4) **Repeat** the titration a few times, until you have at least three results that are **concordant** (very similar).
5) Use the results from each repeat to calculate the **mean** volume of acid used. Remember to leave out any **anomalous results** when calculating your mean — they can distort your answer.

There's more stuff about how to record and handle the data you get from experiments like this in the Practical Skills section. Have a look if you want to know more about means, anomalous results, precision and experimental error.

Choppy seas made it difficult for Captain Cod to read the burette accurately.

Indicators Show you when the Reaction's **Just Finished**

In titrations, indicators that change colour quickly over a **very small pH range** are used so you know **exactly** when the reaction has ended.

The main two indicators for **acid/alkali reactions** are:

1) **methyl orange** — this is **red** in acid and **yellow** in alkali.
2) **phenolphthalein** — this is **colourless** in acid and **pink** in alkali.

Universal indicator is no good here — its colour change is too gradual.

It's a good idea to stand your flask on a white tile — it'll make it easier to see exactly when the end point is.

You can Calculate **Concentrations** from Titrations

The next step is to use the **mean volume** from your titrations to find the **concentration** of the solution in the flask.

Example: In a titration experiment, 25.0 cm^3 of 0.500 mol dm^{-3} HCl neutralised 35.0 cm^3 of NaOH solution. Calculate the concentration of the sodium hydroxide solution in mol dm^{-3}.

First write a **balanced equation** and decide **what you know** and what you **need to know**:

HCl +	NaOH → NaCl + H_2O
25.0 cm^3	**35.0 cm^3**
0.500 mol dm^{-3}	**?**

Now work out how many **moles of HCl** you have:

Number of moles HCl = concentration × volume (dm^3) = $0.500 \times \frac{25.0}{1000}$ = 0.0125 moles

This is the formula from page 14.

From the equation, you know 1 mole of HCl neutralises 1 mole of NaOH.
So 0.0125 moles of HCl must neutralise **0.0125** moles of NaOH.

Now it's a doddle to work out the **concentration of NaOH**:

Concentration of NaOH = moles ÷ volume (dm^3) = $0.0125 \div \frac{35.0}{1000}$ = **0.357 mol dm^{-3}**

Titrations

You use a Pretty Similar Method to Calculate Volumes for Reactions

Example: 20.4 cm^3 of a 0.500 mol dm^{-3} solution of sodium carbonate reacts with 1.50 mol dm^{-3} nitric acid. Calculate the volume of nitric acid required to neutralise the sodium carbonate.

Like before, first write a **balanced equation** for the reaction and decide **what you know** and what you **want to know**:

$$Na_2CO_3 + 2HNO_3 \rightarrow 2NaNO_3 + H_2O + CO_2$$

Na_2CO_3	$2HNO_3$
20.4 cm^3	?
0.500 mol dm^{-3}	**1.50 mol dm^{-3}**

Now work out how many **moles** of Na_2CO_3 you've got:

Number of moles of Na_2CO_3 = concentration × volume (dm^3) = $0.500 \times \frac{20.4}{1000}$ = 0.0102 moles.

1 mole of Na_2CO_3 neutralises 2 moles of HNO_3, so 0.0102 moles of Na_2CO_3 neutralises **0.0204 moles of HNO_3**.

Now you know the number of moles of HNO_3 and the concentration, you can work out the **volume**:

Volume of $HNO_3 = \frac{\text{number of moles}}{\text{concentration}} = \frac{0.0204}{1.50}$ = **0.0136 dm^3**

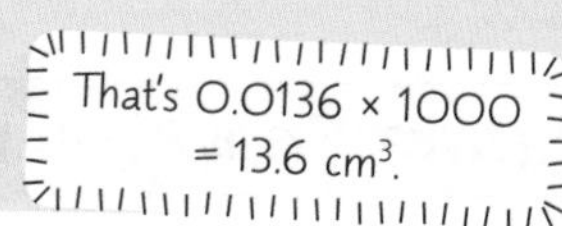
That's 0.0136 × 1000
= 13.6 cm³.

Practice Questions

Q1 What is a standard solution? Describe how to make one.

Q2 When you're doing a titration, why do you add the acid dropwise when you're getting near the end point?

Q3 Write down the formula for calculating number of moles from the concentration and volume of a solution. Rearrange it so that you could use it to calculate concentration. Then do the same for volume.

Exam Questions

Q1 Calculate the concentration in mol dm^{-3} of a solution of ethanoic acid (CH_3COOH) if 25.4 cm^3 of it is neutralised by 14.6 cm^3 of 0.500 mol dm^{-3} sodium hydroxide solution. The equation for this reaction is: $CH_3COOH + NaOH \rightarrow CH_3COONa + H_2O$. [3 marks]

Q2 You are supplied with 0.750 g of calcium carbonate and a solution of 0.250 mol dm^{-3} sulfuric acid. What volume of acid will be needed to neutralise the calcium carbonate? The equation for this reaction is: $CaCO_3 + H_2SO_4 \rightarrow CaSO_4 + H_2O + CO_2$. [4 marks]

Q3 50.0 cm^3 of nitric acid was titrated with 0.400 mol dm^{-3} sodium hydroxide solution. The equation for this reaction is: $HNO_3 + NaOH \rightarrow NaNO_3 + H_2O$. The results of the titration are shown in the table on the right.

a) Identify any anomalous results, explaining your reasoning. [1 mark]

b) Calculate the mean titre, ignoring anomalous results. [1 mark]

c) Calculate the concentration of the nitric acid. [3 marks]

Titration	Volume NaOH (cm^3)
1	45.00
2	45.10
3	42.90
4	44.90

Burettes and pipettes — big glass things, just waiting to be dropped...

Titrations are fiddly. But you do get to use big, impressive-looking equipment and feel like you're doing something important. Then there are the results to do calculations with. The best way to start is always to write out the balanced equation and put what you know about each substance underneath it. Then think about what you're trying to find out.

Formulas, Yield and Atom Economy

Here's another topic that's piled high with numbers — it's all just glorified maths really.

Empirical and Molecular Formulas are Ratios

You have to know what's what with empirical and molecular formulas, so here goes...

1) The **empirical formula** gives just the simplest whole number ratio of atoms of each element in a compound.
2) The **molecular formula** gives the **actual** numbers of atoms of each element in a compound.
3) The **molecular formula** is made up of a whole **number** of empirical units.

Example: A molecule with M_r = 166.0 has the empirical formula $C_4H_3O_2$. Find its molecular formula.

First find the **empirical mass** (that's just the total mass of all the atoms in the empirical formula):

$$(4 \times 12.0) + (3 \times 1.0) + (2 \times 16.0) = 48.0 + 3.0 + 32.0 = 83.0$$

Now compare the empirical mass with the **molecular mass**: M_r = 166, so there are $\frac{166.0}{83.0}$ = 2 empirical units in the molecule.

The molecular formula must be the **empirical formula × 2**, so the molecular formula is $C_8H_6O_4$.

Empirical Formulas can be Calculated from Percentage Composition

You can work out the empirical formula of a compound from the **percentages** of the different elements it contains.

Example: A compound is found to have percentage composition 56.5% potassium, 8.70% carbon and 34.8% oxygen by mass. Find its empirical formula.

Use $n = \frac{\text{mass}}{M_r}$

In **100 g** of the compound there would be:

$\frac{56.5}{39.1} = 1.445$ moles of K $\frac{8.70}{12.0} = 0.725$ moles of C $\frac{34.8}{16.0} = 2.175$ moles of O

Divide each number of moles by the **smallest** of these **numbers** — in this case it's 0.725.

K: $\frac{1.445}{0.725} = 2.00$ C: $\frac{0.725}{0.725} = 1.00$ O: $\frac{2.175}{0.725} = 3.01$

The ratio of K : C : O is 2 : 1 : 3. So the empirical formula's got to be K_2CO_3.

Empirical Formulas can be Calculated from Experiments

You need to be able to work out empirical formulas using **masses** from **experimental results** too.

Example: When a hydrocarbon is burnt in excess oxygen, 4.40 g of carbon dioxide and 1.80 g of water are made. What is the empirical formula of the hydrocarbon?

First work out how many moles of the products you have.

No. of moles of $CO_2 = \frac{\text{mass}}{M_r} = \frac{4.40}{12.0 + (16.0 \times 2)} = \frac{4.40}{44.0} = 0.100$ moles

1 mole of CO_2 contains 1 mole of carbon atoms, so the original hydrocarbon must have contained **0.100 moles** of **carbon atoms**.

No. of moles of $H_2O = \frac{\text{mass}}{M_r} = \frac{1.80}{(2 \times 1.0) + 16.0} = \frac{1.80}{18.0} = 0.100$ moles

1 mole of H_2O contains 2 moles of hydrogen atoms, so the original hydrocarbon must have contained **0.200 moles** of **hydrogen atoms**.

This works because the only place the carbon in the carbon dioxide and the hydrogen in the water could have come from is the hydrocarbon.

Ratio C : H = 0.100 : 0.200. Now divide both numbers by the smallest — here it's 0.100.

Ratio **C : H = 1 : 2**. So the empirical formula must be CH_2.

Formulas, Yield and Atom Economy

Percentage Yield Is Never 100%

1) The **theoretical yield** is the **mass of product** that **should** be formed in a chemical reaction. It assumes **no** chemicals are '**lost**' in the process. You can use the **masses of reactants** and a **balanced equation** to calculate the theoretical yield for a reaction.

Example: 1.40 g of iron filings reacts with ammonia and sulfuric acid to make hydrated ammonium iron(II) sulfate.

$Fe_{(s)} + 2NH_{3\,(aq)} + 2H_2SO_{4\,(aq)} + 6H_2O_{(l)} \rightarrow (NH_4)_2Fe(SO_4)_2.6H_2O_{(s)} + H_{2\,(g)}$

Calculate the theoretical yield of the reaction.

You started with 1.40 g of iron filings and iron has a relative atomic mass of 55.8, so:

Number of moles of iron (A_r = 55.8) reacted = $\frac{\text{mass}}{M_r} = \frac{1.40}{55.8} = 0.0251$

From the equation, **moles of iron : moles of hydrated ammonium iron(II) sulfate** is **1 : 1**, so 0.0251 moles of hydrated ammonium iron(II) sulfate should form.

M_r of $(NH_4)_2Fe(SO_4)_2.6H_2O_{(s)}$ = 392.0
Theoretical yield = moles × M_r = 0.0251 × 392.0 = **9.84 g**

2) For any reaction, the **actual** mass of product (the **actual yield**) will always be **less** than the theoretical yield. There are many reasons for this. For example, sometimes not all the 'starting' chemicals react fully. And some chemicals are always 'lost', e.g. some solution gets left on filter paper, or is lost during transfers between containers.
3) Once you know the **theoretical yield** and the **actual yield**, you can use them to work out the **percentage yield**.

$$\text{Percentage Yield} = \frac{\text{Actual Yield}}{\text{Theoretical Yield}} \times 100$$

4) So, in the ammonium iron(II) sulfate example above, the theoretical yield was 9.84 g. Say you weighed the hydrated ammonium iron(II) sulfate crystals that you had produced and found the actual yield was **5.22 g**. Now you can just pop these numbers into the formula to find the percentage yield:

Percentage yield = (5.22 ÷ 9.84) × 100 = **53.0%**

Here's another example of calculating the percentage yield:

Example: 0.475 g of CH_3Br reacts with excess NaOH in the following reaction:

$CH_3Br + NaOH \rightarrow CH_3OH + NaBr$

0.153 g of CH_3OH is produced. What is the percentage yield?

Find the number of moles of **CH_3Br** that you started off with:

Number of moles of CH_3Br = $\frac{\text{mass}}{M_r} = \frac{0.475}{(12.0 + (3 \times 1.0) + 79.9)} = \frac{0.475}{94.9} = 0.00501$ mol

From the equation, **moles of CH_3Br : CH_3OH** is **1 : 1**, so 0.00501 moles of CH_3OH should form.

M_r of CH_3OH = 32.0
Theoretical yield = 0.00501 × 32.0 = **0.160 g.**

Now put these numbers into the percentage yield formula:

Percentage yield = $\frac{\text{Actual Yield}}{\text{Theoretical Yield}} \times 100 = \frac{0.153\text{ g}}{0.160\text{ g}} \times 100 =$ **95.6%**

Formulas, Yield and Atom Economy

Atom Economy is a Measure of the *Efficiency* of a Reaction

1) **Percentage yield** tells you how wasteful a **process** is — it's based on how much of the product is **lost** during the process (see previous page).
2) But percentage yield doesn't measure how wasteful the **reaction** itself is. A reaction with a 100% yield could still be wasteful if a lot of the atoms from the **reactants** wind up in **by-products** rather than the **desired product**.
3) **Atom economy** is a measure of the proportion of reactant **atoms** that become part of the desired product (rather than by-products) in the **balanced** chemical equation. It's calculated using this formula:

$$\%\text{ atom economy} = \frac{\text{molecular mass of desired product}}{\text{sum of molecular masses of all reactants}} \times 100$$

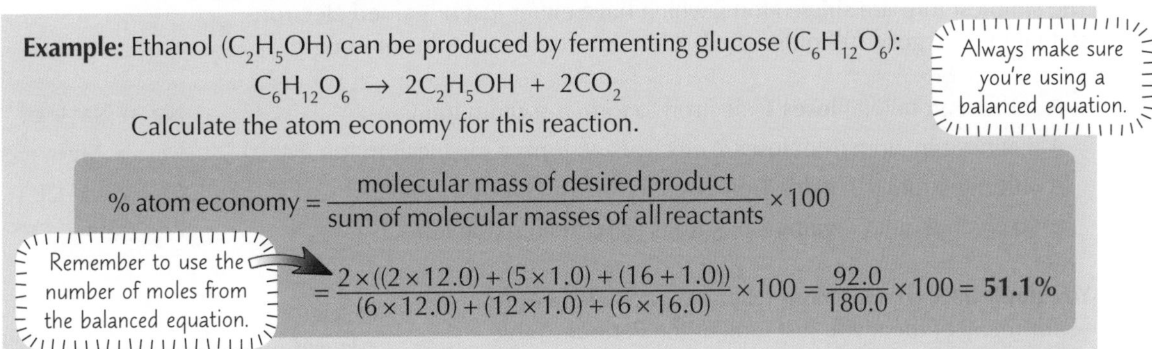

Example: Ethanol (C_2H_5OH) can be produced by fermenting glucose ($C_6H_{12}O_6$):

$$C_6H_{12}O_6 \rightarrow 2C_2H_5OH + 2CO_2$$

Calculate the atom economy for this reaction.

$$\%\text{ atom economy} = \frac{\text{molecular mass of desired product}}{\text{sum of molecular masses of all reactants}} \times 100$$

$$= \frac{2 \times ((2 \times 12.0) + (5 \times 1.0) + (16 + 1.0))}{(6 \times 12.0) + (12 \times 1.0) + (6 \times 16.0)} \times 100 = \frac{92.0}{180.0} \times 100 = \mathbf{51.1\%}$$

Always make sure you're using a balanced equation.
Remember to use the number of moles from the balanced equation.

4) Wherever possible, companies in the chemical industry try to use processes with **high atom economies**.
5) Processes with high atom economies are better for the **environment** because they produce less **waste**. Any waste that's made needs to be **disposed of safely** so the less that's made, the better.
6) They make more efficient use of **raw materials**, so they're more **sustainable** (they use up natural resources more slowly).
7) They're also **less expensive**. A company using a process with a high atom economy will spend less on separating the desired product from the waste products and also less on treating waste.

Practice Questions

Q1 Define 'empirical formula'.

Q2 What is the difference between a molecular formula and an empirical formula?

Q3 Write down the formula for calculating percentage yield.

Q4 Write down the formula for calculating atom economy.

Q5 Explain why it is important for chemical companies to develop processes that have high atom economies

Exam Questions

Q1 Hydrocarbon X has a relative molecular mass of 78.0. It is found to have 92.3% carbon and 7.70% hydrogen by mass. Find the empirical and molecular formulae of X. [4 marks]

Q2 Phosphorus trichloride (PCl_3) can react with chlorine to give phosphorus pentachloride (PCl_5). This is the equation for this reaction: $PCl_3 + Cl_2 \rightarrow PCl_5$

a) If 0.275 g of PCl_3 (M_r = 137.5) reacts with chlorine, what is the theoretical yield of PCl_5? [3 marks]

b) When this reaction is performed 0.198 g of PCl_5 is collected. Calculate the percentage yield. [1 mark]

c) State the atom economy of this reaction. Explain your answer. [2 marks]

The Empirical Strikes Back...

This is the kind of topic where it isn't enough to just learn the facts — you have to know how to do the calculations too. That takes practice — so make sure you understand all the examples on these pages, then test yourself on the questions.

Ionic Bonding

Every atom's aim in life is to have a full outer shell of electrons. Once they've managed this, they're happy.

***Compounds** are **Atoms** of Different Elements **Bonded** Together*

1) When different elements join or bond together, you get a **compound**.
2) There are two main types of bonding in compounds — **ionic** and **covalent**. You need to make sure you've got them **both** totally sussed.

E.g. when the elements hydrogen and oxygen combine, the compound water (H_2O) is formed.

Ionic Bonding** is when **Ions** are Held Together by **Electrostatic Attraction

1) Ions are formed when one or more electrons are **transferred** from one atom to another.
2) The simplest ions are single atoms which have either lost or gained electrons so that they've got a **full outer shell**. Here are some examples of ions:

A sodium atom (Na) **loses** 1 electron to form a sodium ion (Na^+)	$Na \rightarrow Na^+ + e^-$
A magnesium atom (Mg) **loses** 2 electrons to form a magnesium ion (Mg^{2+})	$Mg \rightarrow Mg^{2+} + 2e^-$
A chlorine atom (Cl) **gains** 1 electron to form a chloride ion (Cl^-)	$Cl + e^- \rightarrow Cl^-$
An oxygen atom (O) **gains** 2 electrons to form an oxide ion (O^{2-})	$O + 2e^- \rightarrow O^{2-}$

3) You **don't** have to remember what ion **each element** forms — for many of them you just look at the Periodic Table. Elements in the same **group** all have the same number of **outer electrons**. So they have to **lose or gain** the same number to get the full outer shell that they're aiming for. And this means that they form ions with the **same charges**.
4) **Electrostatic attraction** holds positive and negative ions together — it's **very** strong. When atoms are held together like this, it's called **ionic bonding**.

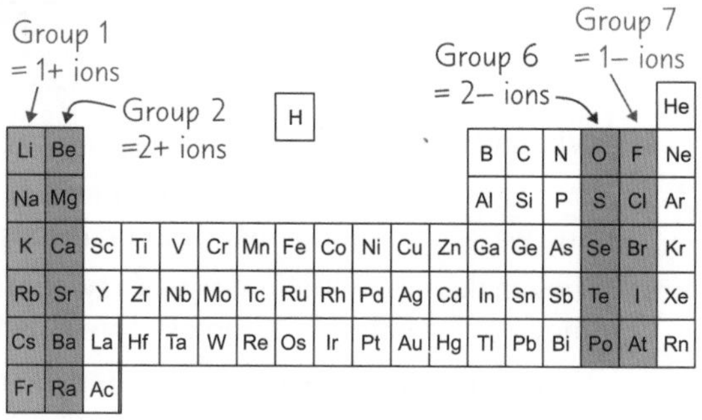

Group 1 = 1+ ions
Group 2 =2+ ions
Group 6 = 2– ions
Group 7 = 1– ions

H																	He
Li	Be											B	C	N	O	F	Ne
Na	Mg											Al	Si	P	S	Cl	Ar
K	Ca	Sc	Ti	V	Cr	Mn	Fe	Co	Ni	Cu	Zn	Ga	Ge	As	Se	Br	Kr
Rb	Sr	Y	Zr	Nb	Mo	Tc	Ru	Rh	Pd	Ag	Cd	In	Sn	Sb	Te	I	Xe
Cs	Ba	La	Hf	Ta	W	Re	Os	Ir	Pt	Au	Hg	Tl	Pb	Bi	Po	At	Rn
Fr	Ra	Ac															

*Not all **Ions** are Made From **Single Atoms***

There are lots of ions that are made up of **groups** of atoms with an **overall charge**. These are called **compound ions**. You need to remember the formulas of these ones:

Sulfate	Hydroxide	Nitrate	Carbonate	Ammonium
SO_4^{2-}	OH^-	NO_3^-	CO_3^{2-}	NH_4^+

*Look at **Charges** to Work Out the **Formula** of an **Ionic Compound***

1) Ionic compounds are made up of a **positively charged** part and a **negatively charged** part.
2) The overall charge of any compound is **zero**. So all the negative charges in the compound must **balance** all the positive charges.
3) You can use the charges on the individual ions present to work out the **formula** of an ionic compound:

Sodium nitrate contains **Na^+ (1+)** and **NO_3^- (1–)** ions. The charges are balanced with one of each ion, so the formula of sodium nitrate is **$NaNO_3$**.

Magnesium chloride contains **Mg^{2+} (2+)** and **Cl^- (1–)** ions.
Because a chloride ion only has a **1–** charge we will need **two** of them to balance out the **2+** charge of a magnesium ion. This gives the formula **$MgCl_2$**.

Ionic Bonding

Sodium Chloride has a **Giant Ionic Lattice** Structure

1) Ionic crystals are giant lattices of ions. A **lattice** is just a **regular structure**.
2) The structure's called '**giant**' because it's made up of the same basic unit repeated over and over again.
3) In **sodium chloride**, the Na^+ and Cl^- ions are packed together. The sodium chloride lattice is **cube** shaped — different ionic compounds have different shaped structures, but they're all still giant lattices.

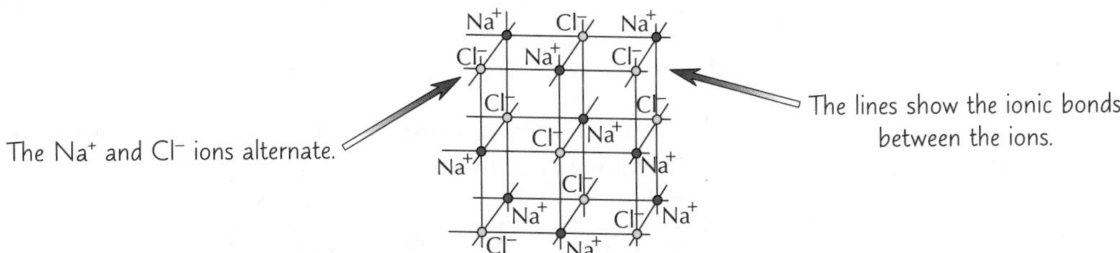

The Na^+ and Cl^- ions alternate.
The lines show the ionic bonds between the ions.
Na^+
Cl^-

The structure of ionic compounds determines their **physical properties**...

Ionic Structure Explains the **Behaviour** of Ionic Compounds

1) **Ionic compounds conduct electricity when they're molten or dissolved — but not when they're solid.**
   The ions in a liquid are free to move (and they carry a charge).
   In a solid the ions are fixed in position by strong ionic bonds.
2) **Ionic compounds have high melting points.**
   Giant ionic lattices are held together by strong electrostatic forces. It takes loads of energy to overcome these forces, so melting points are very high (for example, 801 °C for sodium chloride).
3) **Ionic compounds tend to dissolve in water.**
   Water molecules are polar — part of the molecule has a small negative charge and other bits have small positive charges (see page 30). These charged parts pull ions away from the lattice, causing it to dissolve.

Practice Questions

Q1 What's a compound?

Q2 What type of force holds ionic substances together?

Q3 Sulfur is in group 6 of the periodic table. What will the charge on a sulfide ion be?

Q4 Do ionic compounds tend to dissolve in water? Why?

Exam Questions

Q1 a) Draw a labelled diagram to show the structure of sodium chloride. Your diagram should show at least eight ions. [3 marks]

b) What is the name of this type of structure? [1 mark]

c) Would you expect sodium chloride to have a high or a low melting point? Explain your answer. [3 marks]

Q2 What is the formula of the ionic compound magnesium carbonate? [1 mark]

Q3 Solid lead(II) bromide does not conduct electricity, but molten lead(II) bromide does. Explain this with reference to ionic bonding. [3 marks]

Atom 1 says, "I think I lost an electron." Atom 2 replies, "are you positive?"...

Make sure that you can explain why ionic compounds do what they do. Their properties are down to the fact that ionic crystals are made up of oppositely charged ions attracted to each other. Ionic bonding ONLY happens between a metal and a non-metal. If you've got two non-metals or two metals, they'll do different sorts of bonding — keep reading...

Covalent Bonding

And now for covalent bonding — this is when atoms share electrons with one another so they've all got full outer shells.

Molecules are Groups of Atoms **Bonded** Together

1) Molecules form when **two or more** atoms bond together — it doesn't matter if the atoms are the **same** or **different**. Chlorine gas (Cl_2), carbon monoxide (CO), water (H_2O) and ethanol (C_2H_5OH) are all molecules.
2) Molecules are held together by strong **covalent bonds**.
3) A single covalent bond contains a **shared pair** of electrons.

Covalent bonding happens between non-metals.

In covalent bonding, two atoms **share** electrons, so they've **both** got **full outer shells** of electrons. Both the positive nuclei are attracted **electrostatically** to the shared electrons.

E.g. two iodine atoms bond covalently to form a molecule of iodine (I_2).

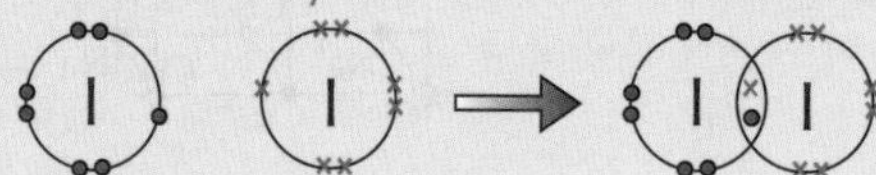

These diagrams don't show all the electrons, just the ones in the outer shells.

4) Here are a few more examples:

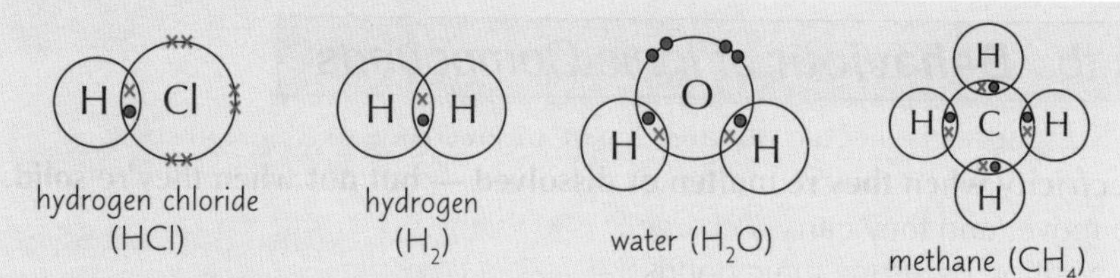

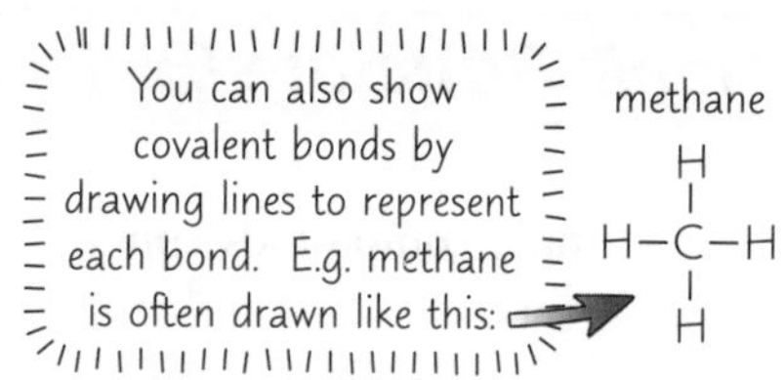

You can also show covalent bonds by drawing lines to represent each bond. E.g. methane is often drawn like this:
methane
H
H—C—H
H

5) The **typical properties** of simple covalent molecules are covered on page 34.

Some Molecules have **Double** or **Triple Bonds**

1) Atoms don't just form single bonds — **double** or even **triple covalent bonds** can be formed between atoms too.
2) These multiple bonds contain **multiple shared pairs** of electrons.

Double bond:

O C O

carbon dioxide (CO_2)

Triple bond:

N N

nitrogen (N_2)

Multiple covalent bonds can be shown using multiple lines, e.g. you can draw N_2 like this: N≡N.

There are **Giant Covalent Structures** Too

1) **Giant covalent** structures have a huge network of **covalently** bonded atoms. (They're sometimes called **macromolecular structures**.)
2) **Carbon** atoms can form this type of structure because they can each form **four** strong, covalent bonds. There are two types of giant covalent carbon structure you need to know about — **graphite** and **diamond**.

Graphite — Sheets of Hexagons with **Delocalised Electrons**

The carbon atoms are arranged in sheets of flat hexagons covalently bonded with three bonds each. The fourth outer electron of each carbon atom is delocalised.

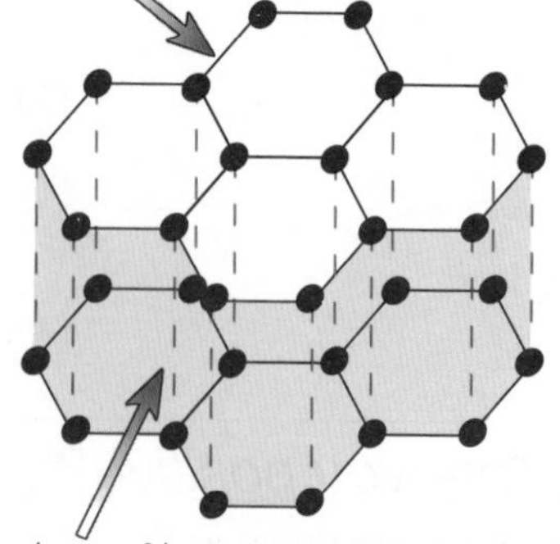

The sheets of hexagons are bonded together by weak van der Waals forces (see page 31).

The **structure** of graphite explains its **properties**:

1) The weak bonds **between** the layers in graphite are easily broken, so the sheets can slide over each other — graphite feels **slippery** and is used as a **dry lubricant** and in **pencils**.
2) The '**delocalised**' electrons in graphite aren't attached to any particular carbon atoms and are **free to move** along the sheets carrying a **charge**. So graphite is an **electrical conductor**.
3) The layers are quite **far apart** compared to the length of the covalent bonds, so graphite has a **low density** and is used to make **strong, lightweight** sports equipment.
4) Because of the **strong covalent bonds** in the hexagon sheets, graphite has a **very high melting point** (it sublimes at over 3900 K).
5) Graphite is **insoluble** in any solvent. The covalent bonds in the sheets are **too strong** to break.

'Sublime' means to change straight from a solid to a gas

Covalent Bonding

Diamond is the Hardest Known Substance

Diamond is also made up of **carbon atoms**. Each carbon atom is **covalently bonded** to **four** other carbon atoms. The atoms arrange themselves in a **tetrahedral** shape.

Because of its **strong covalent** bonds:

1) Diamond has a **very high melting point** — it also sublimes at over 3900 K.
2) Diamond is extremely **hard** — it's used in diamond-tipped drills and saws.
3) **Vibrations** travel easily through the stiff lattice, so it's a **good thermal conductor**.
4) It **can't conduct** electricity — all the outer electrons are held in localised bonds.
5) Like graphite, diamond won't dissolve in **any** solvent.
6) You can 'cut' diamond to form **gemstones**.
   Its structure makes it **refract light** a lot, which is why it sparkles.

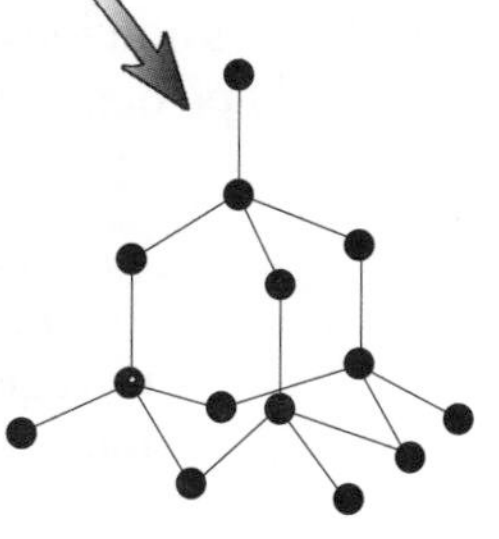

Diamond

Dative Covalent Bonding is where Both Electrons come from One Atom

The **ammonium ion** (NH_4^+) is a classic example of dative covalent (or coordinate) bonding.
It forms when the nitrogen atom in an ammonia molecule **donates a pair of electrons** to a proton (H^+):

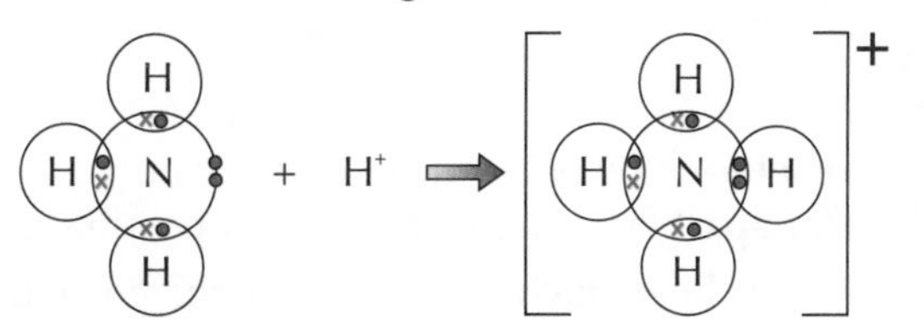

or

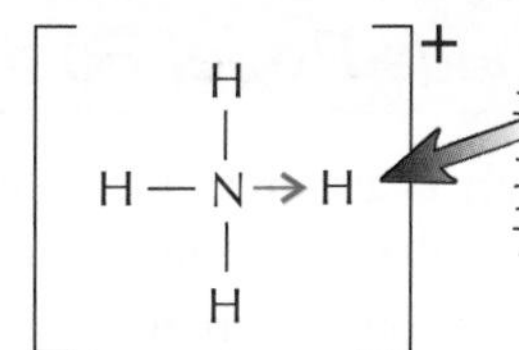

Dative covalent bonding is shown in diagrams by an arrow, pointing away from the 'donor' atom.

Practice Questions

Q1 Describe how atoms are held together in covalent molecules.

Q2 Draw a diagram to show the arrangement of the outer electrons in a molecule of iodine, I_2.

Q3 How are the sheets of carbon atoms in graphite held together?

Q4 In diamond, how many other carbons is each carbon atom bonded to?

Exam Questions

Q1 Ethene, C_2H_4, is an covalently bonded organic compound.
It contains four carbon-hydrogen single bonds and one carbon-carbon double bond.

a) What is a covalent bond? [1 mark]

b) Explain how a single covalent bond differs from a double covalent bond. [1 mark]

Q2 a) What type of bond is formed when an ammonia molecule (NH_3) reacts with a hydrogen ion (H^+)? [1 mark]

b) Explain how this type of bonding occurs. [1 mark]

Q3 Carbon can be found as diamond and as graphite.

a) What type of structure do diamond and graphite display? [1 mark]

b) Draw diagrams to illustrate the structures of diamond and graphite. [2 marks]

c) Compare and explain the electrical conductivities of diamond and graphite in terms of their structure and bonding. [4 marks]

Carbon is a girl's best friend...

OK, so first things first: make sure you've got the hang of what covalent bonding is. You need to know the structures of graphite and diamond too, and why the differences in their structures gives them such different properties. Pretty amazing isn't it — they're both just carbon, but one's crumbly and grey and the other's hard and shiny. And expensive...

Shapes of Molecules

Chemistry would be heaps more simple if all molecules were flat. But they're not.

Molecular Shape depends on Electron Pairs around the Central Atom

Molecules and ions come in loads of **different shapes**.
Their shape depends on the **number of pairs** of electrons in the outer shell of the central atom.

For example, in **ammonia** the outermost shell of the **nitrogen atom** contains **four** pairs of electrons.

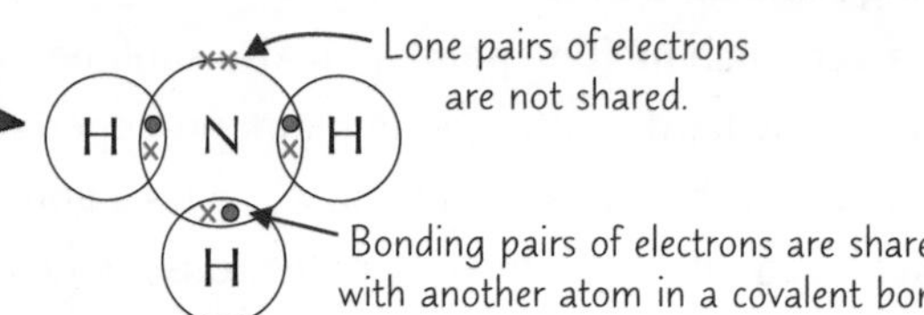

Lone pairs of electrons are not shared.
H
N
H
H
Bonding pairs of electrons are shared with another atom in a covalent bond.

A lone pear

Electron Pairs exist as Charge Clouds

Bonding pairs and lone pairs of electrons exist as **charge clouds**.
A charge cloud is an area where you have a really **big chance** of finding an electron pair. The electrons don't stay still — they **whizz around** inside the charge cloud.

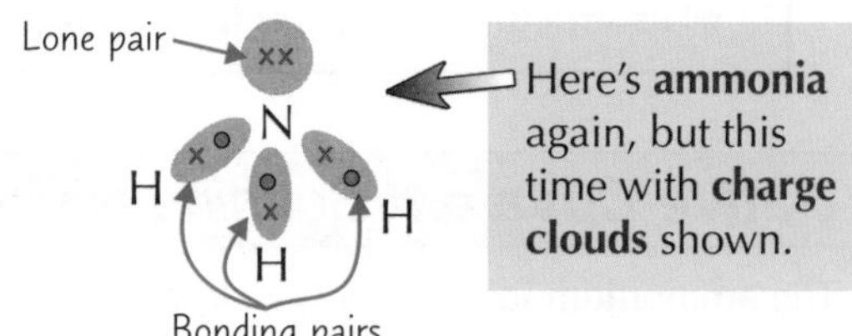

Here's **ammonia** again, but this time with **charge clouds** shown.

Electron Charge Clouds Repel Each Other

1) Electrons are all **negatively charged**, so the charge clouds will **repel** each other as much as they can. So the **pairs of electrons** in the outer shell of an atom will sit as **far apart** from each other as they possibly can.
2) This sounds straightforward, but the **shape** of the charge cloud affects **how much** it repels other charge clouds. Lone-pair charge clouds repel **more** than bonding-pair charge clouds.
3) So, the **greatest** angles are between **lone pairs** of electrons, and bond angles between bonding pairs are often **reduced** because they are pushed together by lone-pair repulsion.

Lone-pair/lone-pair angles are the biggest.	**Lone-pair/bonding-pair angles are the second biggest.**	**Bonding-pair/bonding-pair angles are the smallest.**

The central atoms in these molecules all have **four pairs** of electrons in their outer shells, but they're all **different shapes**:

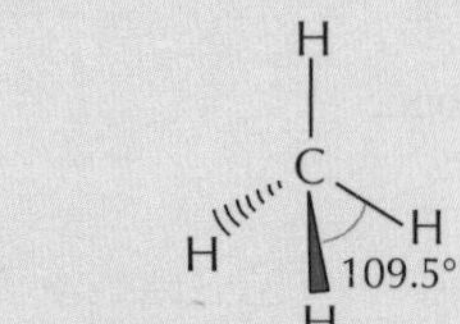

Methane — no lone pairs

The lone pair repels the bonding pairs

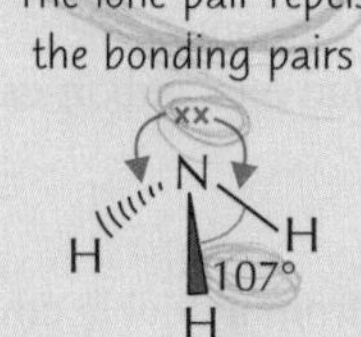

Ammonia — 1 lone pair

2 lone pairs reduce the bond angle even more

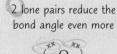

O
H 104.5° H

Water — 2 lone pairs

Wedges (▬) show bonds that are sticking out of the page towards you. Broken lines (⫶⫶⫶) show bonds that go into the page.

4) This is sometimes known by the long-winded name **'Valence-Shell Electron-Pair Repulsion Theory'**.

Use the Number of Electron Pairs to Predict the Shape of a Molecule

To predict the shape of a molecule, you'll need to know how many **bonding** and **lone electron pairs** there are on the central atom of the molecule. Here's how:

1) First work out which one is the **central atom** (that's the one all the other atoms are bonded to).
2) Use the periodic table to work out the **number of electrons** in the **outer shell** of the central atom.
3) **Add one** to this number for every atom that the central atom is **bonded** to.
4) **Divide by 2** to find the number of electron pairs on the central atom.
5) **Compare** the number of **electron pairs** to the number of **bonds** to find the number of lone pairs and the number of bonding pairs on the cental atom.

Now you can use this information to work out the shape of the molecule...

If you're dealing with an ion, you need to take its charge into account too. After step 3, add 1 for each negative charge on the ion (or subtract 1 for each positive charge).

Shapes of Molecules

*Molecules With Different Numbers of **Electron Pairs Have Different Shapes***

Here are the **shapes** that molecules with different numbers of electron pairs will take (and some handy examples):

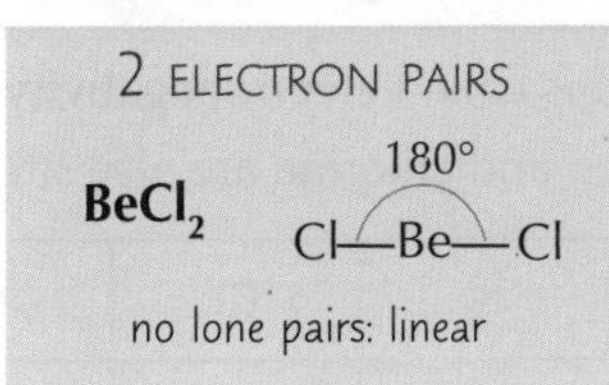

3 ELECTRON PAIRS

BF_3

F, F, B, F, 120°

no lone pairs: trigonal planar

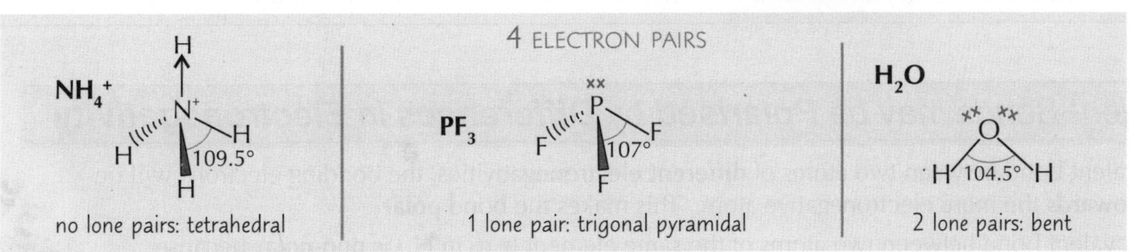

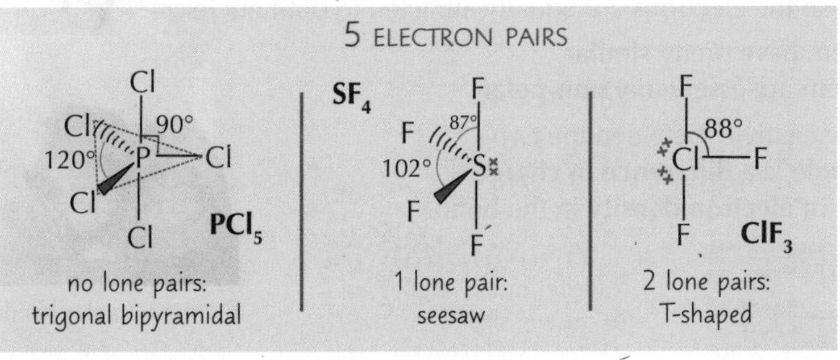

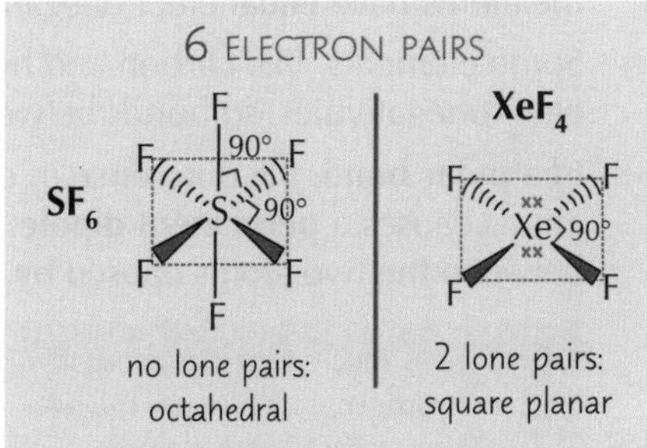

Molecules with 5 electron pairs and 1 lone pair are pretty rare — they have a shape like SF_6, but with the bottom F replaced by the lone pair.

***Example**: Predicting the Shape of the Molecule H_2S*

1) The central atom is **sulfur**.
2) Sulfur is in Group 6, so it has **6 electrons** in its outer shell to start with.
3) The sulfur atom is bonded to 2 hydrogen atoms, so it has (6 + 2) = **8 electrons** in its outer shell in H_2S.
4) The number of electron pairs on the central sulfur atom is 8 ÷ 2 = **4 pairs**.
5) The sulfur atom has **4 electron pairs** and has made **2 bonds** — so it has **2 bonding pairs** and **2 lone pairs**.

This means that H_2S will have a **bent** shape (like water).

S

H H

Practice Questions

Q1 What is a lone pair of electrons?

Q2 Write down the order of the strength of repulsion between different kinds of electron pair.

Q3 Draw an example of a tetrahedral molecule.

Exam Question

Q1 Nitrogen and boron can form the chlorides NCl_3 and BCl_3.

a) Draw the shape of NCl_3. Show the approximate value of the bond angles name the shape. [3 marks]

b) Draw the shape of BCl_3. Show the approximate value of the bond angles name the shape. [3 marks]

c) Explain why the shapes of NCl_3 and BCl_3 are different. [3 marks]

These molecules ain't square...

Don't panic if you get asked to predict the shape of a molecule that you don't know — it'll be just like one you do know (e.g. PH_3 is like NH_3). So learn the shapes on this page — and make sure you remember the bond angles too.

Polarisation and Intermolecular Forces

Intermolecular forces hold molecules together. They're pretty important, cos we'd all be gassy clouds without them. Some of these intermolecular forces are down to polarisation. So you best make sure you know about that first...

Some Atoms **Attract** Bonding Electrons More than Other Atoms

1) An atom's ability to attract the electron pair in a covalent bond is called **electronegativity**.
2) **Fluorine** is the most electronegative element. Oxygen, nitrogen and chlorine are also strongly electronegative.

Element	H	C	N	Cl	O	F
Electronegativity (Pauling Scale)	2.2	2.6	3.0	3.0	3.4	4.0

Covalent Bonds may be **Polarised** by **Differences** in **Electronegativity**

In a covalent bond between two atoms of **different** electronegativities, the bonding electrons will be **pulled towards** the more electronegative atom. This makes the bond **polar**.

1) A covalent bond between two atoms of the same element (e.g. in H_2) is **non-polar** because the atoms have **equal** electronegativities, so the electrons are equally attracted to both nuclei.

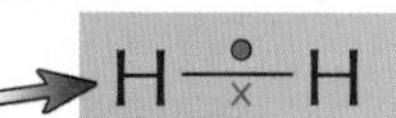

2) Some elements, like carbon and hydrogen, have pretty **similar** electronegativities, so bonds between them are essentially **non-polar**.
3) In a **polar bond**, the difference in electronegativity between the two atoms causes a **permanent dipole**. A dipole is a **difference in charge** between the two atoms caused by a shift in **electron density** in the bond.

Permanent polar bonding

Chlorine is much more electronegative than hydrogen, so hydrogen chloride has a permanent dipole.

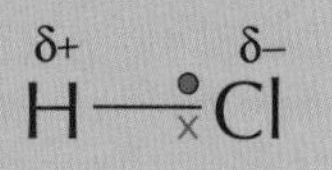

'δ' (delta) means 'slightly', so 'δ+' means 'slightly positive'.

4) The greater the **difference** in electronegativity between the atoms, the **more polar** the bond.

Whole Molecules can be **Polar** Too

1) If you have a molecule that contains polar bonds, you can end up with an uneven distribution of charge across the whole molecule. When this happens, the molecule is **polar**.
2) Not all molecules that contain polar bonds are polar though. If the polar bonds are arranged **symmetrically** in the molecule, then the charges **cancel out** and there is no permanent dipole

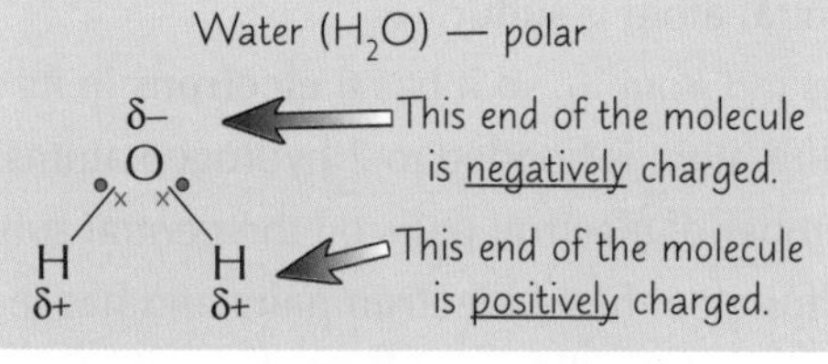

Carbon dioxide (CO_2) — non-polar
δ– δ+ δ–
O=C=O
The positive and negative charges are spread out evenly across the molecule.

Polar Molecules have **Permanent Dipole-Dipole** Forces

In a substance made up of molecules that have **permanent dipoles**, there will be **weak electrostatic forces** of attraction **between** the δ+ and δ– charges on neighbouring molecules.

E.g. hydrogen chloride gas has polar molecules:

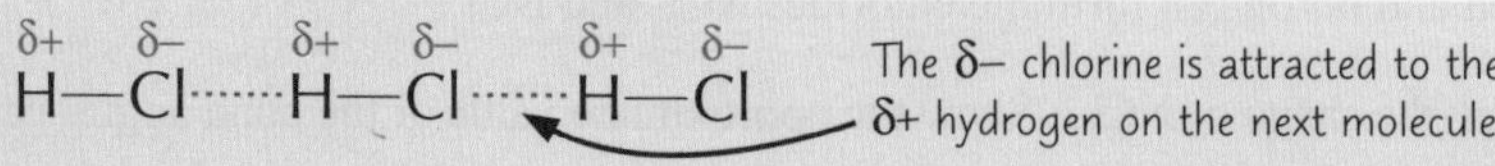

If you put a **charged rod** next to a jet of a polar liquid, like water, the liquid will **move** towards the rod. It's because **polar liquids** contain molecules with **permanent dipoles**. It doesn't matter if the rod is **positively** or **negatively** charged. The polar molecules in the liquid can **turn around** so the oppositely charged end is attracted towards the rod.

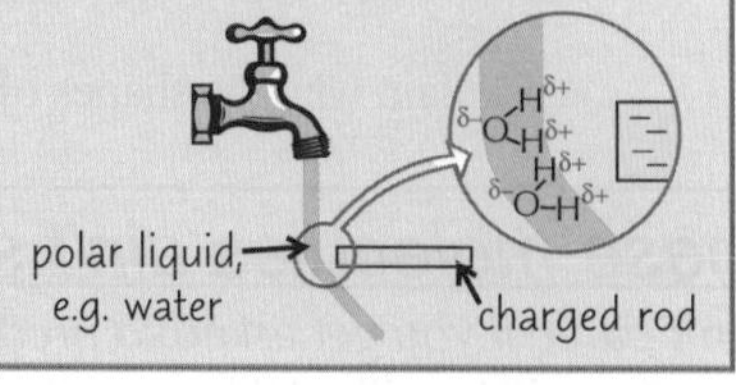

Polarisation and Intermolecular Forces

Intermolecular Forces are **Very Weak**

Intermolecular forces are forces **between** molecules. They're much **weaker** than covalent, ionic or metallic bonds. There are three types you need to know about:

1) **Induced dipole-dipole** or **van der Waals** forces.
2) **Permanent dipole-dipole forces** (these are the ones caused by polar molecules — see previous page).
3) **Hydrogen bonding**.

Intermolecular forces are important because they affect the **physical properties** of a compound.

Van der Waals Forces are Found Between **All** Atoms and Molecules

Van der Waals forces cause **all** atoms and molecules to be **attracted** to each other.

1) **Electrons** in charge clouds are always **moving** really quickly. At any moment, the electrons in an atom are likely to be more to one side than the other. At this moment, the atom would have a **temporary dipole**.
2) This dipole can cause **another** temporary dipole in the opposite direction on a neighbouring atom. The two dipoles are then **attracted** to each other.

3) The second dipole can cause yet another dipole in a **third atom**. It's kind of like a domino effect.
4) Because the electrons are constantly moving, the dipoles are being **created** and **destroyed** all the time. Even though the dipoles keep changing, the **overall effect** is for the atoms to be **attracted** to each another.

Van der Waals Forces Can Hold Molecules in a **Lattice**

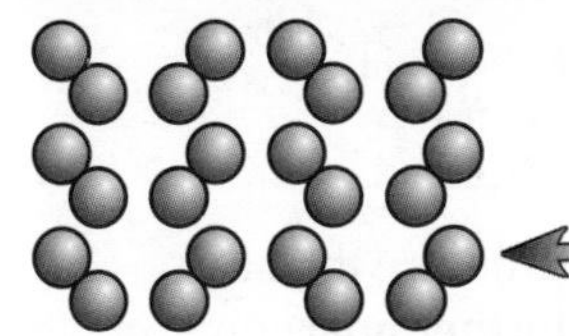

Iodine (**I_2**) is a **solid** at room temperature. It's the **Van der Waals forces** between the iodine molecules that are responsible for holding them together in a **lattice**:

1) Iodine atoms are held together in pairs by **strong** covalent bonds to form I_2 molecules.
2) But the molecules are then held together in a **molecular lattice** arrangement by **weak** van der Waals attractions.

Stronger **Van der Waals Forces** mean **Higher Boiling Points**

1) Not all van der Waals forces are the same strength — larger molecules have **larger electron clouds**, meaning **stronger** van der Waals forces.
2) The **shape** of molecules also affects the strength of Van der Waals forces. Long, straight molecules can lie closer together than branched ones — the **closer** together two molecules get, the **stronger** the forces between them are.
3) When you **boil** a liquid, you need to **overcome** the intermolecular forces, so that the particles can **escape** from the liquid surface. It stands to reason that you need **more energy** to overcome **stronger** intermolecular forces, so liquids with stronger van der Waals forces will have **higher boiling points**.

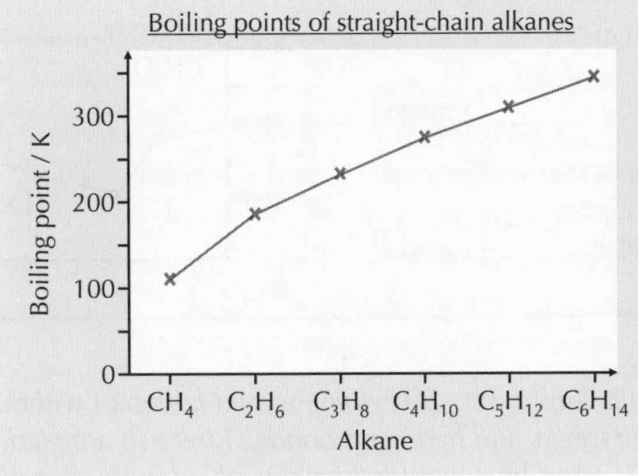

As the alkane chains get **longer**, the **number of electrons** in the molecules increases. This means the van der Waals forces are **stronger**, and so the boiling points **increase**.

Van der Waals forces affect other physical properties too, such as melting point and viscosity.

Polarisation and Intermolecular Forces

***Hydrogen Bonding** is the **Strongest** Intermolecular Force*

1) Hydrogen bonding **only** happens when **hydrogen** is covalently bonded to **fluorine**, **nitrogen** or **oxygen**.
2) Fluorine, nitrogen and oxygen are very **electronegative**, so they draw bonding electrons away from the hydrogen atom. The bond is so **polarised**, and hydrogen has such a **high charge density** (because it's so small), that the hydrogen atoms form weak bonds with **lone pairs of electrons** on the fluorine, nitrogen or oxygen atoms of **other molecules**.
3) Molecules which have hydrogen bonding usually contain **-OH** or **-NH** groups. **Water** and **ammonia** both have hydrogen bonding.

Water:

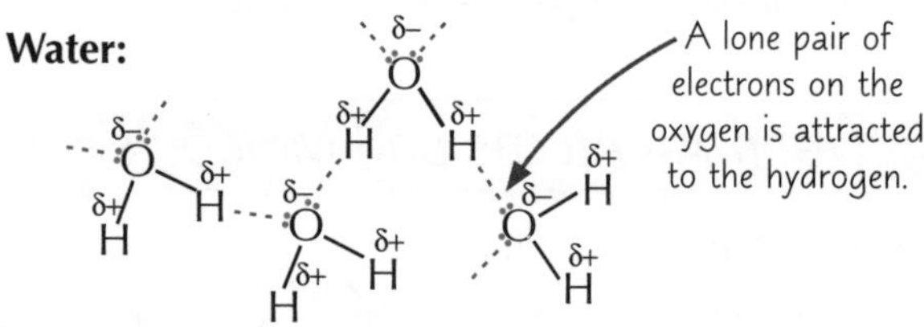

Ammonia:

A lone pair of electrons on the nitrogen is attracted to the hydrogen.

4) Hydrogen bonding has a **huge effect** on the properties of substances:

Substances with hydrogen bonds have **higher boiling** and **melting points** than other similar molecules because of the **extra energy** needed to break the hydrogen bonds.

This is the case with **water** and also **hydrogen fluoride**, which has a **much higher boiling point** than other hydrogen halides:

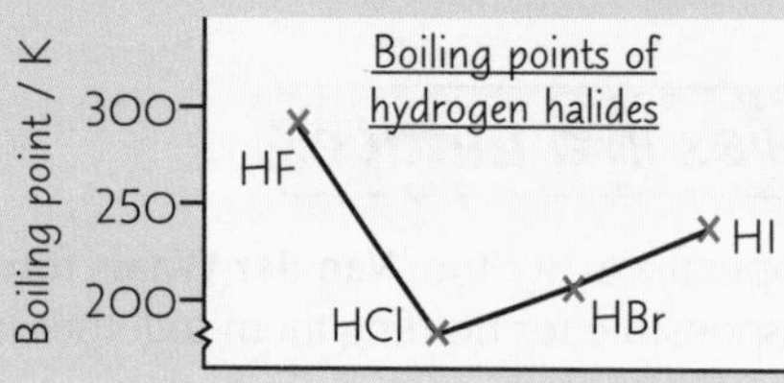

As liquid water cools to form **ice**, the molecules make **more hydrogen bonds** and arrange themselves into a regular **lattice** structure:

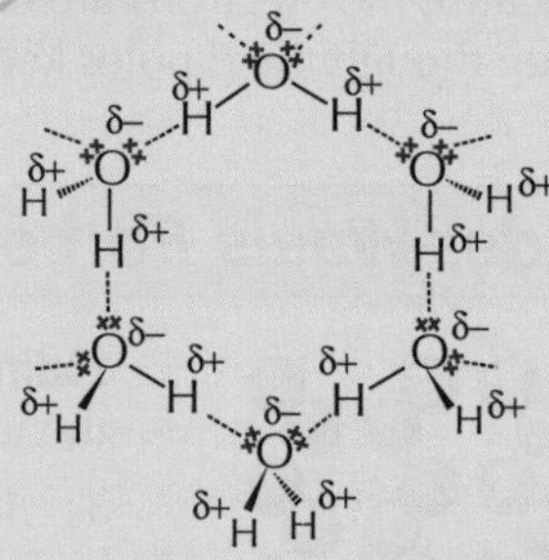

In this regular structure the H_2O molecules are **further apart** on average than the molecules in liquid water — so ice is **less dense** than liquid water.

Practice Questions

Q1 Define the term electronegativity.

Q2 What is a dipole?

Q3 Write down the three types of intermolecular force.

Exam Questions

Q1 Predict whether octene (C_8H_{16}) or decene ($C_{10}H_{20}$) will have a higher boiling point. Explain your answer. [3 marks]

Q2 a) State whether the C–Cl covalent bond is polar. Explain your answer. [1 mark]

b) The molecule CCl_4 has a tetrahedral shape. Will CCl_4 be polar? Explain your answer. [2 marks]

Q3 a) Draw a labelled diagram to show the intermolecular forces that exist between two molecules of water. Include all lone pairs and partial charges in your diagram. [3 marks]

b) The graph on the right shows the boiling points of some of the group 6 hydrides. Explain why water's boiling point is higher than expected in comparison to the other Group 6 hydrides. [3 marks]

Boiling point / K: 400, 300, 200 — H_2O, H_2S, H_2Se, H_2Te

Van der Waal — a German hit for Oasis...

Intermolecular forces are a bit wimpy and weak, but they're really important. Everything would fall apart without them. Learn the three types — van der Waals, permanent dipole-dipole forces and hydrogen bonds. I bet fish are glad that water forms hydrogen bonds. If it didn't, their water would boil. (And they wouldn't have evolved in the first place.)

Metallic Bonding and Properties of Materials

A bit of a mish-mash of a topic now to finish off the section — starting with a bit about bonding in metals.

Metals *have Giant Structures*

Metal elements exist as **giant metallic lattice structures.**

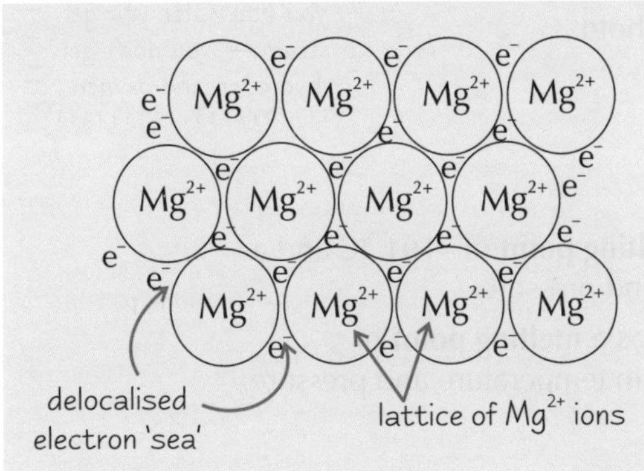

1) The outermost shell of electrons of a metal atom is **delocalised** — the electrons are free to move about the metal. This leaves a **positive metal ion**, e.g. Na^+, Mg^{2+}, Al^{3+}.
2) The positive metal ions are **attracted** to the delocalised negative electrons. They form a lattice of closely packed positive ions in a **sea** of delocalised electrons — this is **metallic bonding**.

Metallic Bonding *Explains the* ***Properties*** *of Metals*

1) Metals have **high melting points** because of the strong **electrostatic attraction** between the positive metal ions and the delocalised sea of electrons.
2) The **number of delocalised electrons per atom** affects the melting point. The **more** there are, the **stronger** the bonding will be and the **higher** the melting point. For example, Mg^{2+} has **two** delocalised electrons per atom, so it's got a **higher melting point** than Na^+, which only has **one**.
3) The delocalised electrons can pass **kinetic energy** to each other, making metals **good thermal conductors**.
4) Metals are **good electrical conductors** because the **delocalised electrons** can move and carry a **current**.
5) Metals are **insoluble** (except in **liquid metals**), because of the **strength** of the metallic bonds.

Right, well that's all you need to know about metals. So, a change of topic now — it's the properties of materials...

The ***Physical Properties*** *of Solids, Liquids and Gases Depend on* ***Particles***

1) A typical **solid** has its particles very **close** together. This gives it a high density and makes it **incompressible**. The particles **vibrate** about a **fixed point** and can't move about freely.
2) A typical **liquid** has a similar density to a solid and is virtually **incompressible**. The particles move about **freely** and **randomly** within the liquid, allowing it to flow.
3) In **gases**, the particles have **loads more** energy and are much **further apart**. So the density is generally pretty low and it's **very compressible**. The particles move about **freely**, with not a lot of attraction between them, so they'll quickly **diffuse** to fill a container.
4) In order to **change** from a solid to a liquid or a liquid to a gas, you need to **break** the forces that are holding the particles together. To do this you need to give the particles more **energy,** e.g. by **heating** them.

Taylor's demonstration of how the particles in a liquid behave had got a bit out of hand.

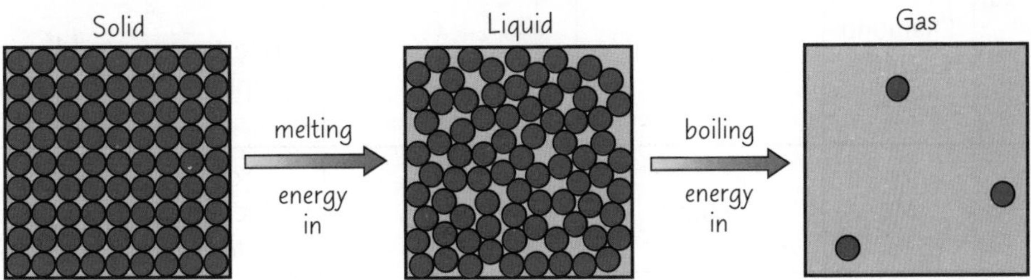

Metallic Bonding and Properties of Materials

Covalent Bonds Don't Break during Melting and Boiling*

**Except for giant covalent substances, like diamond.*

This is something that confuses loads of people — prepare to be enlightened...

1) To **melt** or **boil** a simple covalent compound you only have to overcome the **intermolecular forces** that hold the molecules together.
2) You **don't** need to break the much stronger covalent bonds that hold the atoms together in the molecules.
3) That's why simple covalent compounds have relatively **low melting** and **boiling points**. For example:

It might help to remember that when you boil water, you get steam — you don't get hydrogen and oxygen.

- **Chlorine**, Cl_2, is a **simple covalent** substance. It has a **melting point** of **–101 °C** and a **boiling point** of **–34 °C** — it's a **gas** at room temperature and pressure.
- **Pentane**, C_5H_{12}, is also a **simple covalent** compound. It has a **melting point** of **–130 °C** and a **boiling point** of **36 °C** — it's a **liquid** at room temperature and pressure.

- By contrast, **diamond**, is a **giant covalent** substance, so you **do** have to break the covalent bonds between atoms to turn it into a liquid or a gas. It never really melts, but **sublimes** at over **3600 °C**.

The Physical Properties of a Solid Depend on the Nature of its Particles

Here are a just a few examples of the ways in which the particles that make up a substance affect it properties:

1) The **melting** and **boiling points** of a substance are determined by the strength of the **attraction** between its particles.
2) A substance will only **conduct electricity** if it contains **charged particles** that are **free to move**.
3) How **soluble** a substance is in **water** depends on the **type** of particles that it contains. Water is a **polar solvent**, so substances that are **polar** or **charged** will dissolve in it well, whereas **non-polar** or **uncharged** substances won't.

Learn the Properties of the Main Substance Types

Make sure you know this stuff like the back of your hand:

Bonding	Examples	Melting and boiling points	Typical state at room temperature and pressure	Does solid conduct electricity?	Does liquid conduct electricity?	Is it soluble in water?
Ionic	NaCl $MgCl_2$	High	Solid	No (ions are held in place)	Yes (ions are free to move)	Yes
Simple covalent (molecular)	CO_2 I_2 H_2O	Low (involves breaking intermolecular forces but not covalent bonds)	May be solid (like I_2) but usually liquid or gas.	No	No	Depends on how polarised the molecule is
Giant covalent (macromolecular)	Diamond Graphite SiO_2	High	Solid	No (except graphite)	— (sublimes rather than melting)	No
Metallic	Fe Mg Al	High	Solid	Yes (delocalised electrons)	Yes (delocalised electrons)	No

Metallic Bonding and Properties of Materials

You Can Use the **Properties** of a Material to **Predict its Structure**

You need to be able to predict the type of structure from a list of its properties.
Here's a quick example.

Example: Substance X has a melting point of 1045 K. When solid, it is an insulator, but once melted it conducts electricity. Identify the type of structure present in substance X.

1) Substance X **doesn't** conduct electricity when it's **solid**, but **does** conduct electricity once **melted.** So it looks like it's **ionic** — that would fit with the fact that it has a **high melting point** too.
2) You can also tell that it definitely **isn't simple covalent** because it has a **high melting point**, it definitely **isn't metallic** because it **doesn't** conduct electricity when it's **solid**, and it definitely **isn't giant covalent** because it **does** conduct electricity when **melted.**

So substance X must be **ionic.**

Practice Questions

Q1 Describe the structure of a giant metallic lattice.

Q2 Explain why metals have high melting points.

Q3 Describe how the particles are arranged in a typical solid, a typical liquid and a typical gas.

Q4 In which state will the particles of a substance have the least energy — solid, liquid or gas?

Q5 If a substance has a low melting point, what type of structure is it most likely to have?

Q6 Out of the four main types of structure (ionic, simple covalent, giant covalent and metallic), which will conduct electricity when they are liquids?

Exam Questions

Q1 a) Illustrate the structure of magnesium metal using a labelled diagram. [2 marks]

b) Explain why metals are good conductors of electricity. [1 mark]

Q2 The table below describes the properties of four compounds, A, B, C and D.

Substance	Melting point	Electrical conductivity of solid	Electrical conductivity of liquid	Solubility in water
A	high	poor	good	soluble
B	low	poor	poor	insoluble
C	high	good	good	insoluble
D	very high	poor	— (compound sublimes rather than melting)	insoluble

Identify the type of structure present in each substance. [4 marks]

Q3 Explain why iodine, I_2, has a much lower boiling point than graphite. [4 marks]

I never used to like Chemistry, but after this I feel we've truly bonded...

You need to learn the info in the table on page 34. With a quick glance in my crystal ball, I can almost guarantee you'll need a bit of it in your exam... let me look closer and tell you which bit.... hmm.... No, it's clouded over. You'll have to learn the lot. Sorry. Tell you what — close the book and see how much of the table you can scribble out from memory.

Enthalpy Changes

If you just can't get enough of physical chemistry, there's more. If you can get enough — there's still more. Sorry...

Chemical Reactions Usually Have Enthalpy Changes

When a chemical reaction happens, there is usually a **change in energy**.
The souped-up chemistry term for this is **enthalpy change**:

> **Enthalpy change**, ΔH (delta H), is the heat energy transferred in a reaction at **constant pressure**. The units of ΔH are **kJ mol⁻¹**.

You might see the symbol $^\ominus$ by an enthalpy change (like this: $\Delta H^\ominus$). It's telling you that the substances were in their **standard states** and the measurement was made under **standard conditions**. Standard conditions are **100 kPa pressure** and a stated temperature (e.g. ΔH_{298}). In this book, all standard enthalpy changes are measured at 298 K (25 °C).

Reactions can be either Exothermic or Endothermic

> **Exothermic** reactions **give out** energy. ΔH is **negative**.

In exothermic reactions, the temperature usually goes up.

Oxidation reactions are usually exothermic. Here are two examples:

- The **combustion** of a fuel like methane:
  $CH_{4(g)} + 2O_{2(g)} \rightarrow CO_{2(g)} + 2H_2O_{(l)} \quad \Delta_c H = -890 \text{ kJ mol}^{-1}$ **exothermic**
- The oxidation of **carbohydrates**, such as glucose, $C_6H_{12}O_6$, in respiration.

> **Endothermic** reactions **absorb** energy. ΔH is **positive**.

In endothermic reactions, the temperature usually falls.

- The **thermal decomposition** of calcium carbonate is endothermic.
  $CaCO_{3(s)} \rightarrow CaO_{(s)} + CO_{2(g)} \quad \Delta_c H = +178 \text{ kJ mol}^{-1}$ **endothermic**
- The main reactions of **photosynthesis** are also endothermic — sunlight supplies the energy.

Reactions are all about Breaking and Making Bonds

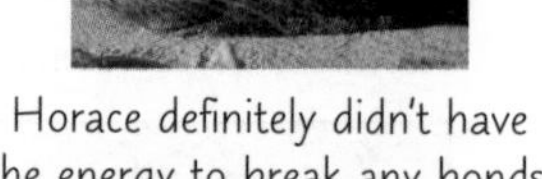
Horace definitely didn't have the energy to break any bonds.

When reactions happen, **reactant bonds** are **broken** and **product bonds** are **formed**.

1) You **need energy** to break bonds, so bond breaking is **endothermic** (ΔH is **positive**). **Stronger** bonds take **more** energy to break.
2) Energy is **released** when bonds are formed, so bond making is **exothermic** (ΔH is **negative**). **Stronger** bonds release **more** energy when they form.
3) The **enthalpy change** for a reaction is the **overall effect** of these two changes. If you need **more** energy to **break** bonds than is released when bonds are made, ΔH is **positive**. If it's less, ΔH is **negative**.

Mean Bond Enthalpies are not Exact

You can look mean bond enthalpies up in data books.

1) **Bond enthalpy** is the energy **required** to **break bonds**.
2) You'd think that every time you broke the **same type of bond** it would require the **same amount of energy**, but annoyingly that's not true — the energy needed to break a bond depends on the environment it's in.
3) In calculations, you use **mean bond enthalpy** — that's the **average energy** needed to break a certain type of bond, over a range of compounds.

> For example, **water** (H_2O) has **two O–H bonds**.
> For the **first** bond, $H\text{–}OH_{(g)}$: $E(\text{H–OH}) = +492$ kJ mol⁻¹
> For the **second** bond, $H\text{–}O_{(g)}$: $E(\text{H–O}) = +428$ kJ mol⁻¹
> **Mean** bond enthalpy = (492 + 428) ÷ 2 = **+460 kJ mol⁻¹**.

The official value for the bond enthalpy for O–H is +463 kJ mol⁻¹. It's a bit different because it's the average for a bigger range of molecules (not just water). For example, it includes the O–H bonds in alcohols and carboxylic acids too.

4) Breaking bonds is **always endothermic**, so mean bond enthalpies are **always positive**.

Enthalpy Changes

Enthalpy Changes Can Be Calculated using **Mean Bond Enthalpies**

In any chemical reaction, energy is **absorbed** to **break bonds** and **given out** during **bond formation**. The difference between the energy absorbed to break bonds and released in making bonds is the overall **enthalpy change of reaction**:

Enthalpy change of reaction = total energy absorbed – total energy released

Example: Calculate the overall enthalpy change for this reaction: $N_{2(g)} + 3H_{2(g)} \rightarrow 2NH_{3(g)}$
Use the mean bond enthalpy values in the table below.

Bonds broken: 1 N≡N bond broken $= 1 \times 945 = 945$ kJ mol^{-1}
3 H–H bonds broken $= 3 \times 436 = 1308$ kJ mol^{-1}
Total Energy Absorbed = 945 + 1308 = **2253 kJ mol^{-1}**
Bonds formed: 6 N–H bonds formed $= 6 \times 391 = 2346$ kJ mol^{-1}
Total Energy Released = **2346 kJ mol^{-1}**
Enthalpy change of reaction = total energy absorbed – total energy released = 2253 – 2346 = **–93 kJ mol^{-1}**

Bond	Mean Bond Enthalpy
N≡N	945 kJ mol^{-1}
H–H	436 kJ mol^{-1}
N–H	391 kJ mol^{-1}

Because they use **average** values, enthalpy changes calculated using mean bond enthalpies **aren't exact** — they are slightly less accurate than enthalpy change value calculated using **Hess's Law** (see page 40).

There are **Different Types** of ΔH

1) **Standard enthalpy of formation**, $\Delta_f H^\ominus$, is the enthalpy change when **1 mole** of a **compound** is formed from its **elements** in their standard states under standard conditions, e.g. $2C_{(s)} + 3H_{2(g)} + \frac{1}{2}O_{2(g)} \rightarrow C_2H_5OH_{(l)}$.
2) **Standard enthalpy of combustion**, $\Delta_c H^\ominus$, is the enthalpy change when **1 mole** of a substance is **completely burned in oxygen** under standard conditions, e.g. $C_2H_{4(g)} + 3O_{2(g)} \rightarrow 2CO_{2(g)} + 2H_2O_{(l)}$.

Practice Questions

Q1 Explain the terms 'exothermic reaction' and 'endothermic reaction'.

Q2 Define 'mean bond enthalpy'.

Q3 Define 'standard enthalpy of formation' and 'standard enthalpy of combustion'.

Exam Questions

Q1 The table on the right shows some mean bond enthalpy values.

Bond	C–H	C=O	O=O	O–H
Mean bond enthalpy (kJ mol^{-1})	435	805	498	464

The complete combustion of methane can be represented by this equation: $CH_{4(g)} + 2O_{2(g)} \rightarrow CO_{2(g)} + 2H_2O_{(l)}$

a) Use the table of mean bond enthalpies to calculate the enthalpy change for this reaction. [3 marks]

b) Is the reaction endothermic or exothermic? Explain your answer. [1 mark]

Q2 Methanol, CH_3OH, when blended with petrol, can be used as a fuel. $\Delta_c H^\ominus [CH_3OH] = -726$ kJ mol^{-1}

a) Write an equation, including state symbols, for the standard enthalpy of combustion of methanol. [1 mark]

b) Write an equation, including state symbols, for the standard enthalpy of formation of methanol. [1 mark]

Q3 Petroleum gas is a fuel that contains propane, C_3H_8.
Explain why the following equation does not represent a standard enthalpy of combustion:

$$2C_3H_{8(g)} + 10O_{2(g)} \rightarrow 6CO_{2(g)} + 8H_2O_{(g)} \qquad \Delta H = -4113 \text{ kJ mol}^{-1}$$

[1 mark]

No Mr bond, I expect you to break endothermically...

So, there you go, breaking bonds needs energy, but making bonds gives out energy. To be honest, all this energy stuff has tired me out. No escaping it though, so just keep ploughing on, brave chemistry warrior. It's worth learning those sneaky ΔH definitions at the end too — I can't help but think that they'll be cropping up again later in the section...

Calorimetry

You can find some enthalpy changes by doing an experiment and then a calculation...

You can find **Enthalpy Changes** *using* **Calorimetry**

1) You use **calorimetry** to find out how much heat is given out by a reaction by measuring a **temperature change**.
2) To find the enthalpy of **combustion** of a **flammable liquid**, you burn it inside some apparatus like this (called a **calorimeter**):
3) As the fuel burns, it heats the water. You can work out the **heat energy** that has been **absorbed** by the water if you know the **mass of the water**, the **temperature change** (ΔT), and the **specific heat capacity of water** (= 4.18 J g^{-1} K^{-1}) — see the next page for the details of how to do this.
4) Ideally, all the heat given out by the fuel as it burns would be **absorbed** by the water — allowing you to work out the enthalpy change of combustion exactly.
5) But in **any** calorimetry experiment, you **always** lose heat **to the surroundings** (however well your calorimeter is **insulated**). This makes it hard to get an **accurate result**.

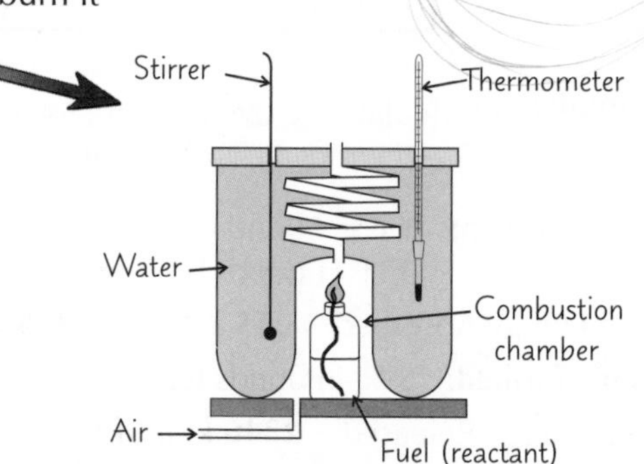

6) Also, when you burn a fuel, some of the combustion may be **incomplete** (meaning **less energy** is given out). Flammable liquids are often quite **volatile** too, so you may lose some fuel to evaporation.

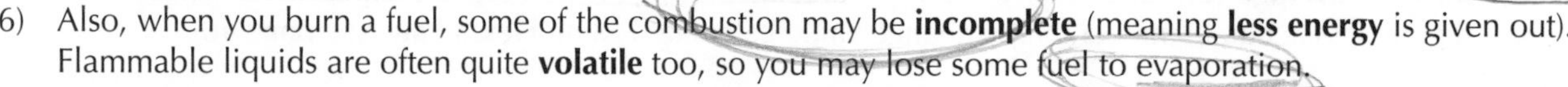

You can also use **Calorimetry** *to* **Measure Enthalpy Changes** *in* **Solution**

1) Calorimetry can also be used to calculate the enthalpy change for reactions that happen **in solution** too, such as **neutralisation**, **dissolution** (**dissolving**) or **displacement**. For example:

> 1) To find the enthalpy change for a **neutralisation reaction**, add a **known volume** of acid to an **insulated container** (e.g. a polystyrene cup) and measure the **temperature**.
> 2) Then add a **known volume** of alkali and record the **temperature** of the mixture at regular intervals over a period of time. (Stir the solution to make sure it's evenly heated.)
> 3) Find the **temperature change** for the experiment. Use it to calculate the **enthalpy change** of the reaction (using the formula on the next page).

Or a known <u>mass</u> if one of the reactants is a solid.

2) You'll need to know the **mass** of the solutions that you've used in order to calculate the enthalpy change of the reaction. You can assume that all solutions have the **same density as water**. Since 1 cm^3 of water has a mass of 1 g, if you have e.g. 50 cm^3 of solution you can assume that it has a mass of 50 g.

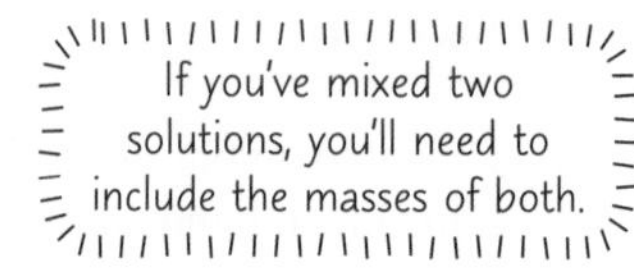

If you've mixed two solutions, you'll need to include the masses of both.

3) If you're trying to find the energy change **per mole** of reactant, you might need the formula **moles = concentration (mol dm^{-3}) × volume (dm^3)** to find the number of moles of a substance in a solution.

You Can Use a **Graph** *to Find an* **Accurate Temperature Change**

1) The most obvious way of finding the **temperature change** in a calorimetry experiment is to subtract the **starting temperature** from the **highest temperature** you recorded. But that **won't** give you a very **accurate** value (because of the heat lost to the surroundings).
2) You can use a **graph** of your results to find an **accurate value**.
3) During the experiment, record the temperature at regular intervals, beginning a couple of minutes **before** you start the reaction.
4) Plot a **graph** of your results. Draw two **lines of best fit**: one through the points **before** the reaction started and one through the points **after** it started.
5) Extend both lines so they **both** pass the time when the reaction started.
6) The **distance between the two lines** at the time the reaction started (before any heat was lost) is the accurate temperature change (**ΔT**) for the reaction.

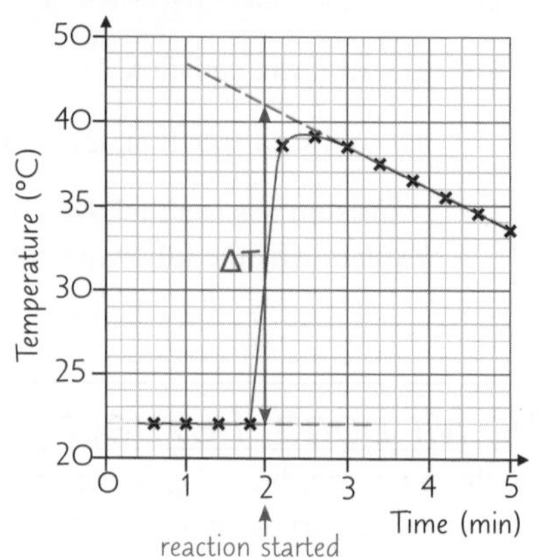

Calorimetry

Calculate **Enthalpy Changes** Using the **Equation q = mcΔT**

Here's the snazzy formula that you need to calculate an **enthalpy change** from a **calorimetry experiment**:

$$q = mc\Delta T$$

where, q = heat lost or gained (in joules). This is the same as the enthalpy change if the pressure is constant.

m = mass of water (or other solution) in the calorimeter (in grams)

c = specific heat capacity of water (4.18 J $g^{-1}K^{-1}$)

ΔT = the change in temperature (in Kelvin) of the water or solution

The specific heat capacity of water is the amount of heat energy it takes to raise the temperature of 1 g of water by 1 K.

You do need to learn the **units** for everything in this formula. (But it's worth knowing that though the official unit for ΔT is **K**, the value is actually the same in **°C** — so if ΔT = 5 °C, then ΔT = 5 K.)

Example: In a laboratory experiment, 1.16 g of an organic liquid fuel were completely burned in oxygen. The heat formed during this combustion raised the temperature of 100 g of water from 295 K to 358 K. Calculate the standard enthalpy of combustion, $\Delta_c H^\circ$, of the fuel. Its M_r is 58.0.

1) First off, you need to calculate the **amount of heat** given out by the fuel using the formula $\boldsymbol{q = mc\Delta T}$:
   $q = 100 \times 4.18 \times (358 - 295) = 26\,334$ J

   Remember — m is the mass of the water, <u>not</u> the mass of fuel.

2) Standard enthalpies of combustion are always given in units of **kJ mol⁻¹**.
   So the next thing to do is to change the **units of *q*** from **joules** to **kilojoules**: (26 334 J ÷ 1000) = 26.334 kJ
3) The standard enthalpy of combustion is the energy produced by burning **1 mole** of fuel. So next you need to find out **how many moles** of fuel produced this much energy, using the old 'number of moles = mass ÷ M_r' formula.
   number of moles of fuel = $\frac{1.16}{58.0} = 0.0200$
4) So the heat produced by 1 mole of fuel ($\Delta_c H^\circ$) = $\frac{-26.334}{0.0200} = -1316.7$ kJ mol^{-1}
   = **−1320 kJ mol⁻¹** (3 s.f.).

You need to add a minus sign here because the reaction's exothermic — you know this because it raised the temperature of the water.

Practice Questions

Q1 Briefly describe an experiment that you could do to find the standard enthalpy change of a combustion reaction.

Q2 Briefly describe an experiment that you could do to find the enthalpy change of a neutralisation reaction.

Q3 Give the formula that you would use to calculate heat change from the results of a calorimetry experiment.

Exam Questions

Q1 The initial temperature of 25.0 cm^3 of 1.00 mol dm^{-3} hydrochloric acid in a polystyrene cup was measured as 19.0 °C. This acid was exactly neutralised by 25 cm^3 of 1.00 mol dm^{-3} sodium hydroxide solution. The maximum temperature of the resulting solution was measured as 25.5 °C.

Calculate the molar enthalpy of neutralisation for the hydrochloric acid.
You may assume the neutral solution formed has a specific heat capacity of 4.18 J K^{-1} g^{-1}. [5 marks]

Q2 A 50.0 cm^3 sample of 0.200 M copper(II) sulfate solution was measured out into a polystyrene beaker. A temperature increase of 2.60 K was recorded when excess zinc powder was stirred in.
The equation for this reaction is: $Zn_{(s)} + CuSO_{4(aq)} \rightarrow Cu_{(s)} + ZnSO_{4(aq)}$

Calculate the enthalpy change when 1 mole of zinc reacts. Assume the specific heat capacity of the solution is 4.18 J $g^{-1}K^{-1}$. Ignore the increase in volume of the solution due to the zinc. [4 marks]

Having trouble with calorimetry? I can enthalpise...

These calculations look a bit nasty, but really they just come down to learning the formula off by heart and getting some practice at using it. Remember to learn all of the units that go with the formula — I know there's a lot of them, but they're really important. There's no point coming out with the answer 8.1 if you don't know what you've got 8.1 of...

Hess's Law

Sometimes you can't work out an enthalpy change by measuring a single temperature change. But there's still a way. You can work it out from the enthalpies of formation or combustion without the need for messy experiments.

*Hess's Law — the Total Enthalpy Change is **Independent** of the Route Taken*

Hess's Law says that:

The **total enthalpy change** of a reaction is **independent** of the **route** taken.

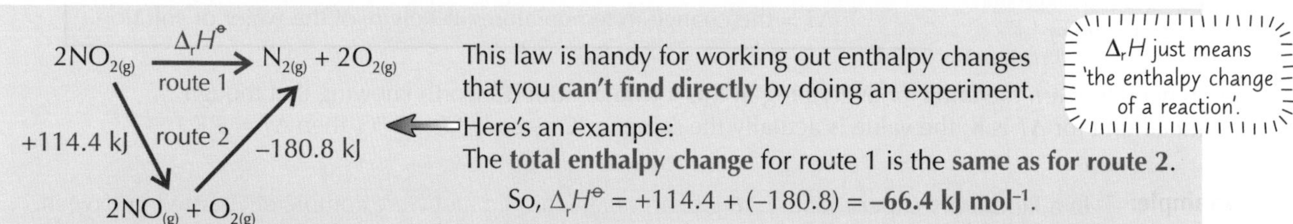

This law is handy for working out enthalpy changes that you **can't find directly** by doing an experiment.

Here's an example:

The **total enthalpy change** for route 1 is the **same as for route 2**.

So, $\Delta_r H^\circ = +114.4 + (-180.8) = \mathbf{-66.4\ kJ\ mol^{-1}}$.

$\Delta_r H$ just means 'the enthalpy change of a reaction'.

*Enthalpy Changes Can be Worked Out From **Enthalpies of Formation***

1) **Enthalpy changes of formation** are useful for calculating enthalpy changes you can't find directly.
2) To do this, you'll need to know $\Delta_f H^\circ$ for **all** the reactants and products that are **compounds**. (Don't panic though — you'll be given the information you need in exam questions.)
3) The value of $\Delta_f H^\circ$ for all **elements** is zero — the element's being 'formed' from the element, so there's no change.

Here's how to calculate $\Delta_r H^\circ$ for this reaction: $SO_{2(g)} + 2H_2S_{(g)} \rightarrow 3S_{(s)} + 2H_2O_{(l)}$

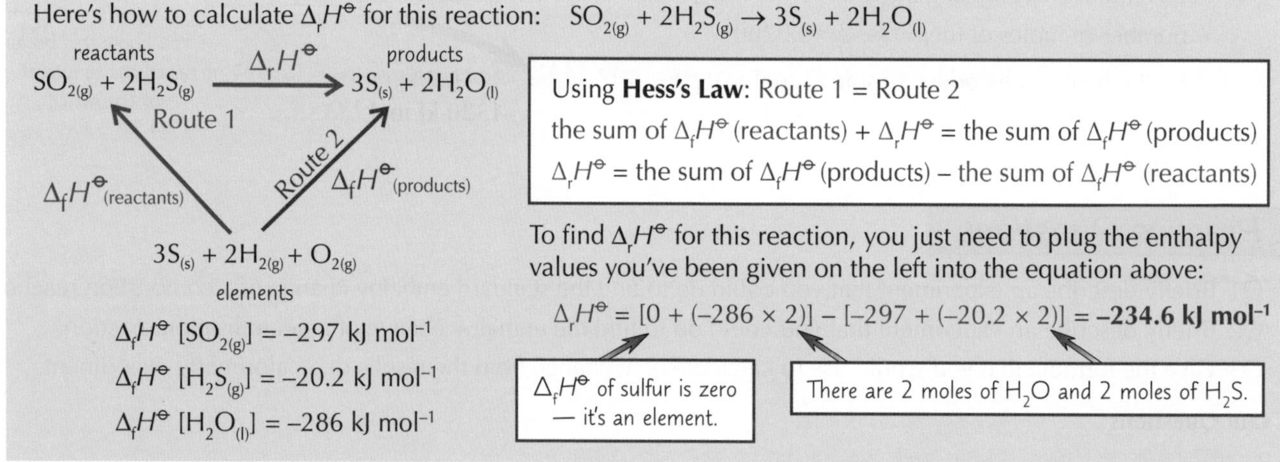

$\Delta_f H^\circ [SO_{2(g)}] = -297\ kJ\ mol^{-1}$

$\Delta_f H^\circ [H_2S_{(g)}] = -20.2\ kJ\ mol^{-1}$

$\Delta_f H^\circ [H_2O_{(l)}] = -286\ kJ\ mol^{-1}$

Using **Hess's Law**: Route 1 = Route 2

the sum of $\Delta_f H^\circ$ (reactants) + $\Delta_r H^\circ$ = the sum of $\Delta_f H^\circ$ (products)

$\Delta_r H^\circ$ = the sum of $\Delta_f H^\circ$ (products) – the sum of $\Delta_f H^\circ$ (reactants)

To find $\Delta_r H^\circ$ for this reaction, you just need to plug the enthalpy values you've been given on the left into the equation above:

$\Delta_r H^\circ = [0 + (-286 \times 2)] - [-297 + (-20.2 \times 2)] = \mathbf{-234.6\ kJ\ mol^{-1}}$

$\Delta_f H^\circ$ of sulfur is zero — it's an element.

There are 2 moles of H_2O and 2 moles of H_2S.

It **always** works, no matter how complicated the reaction:

Here's how to calculate $\Delta_r H^\circ$ for this reaction: $2NH_4NO_{3(s)} + C_{(s)} \rightarrow 2N_{2(g)} + CO_{2(g)} + 4H_2O_{(l)}$

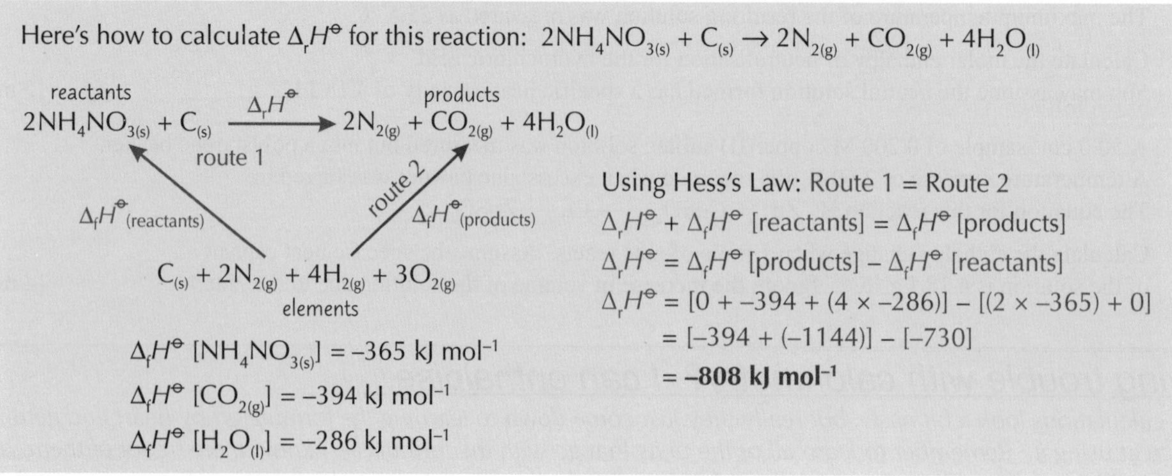

$\Delta_f H^\circ [NH_4NO_{3(s)}] = -365\ kJ\ mol^{-1}$

$\Delta_f H^\circ [CO_{2(g)}] = -394\ kJ\ mol^{-1}$

$\Delta_f H^\circ [H_2O_{(l)}] = -286\ kJ\ mol^{-1}$

Using Hess's Law: Route 1 = Route 2

$\Delta_r H^\circ + \Delta_f H^\circ$ [reactants] = $\Delta_f H^\circ$ [products]

$\Delta_r H^\circ = \Delta_f H^\circ$ [products] – $\Delta_f H^\circ$ [reactants]

$\Delta_r H^\circ = [0 + -394 + (4 \times -286)] - [(2 \times -365) + 0]$

$= [-394 + (-1144)] - [-730]$

$= \mathbf{-808\ kJ\ mol^{-1}}$

Hess's Law

Enthalpy Changes Can be Worked Out From **Enthalpies of Combustion**

You can use a similar method to find an enthalpy change from **enthalpy changes of combustion**.

Here's how to use enthalpy changes of combustion to calculate $\Delta_f H^\ominus$ of **ethanol** (C_2H_5OH):

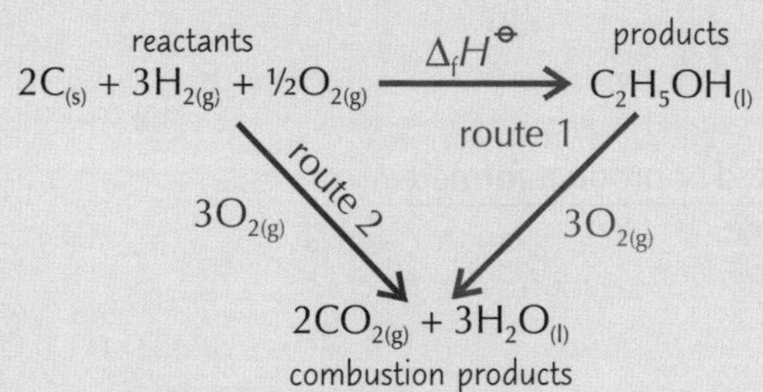

Using Hess's Law: Route 1 = Route 2

$\Delta_f H^\ominus [C_2H_5OH] + \Delta_c H^\ominus [C_2H_5OH] = 2\Delta_c H^\ominus [C] + 3\Delta_c H^\ominus [H_2]$

$\Delta_f H^\ominus [C_2H_5OH] + (-1367) = (2 \times -394) + (3 \times -286)$

$\Delta_f H^\ominus [C_2H_5OH] = -788 + -858 - (-1367)$

$= \mathbf{-279\ kJ\ mol^{-1}}$

$\Delta_c H^\ominus [C_{(s)}] = -394\ kJ\ mol^{-1}$

$\Delta_c H^\ominus [H_{2(g)}] = -286\ kJ\ mol^{-1}$

$\Delta_c H^\ominus [C_2H_5OH_{(l)}] = -1367\ kJ\ mol^{-1}$

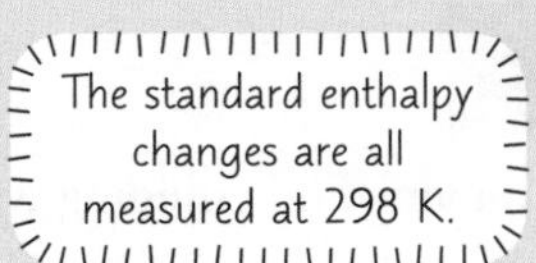

The Loch Hess Monster: Brilliant at chemistry but very, very shy.

If you ever need to go along an arrow **backwards** in a Hess's Law diagram, you **subtract** the enthalpy change that goes with that arrow. For example, you could find the $\Delta_f H^\ominus$ for ethanol using these routes instead:

This time, route 1 is just $\Delta_f H^\ominus [C_2H_5OH]$, while route 2 involves going **forwards** along the step 1 blue arrow and **backwards** along the step 2 blue arrow.

$\Delta_f H^\ominus [C_2H_5OH] = 2\Delta_c H^\ominus [C] + 3\Delta_c H^\ominus [H_2] - \Delta_c H^\ominus [C_2H_5OH]$

$= (2 \times -394) + (3 \times -286) - (-1367) = \mathbf{-279\ kJ\ mol^{-1}}$

$2C_{(s)} + 3H_{2(g)} + \frac{1}{2}O_{2(g)} \xrightarrow{\Delta_f H^\ominus} C_2H_5OH_{(l)}$ route 1

$3O_{2(g)}$ step ① route 2 $3O_{2(g)}$ step ②

$2CO_{2(g)} + 3H_2O_{(l)}$

$\Delta_c H^\ominus [C_{(s)}] = -394\ kJ\ mol^{-1}$

$\Delta_c H^\ominus [H_{2(g)}] = -286\ kJ\ mol^{-1}$

$\Delta_c H^\ominus [C_2H_5OH_{(l)}] = -1367\ kJ\ mol^{-1}$

Practice Questions

Q1 What does Hess's Law state?

Q2 State the value of the standard enthalpy change of formation of any element.

Q3 Use the diagram on the right to find the enthalpy change of formation of methane*.

$C_{(s)} + 2H_{2(g)} \xrightarrow{\Delta_f H^\ominus} CH_{4(g)}$ route 1

$+2O_{2(g)}$ $\quad -2O_{2(g)}$

$\Delta_c H^\ominus(C) + 2\Delta_f H^\ominus(H_2O) = -965\ kJ$ route 2 $-\Delta_c H^\ominus(CH_4) = +890\ kJ$

$CO_{2(g)} + 2H_2O_{(l)}$

*Answer on page 215.

Exam Questions

Q1 Using the facts that (at 298 K) the standard enthalpy change of formation of $Al_2O_{3(s)}$ is $-1676\ kJ\ mol^{-1}$ and the standard enthalpy change of formation of $MgO_{(s)}$ is $-602\ kJ\ mol^{-1}$, calculate the enthalpy change of the following reaction.

$$Al_2O_{3(s)} + 3Mg_{(s)} \rightarrow 2Al_{(s)} + 3MgO_{(s)}$$ [2 marks]

Q2 Calculate the enthalpy change for the hydrogenation of ethene at 298 K: $C_2H_{4(g)} + H_{2(g)} \rightarrow C_2H_{6(g)}$

Use the following standard enthalpies of combustion in your calculations:

$\Delta_c H^\ominus(C_2H_4) = -1400\ kJ\ mol^{-1}$ $\quad \Delta_c H^\ominus(C_2H_6) = -1560\ kJ\ mol^{-1}$ $\quad \Delta_c H^\ominus(H_2) = -286\ kJ\ mol^{-1}$ [2 marks]

The law is an ass — not in this case though...obviously...

To get your head around those Hess diagrams, you're going to have to do more than skim them. You need to be able to use this stuff for any reaction they give you. It'll also help if you know the definitions for those standard enthalpy thingumabobs on page 37. If you didn't bother learning them, have a quick flick back and remind yourself about them.

Reaction Rates

The rate of a reaction is just how quickly it happens. Lots of things can make it go faster or slower.

Reaction Rate is the Amount of Stuff Reacting Divided by Time

1) **Reaction rate** is defined as the **change** in **concentration** (or **amount**) of a reactant or product over **time**.
2) A simple formula for finding the rate of a chemical reaction is:

$$\text{rate of reaction} = \frac{\text{amount of reactant used or product formed}}{\text{time}}$$

Particles Must Collide to React

1) Particles in liquids and gases are **always moving** and **colliding** with **each other**.
2) They **don't** react every time they collide though — only when the **conditions** are right. A reaction **won't** take place between two particles **unless**:

- They collide in the **right direction** — they need to be **facing** each other in the right way.
- They collide with at least a certain **minimum** amount of kinetic (movement) **energy**.

3) This stuff's called **collision theory**.

Particles Must have Enough Energy for a Reaction to Happen

The **minimum amount of kinetic energy** that particles need to react is called the **activation energy**.

The particles must have at least this much energy to **break their bonds** and start the reaction.
To make this a bit clearer, here's an **enthalpy profile diagram**:

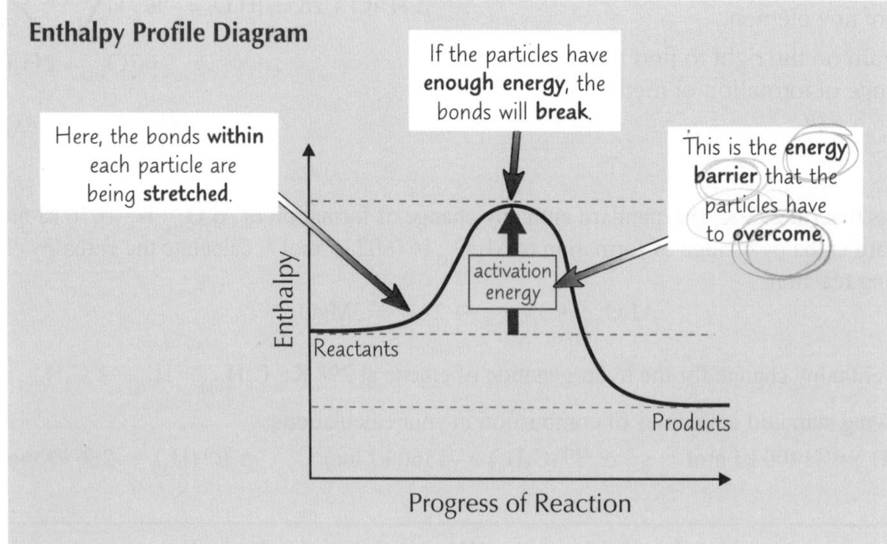

Reactions with **low activation energies** often happen **pretty easily**. But reactions with **high activation energies** don't. You need to give the particles extra energy by **heating** them.

Reaction Rates

*Molecules in a Gas **Don't** all have the **Same Amount of Energy***

1) Imagine looking down on Oxford Street when it's teeming with people. You'll see some people ambling along **slowly**, some hurrying **quickly**, but most of them will be walking with a **moderate speed**.
2) It's the same with the **molecules** in a **gas**. Some **don't have much kinetic energy** and move **slowly**. Others have **loads** of **kinetic energy** and **whizz** along. But most molecules are somewhere **in between**.
3) If you plot a **graph** of the **numbers of molecules** in a **gas** with different **kinetic energies** you get a **Maxwell-Boltzmann distribution**. They look like this:

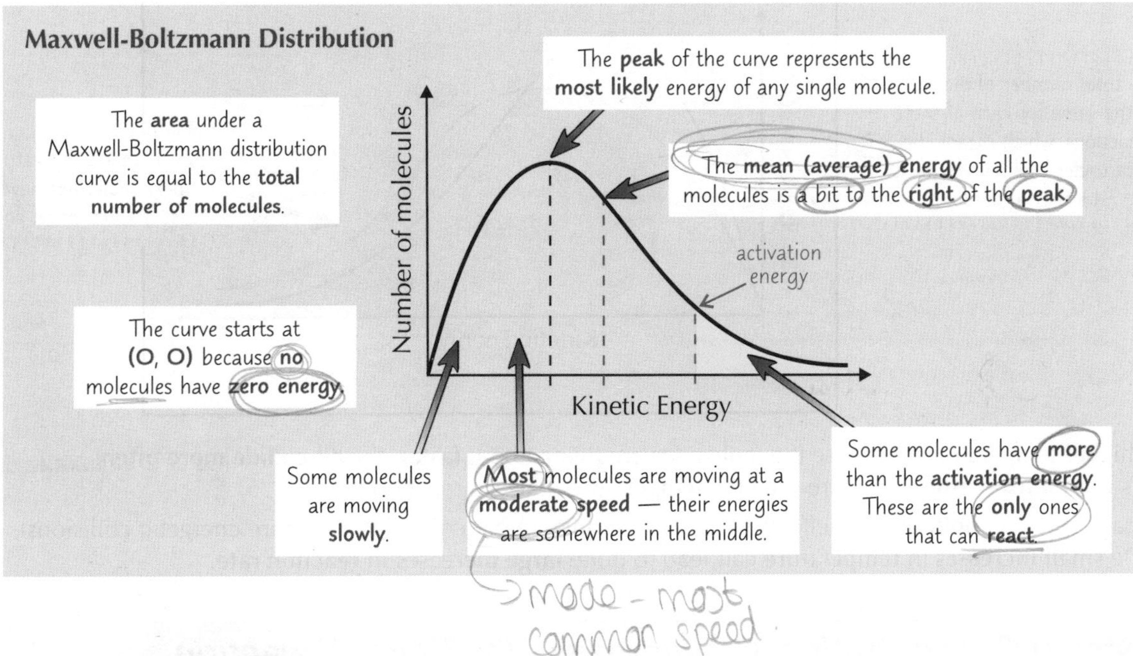

Practice Questions

Q1 Explain what is meant by the term reaction rate.

Q2 Define the term 'activation energy'.

Q3 What does the area under a Maxwell-Boltzmann distribution curve represent?

Exam Questions

Q1 Nitrogen oxide (NO) and ozone (O_3) react to produce nitrogen dioxide (NO_2) and oxygen (O_2). Collisions between the two molecules do not always lead to a reaction, even if the molecules are orientated correctly. Explain why this is. [1 mark]

Q2 The Maxwell-Boltzmann distribution curve on the right shows the distribution of molecular energies in a sample of a gas at 25 °C.

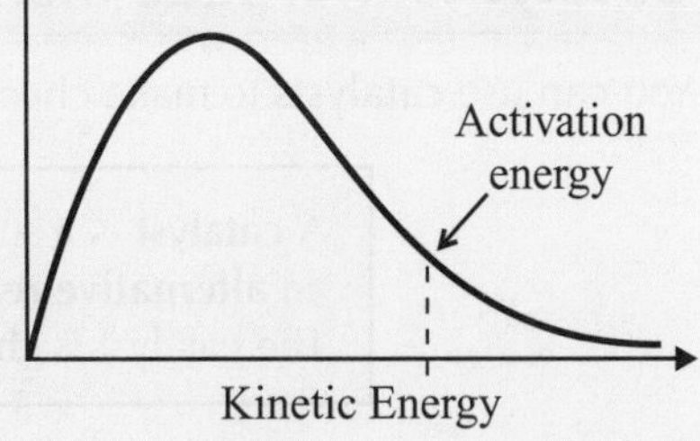

a) Label the *y*-axis. [1 mark]

b) Shade the area of the curve representing molecules with energy greater than the activation energy. [1 mark]

c) Draw a cross on the *x*-axis to mark the position of the most likely energy of a molecule of the gas. [1 mark]

Chemical reactions and waking up — both require activation energy...

Collision theory and activation energy make a lot of sense when you think about them. Particles have to both meet in the first place and have enough energy to react. Make sure you've got the hang of what Maxwell-Boltzmann distributions look like, and where to find the mean and most likely energy — they're easy marks if this topic crops up in the exam...

More on Reaction Rates

Knowing about reaction rates is all well and good, but how do you get a reaction to speed up or slow down? Read on...

Increasing** the **Temperature** makes Reactions **Faster

1) If you increase the **temperature** of a reaction, the particles will on average have more **kinetic energy** and will move **faster**.
2) So, a **greater proportion** of molecules will have at least the **activation energy** and be able to **react**. This changes the **shape** of the **Maxwell-Boltzmann distribution curve** — it pushes it over to the **right**.

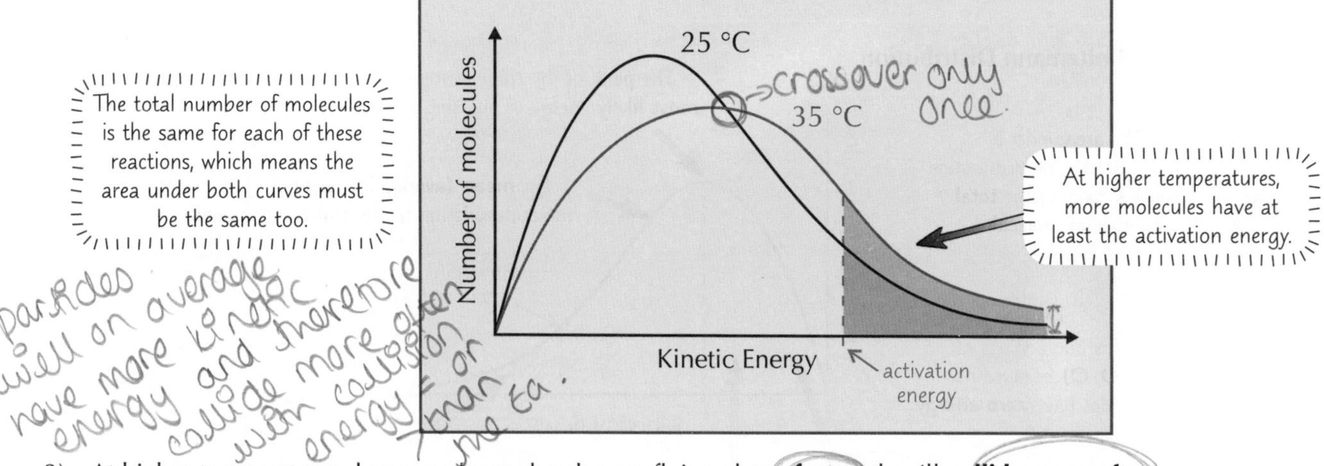

The total number of molecules is the same for each of these reactions, which means the area under both curves must be the same too.
Number of molecules
25 °C
35 °C
Kinetic Energy
activation energy
At higher temperatures, more molecules have at least the activation energy.

3) At higher temperatures, because the molecules are flying about **faster**, they'll **collide more often**. This is **another reason** why increasing the temperature makes a reaction faster.
4) Because you get **both** of these effects happening at once (more collisions and more energetic collisions), quite **small increases** in **temperature** can lead to quite **large increases** in **reaction rate**.

Increasing Concentration** also **Increases** the **Rate** of **Reaction

1) If you increase the **concentration** of reactants in a **solution**, the particles will on average be **closer together**.
2) If they're closer, they'll **collide more often**. If collisions occur **more frequently**, they'll have **more chances** to react. This is why increasing **concentration** increases **reaction rate**.

***Increasing Pressure Increases** the **Rate of Reaction** Too*

1) If a reaction involves gases, increasing the **pressure** works in just the same way as increasing concentration.
2) Raising the pressure pushes all of the gas particles **closer** together, making them **more likely** to collide. So **collisions** take place **more frequently** and the reaction rate **increases**.

***Catalysts Increase** the **Rate** of **Reactions** Too*

You can use **catalysts** to make chemical reactions happen **faster**. Learn this definition:

> A **catalyst** is a substance that increases the **rate** of a reaction by providing an **alternative reaction pathway** with a **lower activation energy**.
> The catalyst is **chemically unchanged** at the end of the reaction.

There's more about this on the next page. I can't wait...

1) Catalysts are **great**. They **don't** get used up in reactions, so you only need a **tiny bit** of catalyst to catalyse a **huge** amount of stuff. They **do** take part in reactions, but they're **remade** at the end.
2) Catalysts are **very fussy** about which reactions they catalyse. They often **only** work on a single reaction.
3) Because catalysts allow you to make the same amount of product **faster** (and often at a **lower temperature** too), they **save heaps of money** in industrial processes.

More on Reaction Rates

Maxwell-Boltzmann Distributions Show Why **Catalysts** Work

If you look at an **energy profile** together with a **Maxwell-Boltzmann distribution**, you can see **why** catalysts work.

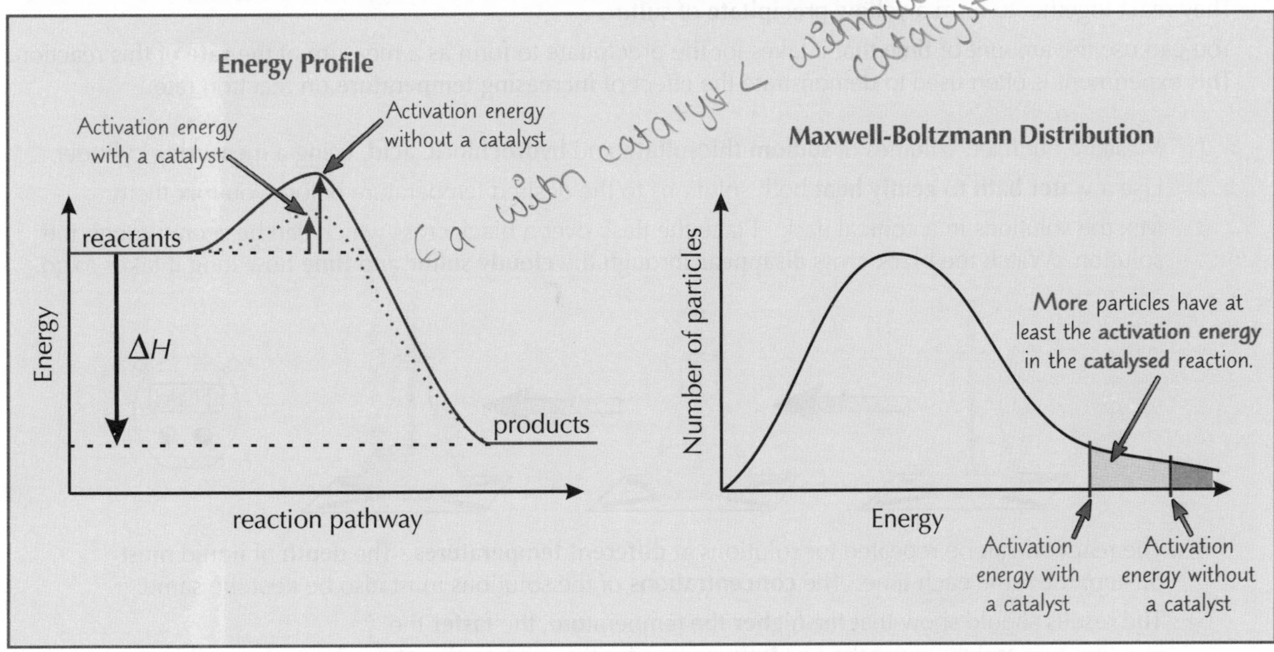

The catalyst **lowers the activation energy**, meaning there's **more particles** with **enough energy** to react when they collide. It does this by allowing the reaction to go **via a different route**. So, in a certain amount of time, **more particles react**.

There are **Different** Ways to **Measure** Reaction Rates

You can follow the rate of a reaction by measuring either how fast the **reactants are used up** or how fast the **products are formed**. Here are three ways to measure reaction rate:

Timing how long a precipitate takes to form.

1) You can use this method when the product is a **precipitate** which clouds a solution.
2) You watch a **mark** through the solution and time how long it takes to be **obscured**.
3) If the **same observer** uses the **same mark** each time, you can compare the rates of reaction, because (roughly) the same amount of precipitate will have been **formed** when the mark becomes obscured.
4) But this method is **subjective** — different people might not agree on the exact moment the mark disappears.

Measuring a decrease in mass.

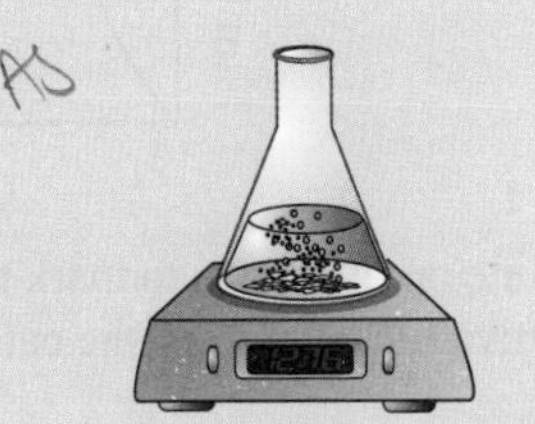

1) When one or more of the products is a **gas**, you can measure the rate of formation using a **mass balance**.
2) As gas is given off, the mass of the reaction mixture decreases.
3) This method is **accurate** and easy to do. But it does release gas into the atmosphere, so it's usually done in a **fume cupboard**.

Measuring the volume of gas given off.

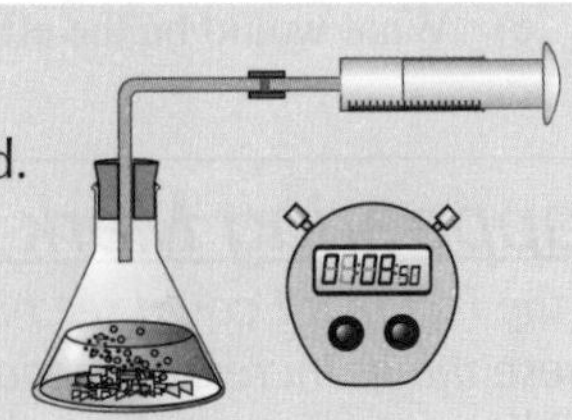

1) This involves using a **gas syringe** to measure the **volume** of gas being produced.
2) You can only use this method when one or more of the products is a **gas**.
3) Gas syringes usually give volumes to the nearest 0.1 cm^3, so this method is **accurate**.

More on Reaction Rates

Example: The Reaction Between **Sodium Thiosulfate** and **Hydrochloric Acid**

Sodium thiosulfate and hydrochloric acid are both **clear**, **colourless solutions**.
They react together to form a **yellow precipitate** of **sulfur**.

You can use the amount of **time** that it takes for the precipitate to form as a measure of the **rate** of this reaction. This experiment is often used to demonstrate the effect of **increasing temperature** on reaction rate:

1) Measure out fixed volumes of **sodium thiosulfate** and **hydrochloric acid**, using a measuring cylinder.
2) Use a **water bath** to **gently heat** both solutions to the desired temperature before you mix them.
3) Mix the solutions in a conical flask. Place the flask over a black cross which can be seen through the solution. Watch the black cross **disappear** through the **cloudy sulfur** and **time** how long it takes to go.

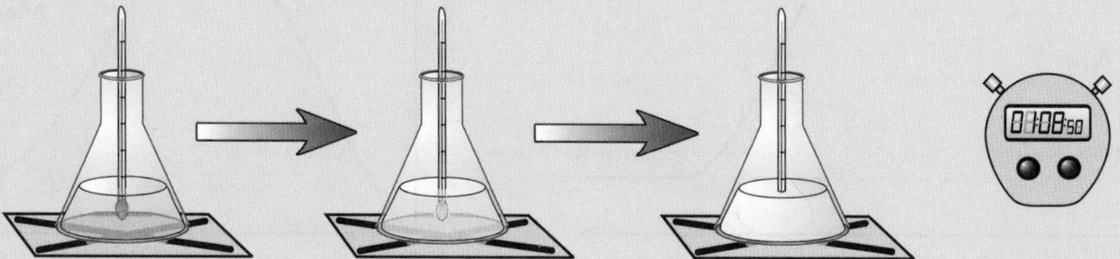

4) The reaction can be repeated for solutions at **different temperatures**. The **depth** of liquid must be kept the same each time. The **concentrations** of the solutions must also be kept the same.
5) The results should show that the **higher** the temperature, the **faster** the reaction rate and therefore the **less time** it takes for the mark to **disappear**.

Practice Questions

Q1 Sketch Maxwell-Boltzmann distribution curves for the molecules in a gas at two different temperatures.

Q2 Explain why increasing the concentration of solutions increases reaction rate.

Q3 Explain what a catalyst is.

Q4 Describe a way to measure the rate of a reaction where the product is a gas.

Exam Question

Q1 The decomposition of hydrogen peroxide, H_2O_2, into water and oxygen is catalysed by manganese(IV) oxide, MnO_2. The Maxwell-Boltzmann distribution for the H_2O_2 molecules at 25 °C is shown below.

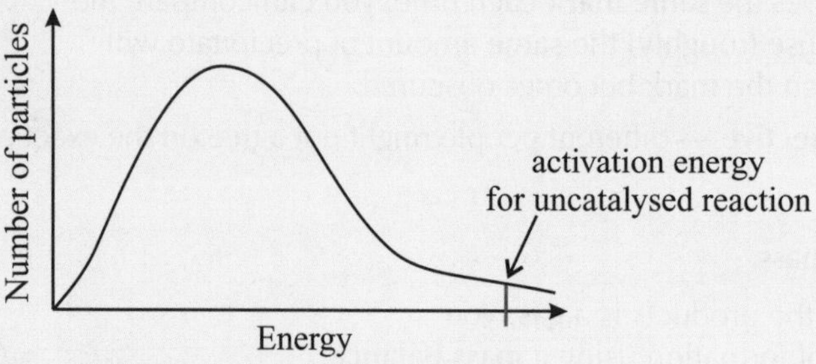

a) Draw a line on the x-axis to mark the approximate position of the activation energy for the catalysed process. [1 mark]

b) Explain how manganese(IV) oxide acts as a catalyst. [1 mark]

c) What would be the effect of raising the temperature on the rate of this reaction? Explain this effect. [4 marks]

Disappearing Mark — the worst magician I've ever seen...

...and the fact that I could see him was the problem to be honest. Talk about an anticlimax. Anyhow, you need to know that these things increase reaction rate: increasing temperature, increasing concentration, increasing pressure (for gases) and adding a catalyst. You need to know how they all do it too, so if you can't remember, now's the time to look back.

Reversible Reactions

There's a lot of to-ing and fro-ing on these pages. Mind your head doesn't start spinning.

Reversible *Reactions Can Reach* ***Dynamic Equilibrium***

1) Lots of chemical reactions are **reversible** — they go **both ways**.
2) To show a reaction's reversible, you stick in a $\rightleftharpoons$. Here's an example:

$$H_{2(g)} + I_{2(g)} \rightleftharpoons 2HI_{(g)}$$

This reaction can go in **either direction**:

forwards $H_{2(g)} + I_{2(g)} \rightleftharpoons 2HI_{(g)}$

...or **backwards** $2HI_{(g)} \rightleftharpoons H_{2(g)} + I_{2(g)}$

3) As the **reactants** get used up, the **forward** reaction **slows down** — and as more **product** is formed, the **reverse** reaction **speeds up**.
4) After a while, the forward reaction will be going at exactly the **same rate** as the backward reaction so the amounts of reactants and products **won't be changing** any more — it'll seem like **nothing's happening**.
5) This is called **dynamic equilibrium**. At equilibrium, the **concentrations** of **reactants** and **products** stay **constant**.
6) A **dynamic equilibrium** can only happen in a **closed system**. This just means nothing can get in or out.

It's a bit like digging a hole while someone else is filling it in at exactly the same speed.

Le Chatelier's Principle *Predicts what will Happen if* ***Conditions are Changed***

If you **change** the **concentration**, **pressure** or **temperature** of a reversible reaction, you're going to **alter** the **position of equilibrium**. This just means you'll end up with **different amounts** of reactants and products at equilibrium.

If the position of equilibrium moves to the **left**, you'll get more **reactants**.

$$H_{2(g)} + I_{2(g)} \rightleftharpoons 2HI_{(g)}$$

lots of H_2 and I_2 — not much HI

If the position of equilibrium moves to the **right**, you'll get more **products**.

$$H_{2(g)} + I_{2(g)} \rightleftharpoons 2HI_{(g)}$$

not much H_2 and I_2 — lots of HI

Mr and Mrs Le Chatelier celebrate another successful year in the principle business.

Le Chatelier's principle tells you how the **position of equilibrium** will change if a **condition changes**:

If a reaction at **equilibrium** is subjected to a change in **concentration**, **pressure** or **temperature**, the position of equilibrium will move to **counteract** the change.

So, basically, if you **raise the temperature**, the position of equilibrium will shift to try to **cool things down**.
And, if you **raise the pressure or concentration**, the position of equilibrium will shift to try to **reduce it again**.

Reversible Reactions

Here's Some **Handy Rules** for Using **Le Chatelier's Principle**

1) You can use Le Chatelier's principle to work out what effect changing the **concentration**, **pressure** or **temperature** will have on the **position of equilibrium**.
2) This only applies to **homogeneous equilibria** — that means reactions where every species is in the **same physical state** (e.g. all liquid or all gas).

CONCENTRATION

1) If you **increase** the **concentration** of a **reactant**, the equilibrium tries to **get rid** of the extra reactant. It does this by making **more product**. So the equilibrium shifts to the **right**.

$2SO_{2(g)} + O_{2(g)} \rightleftharpoons 2SO_{3(g)}$ If you increase the concentration of SO_2 or O_2, the speed of the forward reaction will increase (to use up the extra reactant), moving the equilibrium to the right.

2) If you **increase** the **concentration** of the **product** (SO_3), the equilibrium tries to remove the extra product. This makes the **reverse reaction** go faster. So the equilibrium shifts to the **left**.
3) **Decreasing** the concentrations has the **opposite effect**.

PRESSURE (Changing this only affects **equilibria involving gases**.)

1) **Increasing** the pressure shifts the equilibrium to the side with **fewer** gas molecules. This **reduces** the pressure.

$2SO_{2(g)} + O_{2(g)} \rightleftharpoons 2SO_{3(g)}$ There are 3 moles of gas on the left, but only 2 on the right. So, an increase in pressure shifts the equilibrium to the right.

2) **Decreasing** the pressure shifts the equilibrium to the side with **more** gas molecules. This **raises** the pressure again.

TEMPERATURE

1) **Increasing** the temperature means **adding heat**. The equilibrium shifts in the **endothermic (positive ΔH) direction** to absorb this heat.

$2SO_{2(g)} + O_{2(g)} \rightleftharpoons 2SO_{3(g)}$ $\Delta H = -197$ kJ mol^{-1} This reaction's exothermic in the forward direction. If you increase the temperature, the equilibrium shifts to the left to absorb the extra heat.

2) **Decreasing** the temperature means **removing heat**. The equilibrium shifts in the **exothermic (negative ΔH) direction** to produce more heat, in order to counteract the drop in temperature.
3) If the forward reaction's **endothermic**, the reverse reaction will be **exothermic**, and vice versa.

Catalysts **Don't Affect** The Position of Equilibrium

Catalysts have **NO EFFECT** on the **position of equilibrium**.
They **can't** increase **yield** — but they **do** mean equilibrium is reached **faster**.

Reversible Reactions

In Industry the **Reaction Conditions** Chosen are a **Compromise**

Companies have to think about how much it **costs** to run a reaction and how much money they can make from it. This means they have a few factors to think about when they're choosing the best conditions for a reaction.

For example, **ethanol** can be produced via a reversible exothermic reaction between **ethene** and **steam:**

$$C_2H_{4(g)} + H_2O_{(g)} \rightleftharpoons C_2H_5OH_{(g)} \qquad \Delta H = -46 \text{ kJ mol}^{-1}$$

This reaction is carried out at pressures of **60-70 atmospheres** and a temperature of **300 °C**, with a catalyst of **phosphoric acid.**

1) Because this is an **exothermic reaction, lower** temperatures favour the forward reaction. This means that at lower temperatures **more** ethene and steam is converted to ethanol — you get a better **yield.**
2) But **lower temperatures** mean a **slower rate of reaction.** There's **no point** getting a **very high yield** of ethanol if it takes you 10 years. So 300 °C is a **compromise** between a **reasonable yield** and a **faster reaction.**
3) **High pressure** shifts the equilibrium to the side with **fewer molecules**, which favours the **forward reaction** here. **High pressure** increases the **rate** of this reaction too. So a pressure of **60-70 atmospheres** is used.
4) Cranking up the pressure even higher than that might sound like a great idea. But **very high pressures** are **really expensive** to produce. You need really strong **pipes** and **containers** to withstand high pressures.
5) So the **60-70 atmospheres** is a **compromise** — it gives a **reasonable yield** for the lowest possible **cost.**

Practice Questions

Q1 Explain what the terms 'reversible reaction' and 'dynamic equilibrium' mean.

Q2 If an equilibrium moves to the right, do you get more products or reactants?

Q3 A reaction at equilibrium is endothermic in the forward direction.
What happens to the position of equilibrium as the temperature is increased?

Q4 What effect do catalysts have on equilibrium position?

Exam Questions

Q1 Nitrogen and oxygen gases were reacted together in a closed flask and allowed to reach equilibrium, with the gas nitrogen monoxide formed: $N_{2(g)} + O_{2(g)} \rightleftharpoons 2NO_{(g)}$ $\Delta H = +181 \text{ kJ mol}^{-1}$

a) State Le Chatelier's principle. [1 mark]

b) Explain how the following changes would affect the position of equilibrium of the above reaction:

i) Pressure is increased. [2 marks]

ii) Temperature is reduced. [2 marks]

iii) The concentration of nitrogen monoxide is reduced. [2 marks]

c) State the effect that a catalyst would have on the composition of the equilibrium mixture. [1 mark]

Q2 Ethanol can be manufactured from ethene and steam: $C_2H_{4(g)} + H_2O_{(g)} \rightleftharpoons C_2H_5OH_{(g)}$ $\Delta H = -46 \text{ kJ mol}^{-1}$
Typical conditions are 300 °C and 60-70 atmospheres.

a) Explain why, for this reaction, decreasing the temperature would increase the yield of ethanol. [3 marks]

b) i) What effect would increasing pressure have on the yield of ethanol? [2 marks]

ii) Suggest why a pressure higher than the one quoted is not often used. [1 mark]

Reverse psychology — don't bother learning this rubbish...

Make sure that you know what happens to a reaction at equilibrium if you change the conditions. A quick reminder about pressure — remember that if there are the same number of molecules of gas on each side of the equation, then you can raise the pressure as high as you like and it won't make a blind bit of difference to the position of equilibrium.

The Equilibrium Constant

You don't just need to know what dynamic equilibrium means. It's nice to be able to describe what's going on but you know scientists — they do insist on using mathsy stuff. Here come the numbers...

K_c is the Equilibrium Constant

[X] = the concentration of species X in mol dm^{-3}.

If you know the **molar concentration** of each substance at equilibrium, you can work out the **equilibrium constant, K_c**.
The equilibrium constant can be written as an expression, like this:

$$aA + bB \rightleftharpoons dD + eE, \quad K_c = \frac{[D]^d[E]^e}{[A]^a[B]^b}$$

You can just bung the **equilibrium concentrations** into your expression to work out the **value** for K_c.
The **units** are a bit trickier though — they **vary**, so you have to work them out after each calculation.

Example: Hydrogen gas and iodine gas are mixed in a closed flask. Hydrogen iodide is formed.

$$H_{2(g)} + I_{2(g)} \rightleftharpoons 2HI_{(g)}$$

Calculate the equilibrium constant for the reaction at 640 K. The equilibrium concentrations are: [HI] = 0.80 mol dm^{-3}, [H_2] = 0.10 mol dm^{-3}, and [I_2] = 0.10 mol dm^{-3}.

Just stick the concentrations into the **expression** for K_c:

$$K_c = \frac{[HI]^2}{[H_2][I_2]} = \frac{0.80^2}{0.10 \times 0.10} = 64$$

To work out the **units** of K_c put the units in the expression instead of the numbers:

Units of $K_c = \frac{(\text{mol dm}^{-3})^2}{(\text{mol dm}^{-3})(\text{mol dm}^{-3})}$ — the concentration units cancel, so there are **no units** for K_c.

So K_c is just **64**.

You Might Need to Work Out the Equilibrium Concentrations

You might have to figure out some of the **equilibrium concentrations** before you can find K_c:

Example: 0.20 moles of phosphorus(V) chloride decomposes at 600 K in a vessel of 5.00 dm^3.
The equilibrium mixture is found to contain 0.080 moles of chlorine.
Write the expression for K_c and calculate its value, including units.

$$PCl_{5(g)} \rightleftharpoons PCl_{3(g)} + Cl_{2(g)}$$

First find out how many moles of PCl_5 and PCl_3 there are at equilibrium:

The **equation** tells you that when **1 mole of PCl_5** decomposes, **1 mole of PCl_3** and **1 mole of Cl_2** are formed.
So if 0.080 moles of chlorine are produced at equilibrium, then there will be **0.080 moles** of PCl_3 as well.
0.080 moles of PCl_5 must have decomposed, so there will be **0.12 moles** left (0.20 – 0.080).

Divide each number of moles by the volume of the flask to give the molar concentrations:

[PCl_3] = [Cl_2] = 0.080 ÷ 5.00 = **0.016 mol dm^{-3}**   [PCl_5] = 0.12 ÷ 5.00 = **0.024 mol dm^{-3}**

Put the concentrations in the expression for K_c and calculate it:

$$K_c = \frac{[PCl_3][Cl_2]}{[PCl_5]} = \frac{0.016 \times 0.016}{0.024} = 0.011$$

Now find the units of K_c: Units of $K_c = \frac{(\text{mol dm}^{-3})(\text{mol dm}^{-3})}{(\text{mol dm}^{-3})}$ = **mol dm^{-3}**

So **K_c = 0.011 mol dm^{-3}**

The Equilibrium Constant

K_c can be used to Find Concentrations in an Equilibrium Mixture

Example: When ethanoic acid was allowed to reach equilibrium with ethanol at 25 °C, it was found that the equilibrium mixture contained 2.0 mol dm^{-3} ethanoic acid and 3.5 mol dm^{-3} ethanol. The K_c of the equilibrium is 4.0 at 25 °C. What are the concentrations of the other components?

$$CH_3COOH_{(l)} + C_2H_5OH_{(l)} \rightleftharpoons CH_3COOC_2H_{5(l)} + H_2O_{(l)}$$

Put all the values you know in the K_c expression: $K_c = \frac{[CH_3COOC_2H_5][H_2O]}{[CH_3COOH][C_2H_5OH]} \Rightarrow 4.0 = \frac{[CH_3COOC_2H_5][H_2O]}{2.0 \times 3.5}$

Rearranging this gives: $[CH_3COOC_2H_5][H_2O] = 4.0 \times 2.0 \times 3.5 = 28$

From the equation you know that $[CH_3COOC_2H_5] = [H_2O]$, so: $[CH_3COOC_2H_5] = [H_2O] = \sqrt{28} = 5.3$ mol dm^{-3}

The concentration of $CH_3COOC_2H_5$ and H_2O is **5.3 mol dm^{-3}**

Temperature Changes Alter K_c

Remember: if the temperature rises, the equilibrium shifts in the endothermic (+ve ΔH) direction to absorb the heat. If the temperature falls, the equilibrium shifts in the exothermic (–ve ΔH) direction to replace the heat.

1) The value of K_c is only **valid** for one particular **temperature**.
2) If you change the temperature of the system, you will also change the **equilibrium concentrations** of the products and reactants, so **K_c will change**.
3) If the temperature change means there's **more product** at equilibrium, **K_c will rise**. If it means there's **less product** at equilibrium, **K_c will decrease**. For example:

The reaction on the right is exothermic in the forward direction. If you **increase** the temperature, you favour the **endothermic** reaction. This means that **less product** is formed.

Exothermic ⟹

$$N_{2(g)} + 3H_{2(g)} \rightleftharpoons 2NH_{3(g)} \quad \Delta H = -46.2 \text{ kJ mol}^{-1}$$

⟸ Endothermic

$$K_c = \frac{[NH_3]^2}{[N_2][H_2]^3}$$

← The concentration of NH_3 will be reduced.
← The concentrations of N_2 and H_2 will increase.

As the temperature **increases**, $[NH_3]$ will **decrease** and $[N_2]$ and $[H_2]$ will **increase**, so K_c will **decrease**.

4) Changing the **concentration** of a reactant or product, will **not affect** the value of K_c.
5) **Catalysts** don't affect K_c either — they'll speed up the reaction in both directions by the same amount, so they just help the system to reach equilibrium **faster**.

Practice Question

Q1 If a reversible reaction has an exothermic forward reaction, how will increasing the temperature affect K_c?

Exam Questions

Q1 A sample of pure hydrogen iodide is placed in a sealed flask, and heated to 443 °C. The following equilibrium is established: $2HI_{(g)} \rightleftharpoons H_{2(g)} + I_{2(g)}$ $(K_c = 0.0200)$
At equilibrium, $[I_2] = 0.770$ mol dm^{-3}. Find the equilibrium concentration of HI. [4 marks]

Q2 Nitrogen dioxide dissociates according to the equation $2NO_{2(g)} \rightleftharpoons 2NO_{(g)} + O_{2(g)}$
When 34.5 g of nitrogen dioxide were heated in a vessel of volume 9.80 dm^3 at 500 °C, 7.04 g of oxygen were found in the equilibrium mixture.

a) Calculate: i) the number of moles of nitrogen dioxide originally. [1 mark]

ii) the number of moles of each gas in the equilibrium mixture. [3 marks]

b) Find the value of K_c at this temperature, and give its units. [4 marks].

I still don't like equilibrium positions — so nothing's changed there...

Working out K_c is easy as pie once you've got all the concentrations figured out. Make sure you practise writing down expressions for the equilibrium constant and doing all those mathsy bits — the questions here should help you with that.

Redox Reactions

This double page has more occurrences of "oxidation" than The Beatles "All You Need is Love" features the word "love".

If Electrons are Transferred, it's a ***Redox Reaction***

1) A **loss** of electrons is called **oxidation**.
2) A **gain** in electrons is called **reduction**.
3) Reduction and oxidation happen **simultaneously** — hence the term "**redox**" reaction.
4) An **oxidising agent accepts** electrons and gets reduced.
5) A **reducing agent donates** electrons and gets oxidised.

Example: the formation of **sodium chloride** from sodium and chlorine is a **redox reaction:**

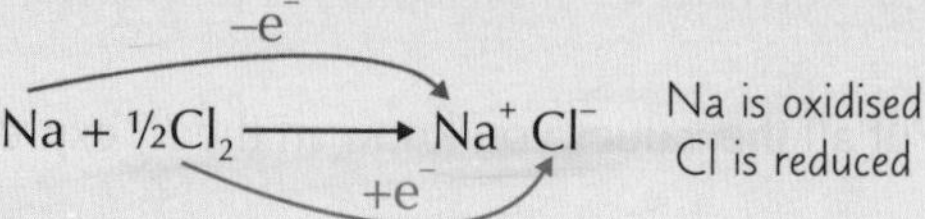

You Need to Know the ***Rules*** *for* ***Assigning Oxidation States****

*Or 'oxidation numbers' if you prefer.

The oxidation state of an element tells you the **total number** of **electrons** it has **donated** or **accepted**. For example, in the redox reaction shown above, the sodium atom has **donated one electron**, so it has an oxidation state of **+1**. The chlorine atom has **accepted one electron**, so it has an oxidation state of **–1**.

There's a set of **rules** that you can use to work out the oxidation state of an atom when it's in a **compound**, in an **ion**, or just **on its own**. Take a deep breath, here we go...

1) Uncombined **elements**, like He and Ar, have an oxidation state of **0**.

2) Elements just bonded to **identical atoms**, like O_2 and H_2, also have an oxidation state of **0**.

3) The oxidation state of a simple **monatomic ion**, like Na^+, is the same as its **charge**.

In this case, that's +1.

4) In **compound ions**, the **overall oxidation state** is just the ion charge.

SO_4^{2-} — **overall oxidation state = –2**,
oxidation state of **O** = **–2** (total = 4 × –2 = –8),
so oxidation state of **S** = **–2 – (–8) = +6**

In a compound ion, the most electronegative element has a negative oxidation state. The other elements have more positive oxidation states.

5) The sum of the oxidation states for a **neutral compound** is 0.

Fe_2O_3 — **overall oxidation state = 0**
oxidation state of **O** = **–2** (total = 3 × –2 = –6),
so oxidation state of **Fe** = **(0 – (–6)) ÷ 2 = +3**

6) Combined **oxygen** is nearly always –2, except in peroxides, where it's –1, (and in the fluorides OF_2, where it's +2, and O_2F_2, where it's +1 and O_2 where it's 0).

In H_2O, oxidation state of **O = –2**, but in H_2O_2, oxidation state of **O = –1**.

7) Combined **hydrogen** is +1, except in metal hydrides where it is –1 (and H_2 where it's 0).

In **HF**, oxidation state of **H = +1**, but in **NaH**, oxidation state of **H = –1**.

Roman Numerals *Show Oxidation States*

Sometimes, oxidation states aren't clear from the formula of a compound.

If you see **Roman numerals** in a chemical name, it's an **oxidation number**.

E.g. iron has oxidation state **+2** in **iron(II) sulfate** ($FeSO_4$), but it has oxidation state **+3** in **iron(III) sulfate** ($Fe_2(SO_4)_3$).

Hands up if you like Roman numerals...

Redox Reactions

You can Write Half-Equations and Combine them into Redox Equations

1) **Ionic half-equations** show oxidation or reduction.
2) You show the **electrons** that are being lost or gained in a half-equation.
For example, this is the half-equation for the **oxidation of sodium**: $Na \rightarrow Na^+ + e^-$
3) You can **combine** half-equations for different oxidising or reducing agents together to make **full equations** for redox reactions.

Here's the electron that the sodium atom has lost.

Magnesium burns in **oxygen** to form **magnesium oxide**.

Oxygen is reduced to O^{2-}:

$$\mathbf{O_2 + 4e^- \rightarrow 2O^{2-}}$$

Magnesium is oxidised to Mg^{2+}:

$$\mathbf{Mg \rightarrow Mg^{2+} + 2e^-}$$

Make sure the atoms and charges balance.

You need both equations to contain the same number of electrons. So double everything in the second equation.

$$\mathbf{2Mg \rightarrow 2Mg^{2+} + 4e^-}$$

Combining the half-equations makes:

$$\mathbf{2Mg + O_2 \rightarrow 2MgO}$$

The electrons aren't included in the full equation. You end up with four on each side — so they cancel.
$2Mg + O_2 + 4e^- \rightarrow 2MgO + 4e^-$

Aluminium reacts with **chlorine** to form **aluminium chloride**.

Aluminium is oxidised to Al^{3+}:

$$\mathbf{Al \rightarrow Al^{3+} + 3e^-}$$

Chlorine is reduced to Cl^-:

$$\mathbf{Cl_2 + 2e^- \rightarrow 2Cl^-}$$

Make sure the atoms and charges balance.

Now make sure the equations each contain the same number of electrons.

$$\mathbf{Al \rightarrow Al^{3+} + 3e^- \xrightarrow{\times 2} 2Al \rightarrow 2Al^{3+} + 6e^-}$$

$$\mathbf{Cl_2 + 2e^- \rightarrow 2Cl^- \xrightarrow{\times 3} 3Cl_2 + 6e^- \rightarrow 6Cl^-}$$

Combining the half-equations makes:

$$\mathbf{2Al + 3Cl_2 \rightarrow 2AlCl_3}$$

Practice Questions

Q1 What is a reducing agent?

Q2 What is the oxidation state of hydrogen in H_2 gas?

Q3 What is the usual oxidation state of oxygen when it's in a compound?

Exam Questions

Q1 Lithium oxide forms when lithium is burned in air. Combustion is a redox reaction.
The equation for the combustion of lithium is: $4Li_{(s)} + O_{2(g)} \rightarrow 2Li_2O_{(s)}$

a) Define oxidation in terms of the movement of electrons. [1 mark]

b) What is the oxidation state of lithium in: i) Li ii) Li_2O [2 marks]

c) State which reactant is this reaction is reduced.
Write a half-equation for this reduction reaction. [2 marks]

Q2 The half-equation for chlorine acting as an oxidising agent is: $Cl_2 + 2e^- \rightarrow 2Cl^-$

a) Define the term oxidising agent in terms of electron movement. [1 mark]

b) Indium is a metal that can be oxidised by chlorine.
Write a balanced half-equation for the oxidation of indium metal to form In^{3+} ions. [1 marks]

c) Use your answer to b) and the equation above to form a balanced equation for the reaction of indium with chlorine by combining half-equations. [2 marks]

Oxidising agent SALE NOW ON — everything's reduced...

Half-equations look evil, with all those electrons flying about. But they're not too bad really. Just make sure you get lots of practice using them. (Oh, look, there are some handy questions up there).

And while we're on the redox page, I suppose you ought to learn the most famous memory aid thingy in the world...

OIL RIG
— **Oxidation Is Loss**
— **Reduction Is Gain**
(of electrons)

Enthalpy Definitions

I'm sure that you remember enthalpies, but here's a quick reminder just in case...

First — A Few Definitions You Should Remember

ΔH is the symbol for **enthalpy change**.
Enthalpy change is the **heat** energy transferred in a reaction at **constant pressure**.
$\Delta H^{\ominus}$ means that the enthalpy change was measured under **standard conditions** (**298 K** and **100 kPa**).
Exothermic reactions have a **negative** ΔH value, because heat energy is given out.
Endothermic reactions have a **positive** ΔH value, because heat energy is absorbed.

There are Different Types of Enthalpy Change

Here are some more definitions: (you're welcome)

Enthalpy change of formation, $\Delta_f H$, is the enthalpy change when **1 mole** of a **compound** is formed from its **elements** in their standard states under standard conditions, e.g. $2C_{(s)} + 3H_{2(g)} + \frac{1}{2}O_{2(g)} \rightarrow C_2H_5OH_{(l)}$.

Bond dissociation enthalpy, $\Delta_{diss}H$, is the enthalpy change when all the **bonds of the same type** in **1 mole** of **gaseous molecules** are broken, e.g. $Cl_{2(g)} \rightarrow 2Cl_{(g)}$.

Enthalpy change of atomisation of an element, $\Delta_{at}H$, is the enthalpy change when **1 mole** of **gaseous atoms** is formed from an element in its **standard state**, e.g. $\frac{1}{2}Cl_{2(g)} \rightarrow Cl_{(g)}$.

Enthalpy change of atomisation of a compound, $\Delta_{at}H$, is the enthalpy change when **1 mole** of a compound in its **standard state** is converted to **gaseous atoms**, e.g. $NaCl_{(s)} \rightarrow Na_{(g)} + Cl_{(g)}$.

First ionisation energy, $\Delta_{ie1}H$, is the enthalpy change when **1 mole** of **gaseous 1+ ions** is formed from **1 mole** of **gaseous atoms**, e.g. $Mg_{(g)} \rightarrow Mg^+_{(g)} + e^-$.

Second ionisation energy, $\Delta_{ie2}H$, is the enthalpy change when **1 mole** of **gaseous 2+ ions** is formed from **1 mole** of **gaseous 1+ ions**, e.g. $Mg^+_{(g)} \rightarrow Mg^{2+}_{(g)} + e^-$.

First electron affinity, $\Delta_{ea1}H$, is the enthalpy change when **1 mole** of **gaseous 1– ions** is formed from **1 mole** of **gaseous atoms**, e.g. $O_{(g)} + e^- \rightarrow O^-_{(g)}$.

Second electron affinity, $\Delta_{ea2}H$, is the enthalpy change when **1 mole** of **gaseous 2– ions** is formed from **1 mole** of **gaseous 1– ions**, e.g. $O^-_{(g)} + e^- \rightarrow O^{2-}_{(g)}$.

Enthalpy change of hydration, $\Delta_{hyd}H$, is the enthalpy change when **1 mole** of **aqueous ions** is formed from **1 mole** of **gaseous ions**, e.g. $Na^+_{(g)} \rightarrow Na^+_{(aq)}$.

Enthalpy change of solution, $\Delta_{solution}H$, is the enthalpy change when **1 mole** of **solute** is dissolved in **enough solvent** that no further enthalpy change occurs on further dilution, e.g. $NaCl_{(s)} \rightarrow NaCl_{(aq)}$.

Practice Questions

Q1 Do exothermic reactions have positive or negative enthalpy changes?

Q2 Give the definition of bond dissociation enthalpy.

Q3 Write the equation representing the enthalpy change of atomisation for $NaCl_{(s)}$.

Q4 Define first ionisation energy and first electron affinity.

My eyes, MY EYES — the definitions make them hurt...

Well, isn't this a lovely way to start off a new section — a barrage of definitions to learn. It's not that bad though — once you're happy with a few of them, the rest will make a lot more sense. And by the way, don't think you won't need to learn this stuff just because there are no exam questions on this page. You WILL need to understand these definitions.

Lattice Enthalpy and Born-Haber Cycles

Now you know all your enthalpy change definitions, here's how to use them... enjoy.

Lattice Enthalpy is a Measure of *Ionic Bond Strength*

Lattice enthalpy, $\Delta_{lattice}H$, can be defined in two ways:

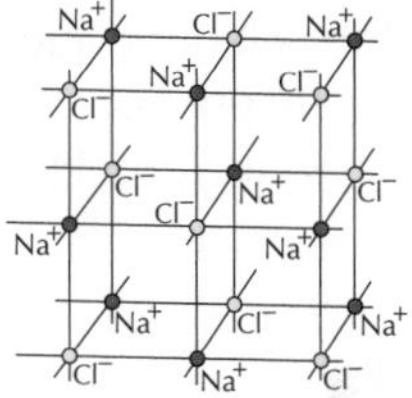

Part of the sodium chloride lattice

1. **Lattice enthalpy of formation**: the enthalpy change when **1 mole** of a **solid ionic compound** is **formed** from its **gaseous ions** under standard conditions.

   Example: $Na^+_{(g)} + Cl^-_{(g)} \rightarrow NaCl_{(s)}$ $\Delta_{lattice}H = -787$ kJ mol^{-1} (**exothermic**)
   $Mg^{2+}_{(g)} + 2Cl^-_{(g)} \rightarrow MgCl_{2(s)}$ $\Delta_{lattice}H = -2526$ kJ mol^{-1} (**exothermic**)

2. **Lattice enthalpy of dissociation**: the enthalpy change when **1 mole** of a **solid ionic compound** is completely **dissociated** into its **gaseous ions** under standard conditions.

   Example: $NaCl_{(s)} \rightarrow Na^+_{(g)} + Cl^-_{(g)}$ $\Delta_{lattice}H = +787$ kJ mol^{-1} (**endothermic**)
   $MgCl_{2(s)} \rightarrow Mg^{2+}_{(g)} + 2Cl^-_{(g)}$ $\Delta_{lattice}H = +2526$ kJ mol^{-1} (**endothermic**)

Notice that lattice formation enthalpy and lattice dissociation enthalpy are exact opposites.

Born-Haber Cycles can be Used to *Calculate Lattice Enthalpies*

Hess's law says that the **total enthalpy change** of a reaction is always the **same**, no matter which route is taken. You can't calculate a lattice enthalpy **directly**, so you have to use a **Born-Haber cycle** to figure out what the enthalpy change would be if you took **another, less direct, route**.

Here's a Born-Haber cycle you could use to calculate the lattice enthalpy of **NaCl**:

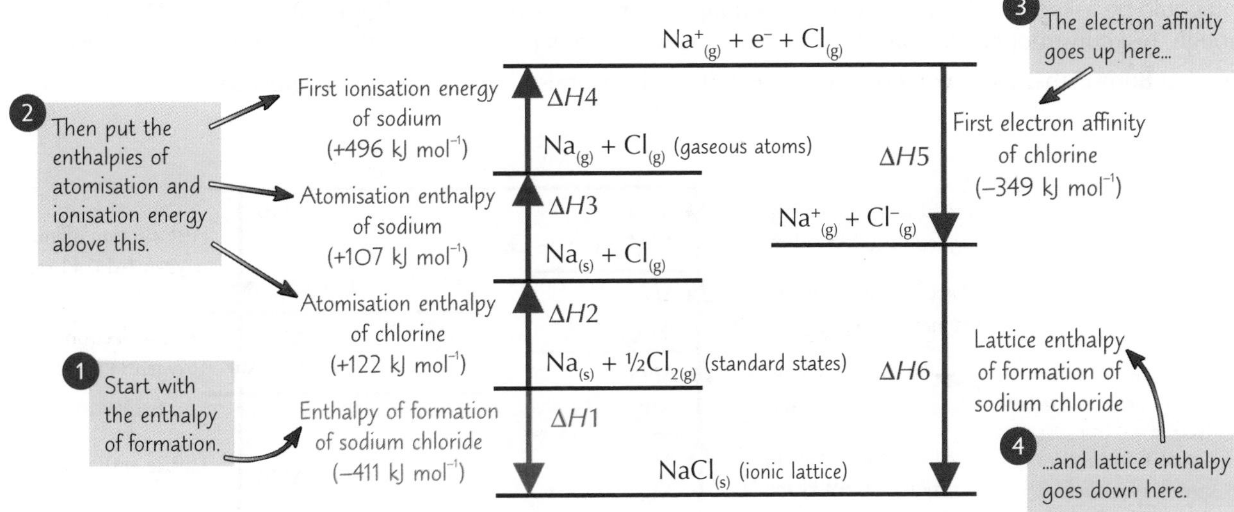

There are **two routes** you can follow to get from the elements in their **standard states** to the **ionic lattice**.
The green arrow shows the **direct route** and the purple arrows show the **indirect route**.
The enthalpy change for each is the **same**.

From **Hess's law**: $\Delta H6 = -\Delta H5 - \Delta H4 - \Delta H3 - \Delta H2 + \Delta H1$
$= -(-349) - (+496) - (+107) - (+122) + (-411) = \mathbf{-787}$ **kJ mol**$^{-1}$

You need a minus sign if you go the wrong way along an arrow.

So the **lattice enthalpy of formation** of sodium chloride is **−787 kJ mol**$^{-1}$.

Lattice Enthalpy and Born-Haber Cycles

Calculations Involving **Group 2 Elements** are a Bit **Different**

Born-Haber cycles for compounds containing **Group 2 elements** have at least one **extra step** compared to the one on the previous page. Make sure you understand what's going on so you can handle whatever compound they throw at you. Here's the Born-Haber cycle for calculating the lattice enthalpy of **magnesium chloride** ($MgCl_2$).

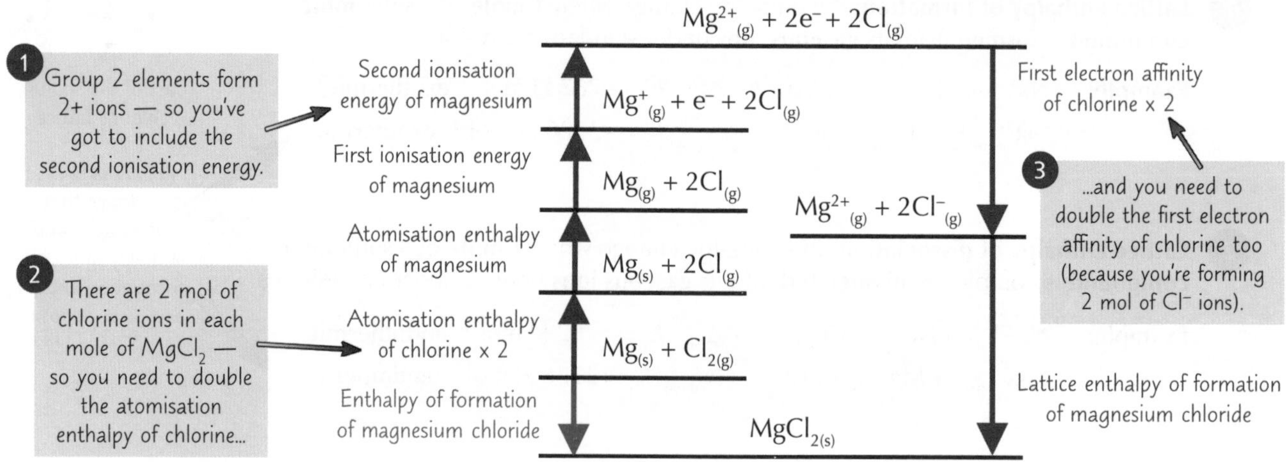

If the formation of a compound involves ions with **charges** of **more than 1**, you just have to add the **extra ionisation energies** and possibly **electron affinities** (see the example below) to the Born-Haber cycle. So you can **adapt** this method to any compound if you have all the information needed.

Born-Haber Cycles can be Used to Calculate **Different Values**

You could be asked to calculate **any** of the **enthalpy or energy values** used in Born-Haber cycles. Don't worry though, just construct the diagram and calculate the unknown value in exactly **the same way** as for lattice enthalpy. Here's a Born-Haber cycle you could use to calculate the **atomisation enthalpy of magnesium**:

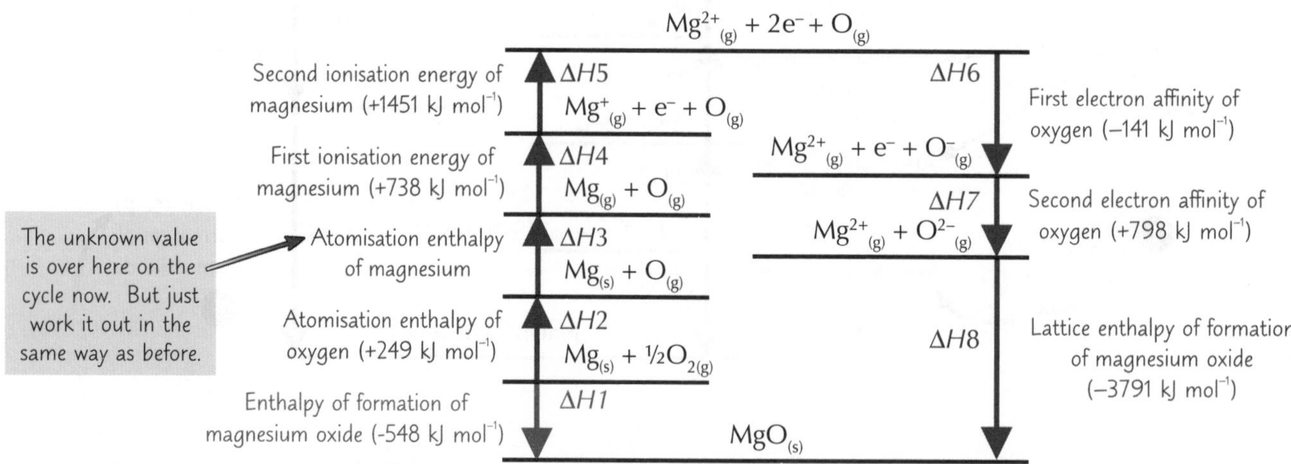

Work your way round the arrows, just like the example on the previous page. Since you're calculating the **atomisation enthalpy of magnesium**, move from [$Mg_{(s)} + O_{(g)}$] to [$Mg_{(g)} + O_{(g)}$] by the **indirect route**.

$\Delta H3 = -\Delta H2 + \Delta H1 - \Delta H8 - \Delta H7 - \Delta H6 - \Delta H5 - \Delta H4$
$= -(+249) + (-548) - (-3791) - (+798) - (-141) - (+1451) - (+738) = \mathbf{+148\ kJ\ mol^{-1}}$

Lattice Enthalpy and Born-Haber Cycles

Theoretical Lattice Enthalpies are Often *Different* from *Experimental Values*

You can work out a **theoretical lattice enthalpy** by doing some calculations based on the **purely ionic model** of a lattice. The purely ionic model of a lattice assumes that all the ions are **spherical**, and have their charge **evenly distributed** around them.

But if you find the **lattice enthalpy experimentally**, the value that you get is often different. This is **evidence** that most ionic compounds have some **covalent character**.

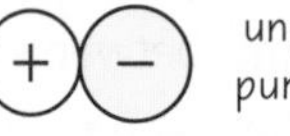

unpolarised ions — purely ionic bonding

The positive and negative ions in a lattice **aren't** usually exactly spherical. Positive ions **polarise** neighbouring negative ions to different extents, and the **more polarisation** there is, the **more covalent** the bonding will be.

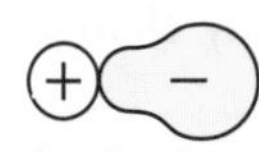

polarised ions — partial covalent bonding

Comparing Lattice Enthalpies Can Tell You '*How Ionic*' an Ionic Lattice Is

Here are experimental and theoretical lattice enthalpy values for some **magnesium halides** and some **sodium halides**.

Compound	Lattice Enthalpy of Formation (kJ mol^{-1})	
	From experimental values	From theory
$MgCl_2$	–2526	–2326
$MgBr_2$	–2440	–2097
MgI_2	–2327	–1944
NaCl	–787	–766
NaBr	–742	–731
NaI	–698	–686

The differences between experimental and theoretical lattice enthalpies are **much bigger** for the **magnesium halides** than for the **sodium halides**. This shows that the bonding in **magnesium halides** is **stronger** than the **ionic model** predicts, so the bonds are **strongly polarised** and have quite a lot of **covalent character.** The bonding in the **sodium halides** is **similar** to the predictions of the **ionic model**, so the compounds are close to being **purely ionic**.

Practice Questions

Q1 Is lattice formation an exothermic or endothermic process?

Q2 Explain why theoretical lattice enthalpies are often different from experimentally determined lattice enthalpies.

Exam Questions

Q1 Using this data:

$\Delta_f H$ [potassium bromide] = –394 kJ mol^{-1} $\Delta_{at} H$ [bromine] = +112 kJ mol^{-1} $\Delta_{at} H$ [potassium] = +89 kJ mol^{-1}
$\Delta_{ie1} H$ [potassium] = +419 kJ mol^{-1} $\Delta_{ea1} H$ [bromine] = –325 kJ mol^{-1}

a) Construct a Born-Haber cycle for potassium bromide (KBr). [3 marks]

b) Use your Born-Haber cycle to calculate the lattice formation enthalpy of potassium bromide. [2 marks]

Q2 Using this data:

$\Delta_f H$ [aluminium chloride] = –706 kJ mol^{-1} $\Delta_{at} H$ [chlorine] = +122 kJ mol^{-1} $\Delta_{at} H$ [aluminium] = +326 kJ mol^{-1}
$\Delta_{ea1} H$ [chlorine] = –349 kJ mol^{-1} $\Delta_{ie2} H$ [aluminium] = +1817 kJ mol^{-1}
$\Delta_{ie3} H$ [aluminium] = +2745 kJ mol^{-1} $\Delta_{lattice} H$ [aluminium chloride] = –5491 kJ mol^{-1}

a) Construct a Born-Haber cycle for aluminium chloride ($AlCl_3$). [3 marks]

b) Use your Born-Haber cycle to calculate the first ionisation energy of aluminium. [2 marks]

Using Born-Haber cycles — it's just like riding a bike...

All this energy going in and out can get a bit confusing. Remember these simple rules: 1) It takes energy to break bonds, but energy is given out when bonds are made. 2) A negative ΔH means energy is given out (it's exothermic). 3) A positive ΔH means energy is taken in (it's endothermic). 4) Never return to a firework once lit.

Enthalpies of Solution

Once you know what's happening when you stir sugar into your tea, your cuppa'll be twice as enjoyable.

Dissolving Involves *Enthalpy Changes*

Ionic lettuce.

When a solid **ionic lattice** dissolves in water, these **two** things happen:

1) The bonds between the ions **break** to give free ions — this is **endothermic**.
2) Bonds between the ions and the water are **made** — this is **exothermic**.
3) The **enthalpy change of solution** is the overall effect on the enthalpy of these two things.

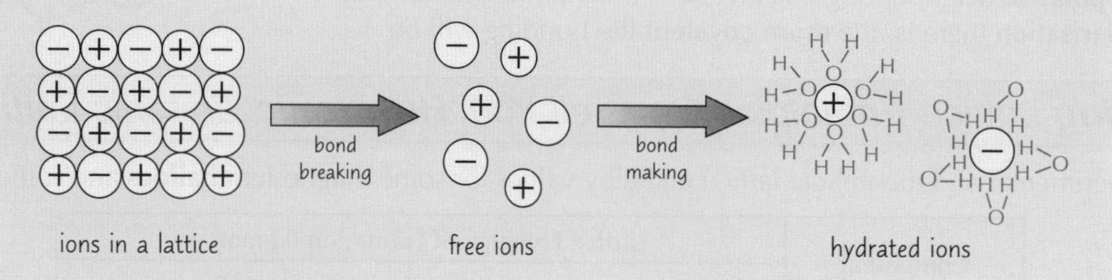
bond breaking
bond making
ions in a lattice
free ions
hydrated ions

4) Water molecules can bond to the ions because oxygen is more **electronegative** than hydrogen, so it draws electrons towards itself, creating a **dipole**. The dipole means the **positively charged hydrogen** atoms can form bonds with **negative ions** and **negatively charged oxygen** atoms can form bonds with **positive ions**.

Substances generally **only** dissolve if the energy released is roughly the same, or **greater than** the energy taken in. So soluble substances tend to have **exothermic** enthalpies of solution.

Enthalpy Change of Solution can be *Calculated*

1) You can work out the **enthalpy change of solution** using a Born-Haber cycle.
2) Imagine that, instead of dissolving the compound directly, you are going to break the lattice into separate **gaseous** ions and then dissolve the gaseous ions in water.
3) Both of these are standard enthalpy changes that you can look up — the **lattice dissociation enthalpy** and the **enthalpies of hydration** of the ions. So you can use them to construct a Born-Haber cycle to find the enthalpy change of solution.

In reality, of course, you don't turn the ions into gases. But the net effect is the same (you start with a solid lattice and end with dissolved ions), so the energy change is the same too.

Here's how to draw the Born-Haber cycle for working out the **enthalpy change of solution** for **sodium chloride**.

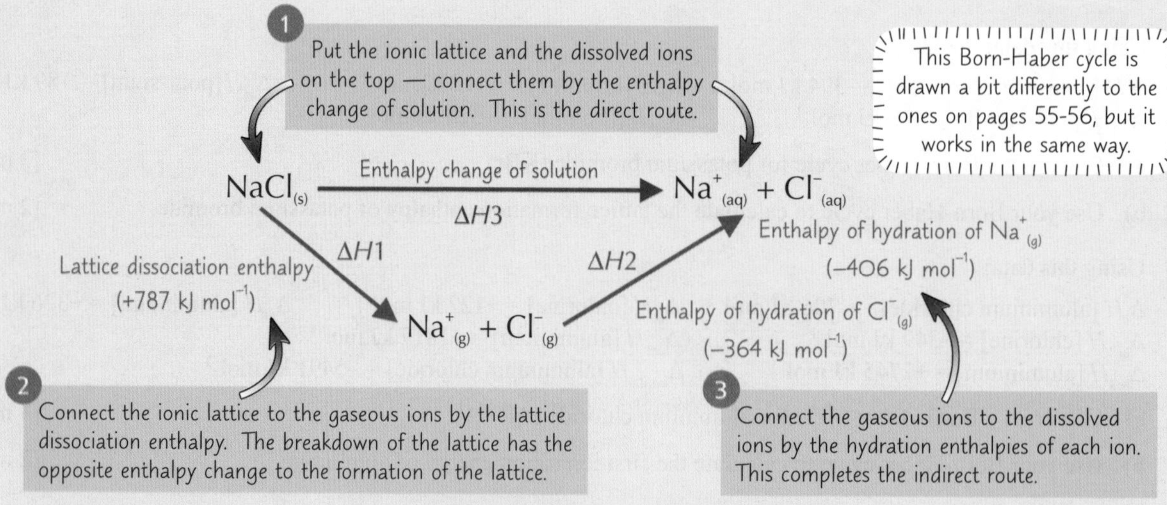
1
Put the ionic lattice and the dissolved ions on the top — connect them by the enthalpy change of solution. This is the direct route.
This Born-Haber cycle is drawn a bit differently to the ones on pages 55-56, but it works in the same way.
NaCl(s)
Enthalpy change of solution
ΔH3
Na+(aq) + Cl−(aq)
Enthalpy of hydration of Na+(g)
(−406 kJ mol−1)
Lattice dissociation enthalpy
(+787 kJ mol−1)
ΔH1
ΔH2
Na+(g) + Cl−(g)
Enthalpy of hydration of Cl−(g)
(−364 kJ mol−1)
2
Connect the ionic lattice to the gaseous ions by the lattice dissociation enthalpy. The breakdown of the lattice has the opposite enthalpy change to the formation of the lattice.
3
Connect the gaseous ions to the dissolved ions by the hydration enthalpies of each ion. This completes the indirect route.

From Hess's law: $\Delta H3 = \Delta H1 + \Delta H2 = +787 + (-406 + -364) = \mathbf{+17\ kJ\ mol^{-1}}$

The enthalpy change of solution is **slightly endothermic**. (But, in case you're wondering, there are other factors at work here that mean that sodium chloride will still dissolve in water — there's more about this on page 60).

Enthalpies of Solution

Here's another. This one's for working out the **enthalpy change of solution for silver chloride:**

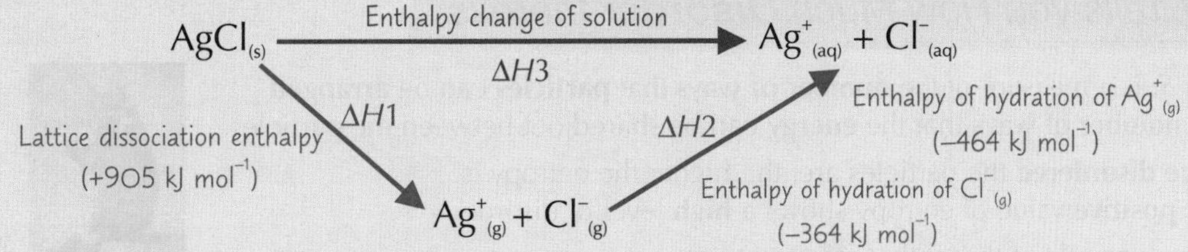

From Hess's law: $\Delta H3 = \Delta H1 + \Delta H2 = +905 + (-464 + -364) =$ **+77 kJ mol⁻¹**

This is much **more endothermic** than the enthalpy change of solution for sodium chloride. As such, silver chloride is **insoluble** in water.

As long as there's only one unknown enthalpy value, you can use these cycles to work out any value on the arrows. For example, if you know the enthalpy change of solution and the enthalpy changes of hydration, you can use those values to work out the lattice dissociation enthalpy.

Practice Questions

Q1 Describe the two steps that occur when an ionic lattice dissolves in water.

Q2 Do soluble substances have exothermic or endothermic enthalpies of solution, in general?

Q3 Sketch a Born-Haber cycle to calculate the enthalpy change of solution of magnesium chloride.

Exam Questions

Q1 a) Draw a Born-Haber cycle for the enthalpy change of solution of $AgF_{(s)}$. Label each enthalpy change. [2 marks]

b) Calculate the enthalpy change of solution for AgF, using the data below. [1 mark]

$\Delta_{lattice}H\,[AgF_{(s)}] = +960\ kJ\ mol^{-1}$, $\Delta_{hyd}H\,[Ag^+_{(g)}] = -464\ kJ\ mol^{-1}$, $\Delta_{hyd}H\,[F^-_{(g)}] = -506\ kJ\ mol^{-1}$.

Q2 Use this Born-Haber cycle to calculate the enthalpy change of solution for SrF_2. [2 marks]

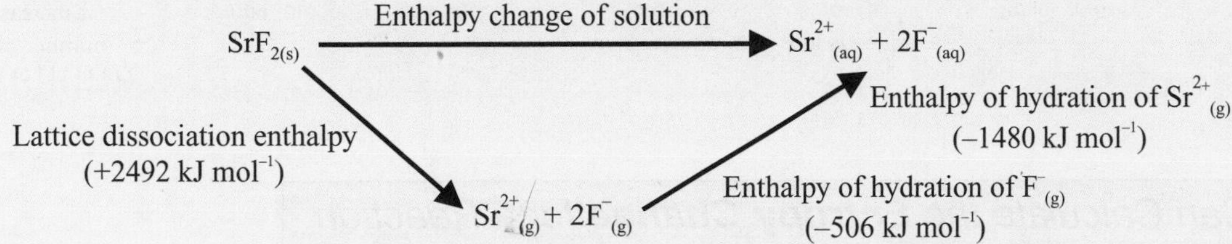

Q3 Show that the enthalpy of hydration of $Cl^-_{(g)}$ is –364 kJ mol⁻¹, given that: [3 marks]

$\Delta_{lattice}H\,[MgCl_{2(s)}] = -2526\ kJ\ mol^{-1}$, $\Delta_{hyd}H\,[Mg^{2+}_{(g)}] = -1920\ kJ\ mol^{-1}$, $\Delta_{sol}H\,[MgCl_{2(s)}] = -122\ kJ\ mol^{-1}$.

N A Chloride & dissociates — attorneys at Hess's Law...

Compared to the ones on pages 55-56, these Born-Haber cycles are an absolute breeze. You've got to make sure the definitions are firmly fixed in your mind though. You only need to know the lattice enthalpy and the enthalpy of hydration of your lattice ions, and you're well on your way to finding out the enthalpy change of solution.

Entropy

Entropy sounds a bit like enthalpy (to start with at least), but they're not the same thing at all. Read on...

***Entropy** Tells you How Much **Disorder** there is*

1) Entropy, S, is a measure of the **number of ways** that **particles** can be **arranged** and the **number of ways** that the **energy** can be shared out between the particles.
2) The more **disordered** the particles are, the higher the entropy is. A **large**, **positive** value of entropy shows a **high** level of disorder.
3) There are a couple of things that affect entropy:

Squirrels do not teach Chemistry. But if they did, this is what a demonstration of increasing entropy would look like.

Physical State affects Entropy

You have to go back to the good old **solid-liquid-gas** particle explanation thingy to understand this.

Solid particles just wobble about a fixed point — there's **hardly any** disorder, so they have the **lowest entropy**.

Gas particles whizz around wherever they like. They've got the most **disordered arrangements** of particles, so they have the **highest entropy**.

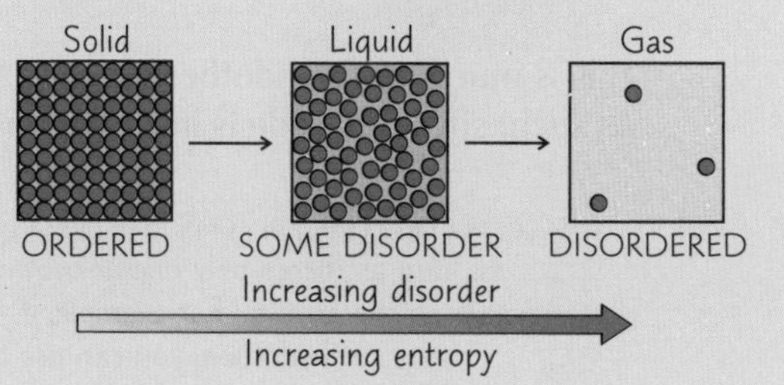

More Particles means More Entropy

It makes sense — the more particles you've got, the **more ways** they and their energy can be **arranged** — so in a reaction like $N_2O_{4(g)} \rightarrow 2NO_{2(g)}$, entropy increases because the **number of moles** increases.

*More **Arrangements** Means More **Stability***

1) Substances always tend towards disorder — they're actually more **energetically stable** when there's more disorder. So particles will move to **increase their entropy**.
2) This is why some reactions are **feasible** (they just happen by themselves — without the addition of energy) even when the enthalpy change is **endothermic**.

Feasible reactions are sometimes called spontaneous reactions.

Example: The reaction of sodium hydrogencarbonate with hydrochloric acid is an **endothermic reaction** — but it is **feasible**. This is due to an **increase in entropy** as the reaction produces carbon dioxide gas and water. Liquids and gases are **more disordered** than solids and so have a **higher entropy**. This increase in entropy **overcomes** the change in enthalpy.

$NaHCO_{3(s)}$	+	$H^+_{(aq)}$	$\rightarrow$	$Na^+_{(aq)}$	+	$CO_{2(g)}$	+	$H_2O_{(l)}$
1 mol solid		1 mol aqueous ions		1 mol aqueous ions		1 mol gas		1 mol liquid

The reaction is also favoured because it increases the number of moles.

*You Can Calculate the **Entropy Change** for a Reaction*

1) During a reaction, there's an entropy change, ΔS, between the **reactants and products**.
2) You can calculate ΔS using this formula:

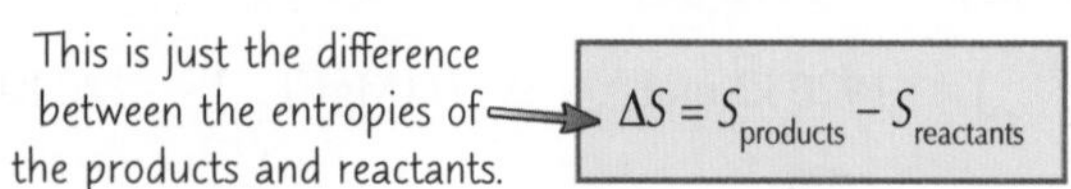

This is just the difference between the entropies of the products and reactants.

$$\Delta S = S_{products} - S_{reactants}$$

The units of entropy are $J\,K^{-1}\,mol^{-1}$.

3) You can find $S_{products}$ by adding up the **standard entropy** of all of the products, and $S_{reactants}$ by adding up the standard entropy of all of the reactants.
4) The **standard entropy** of a substance, S°, is the entropy of 1 mole of that substance under standard conditions (at a pressure of 100 kPa and a temperature of 298 K).

Entropy

Example: Use the data given below to calculate the entropy change of the reaction between ammonia and hydrogen chloride: $\mathbf{NH_{3(g)} + HCl_{(g)} \rightarrow NH_4Cl_{(s)}}$

$S^\circ[NH_{3(g)}] = 192 \text{ J K}^{-1} \text{ mol}^{-1}$, $S^\circ[HCl_{(g)}] = 187 \text{ J K}^{-1} \text{ mol}^{-1}$, $S^\circ[NH_4Cl_{(s)}] = 94.6 \text{ J K}^{-1} \text{ mol}^{-1}$.

$S_{products} = S^\circ[NH_4Cl_{(s)}] = 94.6 \text{ J K}^{-1} \text{ mol}^{-1}$

$S_{reactants} = S^\circ[NH_{3(g)}] + S^\circ[HCl_{(g)}] = 192 + 187 = 379 \text{ J K}^{-1} \text{ mol}^{-1}$

$\Delta S = S_{products} - S_{reactants} = 94.6 - 379 = \mathbf{-284.4 \text{ J K}^{-1} \text{ mol}^{-1}}$

This shows a negative change in entropy. It's not surprising as 2 moles of gas have combined to form 1 mole of solid.

If the balanced equation for the reaction contains more than one mole of a reactant or product, you'll need to multiply the entropy value for that substance by the same number.

Example: Nitrogen gas reacts with hydrogen gas to form ammonia: $\mathbf{3H_{2(g)} + N_{2(g)} \rightarrow 2NH_{3(g)}}$

Use the data given below to calculate the entropy change associated with this reaction.

$S^\circ[NH_{3(g)}] = 192 \text{ J K}^{-1} \text{ mol}^{-1}$, $S^\circ[H_{2(g)}] = 131 \text{ J K}^{-1} \text{ mol}^{-1}$, $S^\circ[N_{2(g)}] = 192 \text{ J K}^{-1} \text{ mol}^{-1}$.

$S_{products} = 2 \times S^\circ[NH_{3(g)}] = 2 \times 192 = 384 \text{ J K}^{-1} \text{ mol}^{-1}$

$S_{reactants} = (3 \times S^\circ[H_{2(g)}]) + S^\circ[N_{2(g)}] = (3 \times 131) + 192 = 585 \text{ J K}^{-1} \text{ mol}^{-1}$

$\Delta S = S_{products} - S_{reactants} = 384 - 585 = \mathbf{-201 \text{ J K}^{-1} \text{ mol}^{-1}}$

An increase in total entropy means that it is **feasible** for a reaction to occur, but it's not a **guarantee** that it will — **enthalpy**, **temperature** and **kinetics** also play a part in whether or not a reaction occurs.

There's more about this on the next page...

Practice Questions

Q1 What does the term 'entropy' mean?

Q2 In each of the following pairs, say which one will have the greater entropy.

a) 1 mole of $NaCl_{(aq)}$ and 1 mole of $NaCl_{(s)}$ b) 1 mole of $Br_{2(l)}$ and 1 mole of $Br_{2(g)}$

c) 1 mole of $Br_{2(g)}$ and 2 moles of $Br_{2(g)}$

Q3 Write down the formula used for finding the entropy change of a reaction.

Exam Questions

Q1 Zinc carbonate breaks down when heated to give zinc oxide and carbon dioxide: $ZnCO_{3(s)} \rightarrow ZnO_{(s)} + CO_{2(g)}$

Would you expect the entropy change for this reaction to be positive or negative? Explain your answer. [2 marks]

Q2 Magnesium burns in oxygen according to the following equation:

$$Mg_{(s)} + \tfrac{1}{2}O_{2(g)} \rightarrow MgO_{(s)}$$

Use the data on the right to calculate the entropy change for this reaction. [2 marks]

Substance	S° (J K^{-1} mol^{-1})
$Mg_{(s)}$	32.7
$O_{2(g)}$	205
$MgO_{(s)}$	26.9

Q3 Hydrogen peroxide decomposes to give water and oxygen gas:

$$2H_2O_{2(l)} \rightarrow 2H_2O_{(l)} + O_{2(g)}$$

Use the data given below to calculate the entropy change for this reaction.

$S^\circ[H_2O_{2(l)}] = 110 \text{ J K}^{-1} \text{ mol}^{-1}$, $S^\circ[O_{2(g)}] = 205 \text{ J K}^{-1} \text{ mol}^{-1}$, $S^\circ[H_2O_{(l)}] = 69.9 \text{ J K}^{-1} \text{ mol}^{-1}$. [2 marks]

Being neat and tidy is against the laws of nature...

Well, there you go. Entropy in all its glory. You haven't seen the back of it yet though, oh no. There's more where this came from. Which is why, if random disorder has left you in a spin, I'd suggest reading it again and making sure you've got your head round this lot before you turn over. You'll thank me for it, you will... Chocolates are always welcome...

Free Energy Change

Free energy — I could definitely do with a bit of that. My gas bill is astronomical.

For a Reaction to be **Feasible** ΔG must be **Negative** or **Zero**

Free energy change, ΔG, is a measure used to predict whether a reaction is **feasible**.
If ΔG is **negative or equal to zero**, then the reaction might happen by itself.
Free energy change takes into account the changes in **enthalpy** and **entropy** in the system. And of course, there's a formula for it:

Even if ΔG shows that a reaction is theoretically feasible, it might have a very high activation energy or be so slow that you wouldn't notice it happening at all.

ΔG = free energy change (in J mol^{-1})

$$\Delta G = \Delta H - T\Delta S$$

ΔH = enthalpy change (in J mol^{-1})
T = temperature (in K)
ΔS = entropy change (in J K^{-1} mol^{-1}) (see pages 60-61)

Example: Calculate the free energy change for the following reaction at 298 K.

$MgCO_{3(g)} \rightarrow MgO_{(s)} + CO_{2(g)}$ $\Delta H^{\ominus} = +117\ 000$ J mol^{-1}, $\Delta S^{\ominus} = +175$ J K^{-1} mol^{-1}

$\Delta G = \Delta H - T\Delta S = +117\ 000 - (298 \times (+175)) = \mathbf{+64\ 900\ J\ mol^{-1}}$ (3 s.f.)

ΔG is positive — so the reaction isn't feasible at this temperature.

The **Feasibility** of Some Reactions Depends on **Temperature**

If a reaction is exothermic (**negative** ΔH) and has a **positive** entropy change, then ΔG is **always negative** since $\Delta G = \Delta H - T\Delta S$. These reactions are feasible at any temperature.
If a reaction is endothermic (**positive** ΔH) and has a **negative** entropy change, then ΔG is **always positive**. These reactions are not feasible at any temperature.
But for other combinations, temperature has an effect.

1) If ΔH is **positive** (endothermic) and ΔS is **positive** then the reaction won't be feasible at some temperatures but will be at a high enough temperature.
For example, the decomposition of calcium carbonate is **endothermic** but results in an **increase in entropy** (the number of molecules increases and CO_2 is a gas).

$$CaCO_{3(s)} \rightarrow CaO_{(s)} + CO_{2(g)}$$

The reaction will only occur when $CaCO_3$ is heated — it isn't feasible at 298 K.

After her surgery, Anne found that a reaction wasn't feasible.

Example: $\Delta H = +10$ kJ mol^{-1}, $\Delta S = +10$ J K^{-1} mol^{-1}

at **300 K** $\Delta G = \Delta H - T\Delta S = +10 \times 10^3 - (300 \times 10) = \mathbf{+7000\ J\ mol^{-1}}$
at **1500 K** $\Delta G = \Delta H - T\Delta S = +10 \times 10^3 - (1500 \times 10) = \mathbf{-5000\ J\ mol^{-1}}$

So a reaction with these enthalpy and entropy changes is feasible at 1500 K, but not at 300 K...

2) If ΔH is **negative** (exothermic) and ΔS is **negative** then the reaction will be feasible at lower temperatures but won't be feasible at higher temperatures.
For example, the process of turning water from a liquid to a solid is **exothermic** but results in a **decrease in entropy** (a solid is more ordered than a liquid), which means it will only occur at certain temperatures (i.e. at 0 °C or below).

Example: $\Delta H = -10$ kJ mol^{-1}, $\Delta S = -10$ J K^{-1}mol^{-1}

at **300 K** $\Delta G = \Delta H - T\Delta S = -10 \times 10^3 - (300 \times -10) = \mathbf{-7000\ J\ mol^{-1}}$
at **1500 K** $\Delta G = \Delta H - T\Delta S = -10 \times 10^3 - (1500 \times -10) = \mathbf{+5000\ J\ mol^{-1}}$

...and this one is feasible at 300 K, but not at 1500 K.

Free Energy Change

You can Calculate the Temperature at which a Reaction Becomes Feasible

When ΔG is **zero**, a reaction is **just feasible.**

You can find the **temperature** when ΔG is zero by rearranging the free energy equation from the previous page.

$\Delta G = \Delta H - T\Delta S$. When $\Delta G = 0$, $T\Delta S = \Delta H$.
So you can find the temperature at which a reaction becomes feasible using the formula:

$$T = \frac{\Delta H}{\Delta S}$$

T = temperature at which a reaction becomes feasible (in K)
ΔH = enthalpy change (in J mol^{-1})
ΔS = entropy change (in J K^{-1} mol^{-1})

Example: Tungsten, W, can be extracted from its ore, WO_3, by reduction using hydrogen:

$$WO_{3(s)} + 3H_{2(g)} \rightarrow W_{(s)} + 3H_2O_{(g)} \qquad \Delta H^{\ominus} = +110 \text{ kJ mol}^{-1}$$

Use the data in the table to find the minimum temperature at which the reaction becomes feasible.

Substance	$S^{\ominus}$/J K^{-1} mol^{-1}
$WO_{3(s)}$	76
$H_{2(g)}$	65
$W_{(s)}$	33
$H_2O_{(g)}$	189

First, convert the **enthalpy change**, ΔH, to joules per mole:

$\Delta H = 110 \times 10^3 = \mathbf{110\ 000\ J\ mol^{-1}}$

Then find the **entropy change**, ΔS:

$\Delta S = S_{products} - S_{reactants} = [33 + (3 \times 189)] - [76 + (3 \times 65)] = \mathbf{+329\ J\ K^{-1}\ mol^{-1}}$

See page 60 for more on the entropy change formula.

Then divide ΔH by ΔS to find the temperature at which the reaction just becomes feasible:

$$T = \frac{\Delta H}{\Delta S} = \frac{110\ 000}{329} = \mathbf{334\ K}$$

Practice Questions

Q1 What does ΔG stand for? What is it used for?

Q2 State whether each of the following reactions are feasible, feasible at certain temperatures or not feasible:
Reaction A: positive ΔH and negative ΔS Reaction B: negative ΔH and positive ΔS

Q3 Write down the formula for calculating the temperature at which a reaction becomes feasible.

Exam Questions

Q1 Magnesium carbonate decomposes to form magnesium oxide and carbon dioxide:

$$MgCO_{3(s)} \rightarrow MgO_{(s)} + CO_{2(g)} \qquad \Delta H = +117 \text{ kJ mol}^{-1} \qquad \Delta S = +175 \text{ J K}^{-1} \text{ mol}^{-1}$$

a) Calculate the free energy change of this reaction at the following temperatures:

i) 500 K ii) 760 K [2 marks]

b) At which of the temperatures in part a) is the reaction feasible? Explain your answer. [2 marks]

Q2 a) Use the equation below and the table on the right to calculate the free energy change for the complete combustion of methane at 298 K. [2 marks]

$$CH_{4(g)} + 2O_{2(g)} \rightarrow CO_{2(g)} + 2H_2O_{(l)} \qquad \Delta H^{\ominus} = -730 \text{ kJ mol}^{-1}$$

b) What is the maximum temperature at which the reaction is feasible? [2 marks]

Substance	$S^{\ominus}$(J K^{-1} mol^{-1})
$CH_{4(g)}$	186
$O_{2(g)}$	205
$CO_{2(g)}$	214
$H_2O_{(l)}$	69.9

The feasibility of revision depends on what's on the telly...

These pages are a bit tricky if you ask me — so make sure you've properly understood them before you move on. The most important bit to learn is the formula for ΔG. If you know that, then you can always work out whether a reaction is feasible even if you can't remember the rules about positive and negative enthalpy and entropy.

Rate Equations

Sorry — this section gets a bit mathsy. Just take a deep breath, dive in, and don't bash your head on the bottom.

Reaction Rate *tells you How Fast* ***Reactants*** *are Converted to* ***Products***

As you saw on page 42, **reaction rate** is the **change in amount** of reactant or product **per unit time** (usually seconds). E.g. if the reactants are in solution, the rate will be **change in concentration per second**. The units will be **mol dm^{-3} s^{-1}**.

You can Work out ***Reaction Rate*** *from the* ***Gradient of a Graph***

If you draw a graph of the **amount of reactant or product** (on the *y* axis) against **time** (on the *x* axis), then the reaction rate is just the **gradient** of the graph. You can work out the gradient of a line using this formula:

gradient = change in y ÷ change in x

Example: The data on this graph came from measuring the volume of gas given off during a chemical reaction.

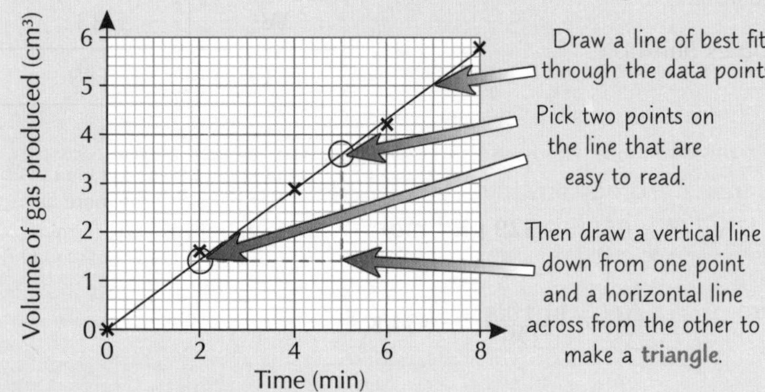
Volume of gas produced (cm³)
Time (min)
Draw a line of best fit through the data points.
Pick two points on the line that are easy to read.
Then draw a vertical line down from one point and a horizontal line across from the other to make a triangle.

change in y = 3.6 – 1.4 = 2.2 cm^3

change in x = 5.0 – 2.0 = 3.0 minutes

gradient = 2.2 ÷ 3.0 = 0.73 cm^3 min^{-1}

So the rate of reaction = **0.73 cm^3 min^{-1}**

You may need to ***Work out*** *the* ***Gradient*** *from a* ***Curved Graph***

When the points on a graph lie in a **curve**, you can't draw a straight line of best fit through them. But you can still work out the gradient, and so the rate, at a **particular point** in the reaction by working out the **gradient of a tangent**.

A tangent is a line that just touches a curve and has the same gradient as the curve does at that point.

1. Find the point on the curve that you need to look at. For example, if you want to find the rate of reaction at 3 minutes, find 3 on the *x*-axis and go up to the curve from there.

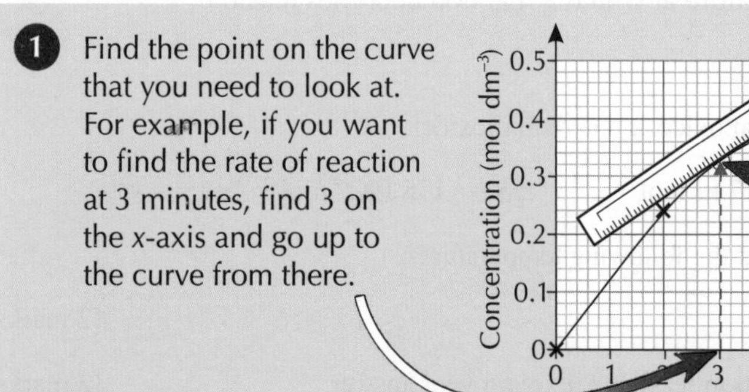
Concentration (mol dm⁻³)
Time (min)

2. Place a ruler at that point so that it's just touching the curve. Position the ruler so that you can see the whole curve.

3. Adjust the ruler until the space between the ruler and the curve is equal on both sides of the point.

Concentration (mol dm^{-3})

Time (min)

Pick two points on the tangent that are easy to read.

4. Draw a line along the ruler to make the tangent. Extend the line right across the graph — it'll help to make your gradient calculation easier as you'll have more points to choose from.

5. Calculate the gradient of the tangent to find the rate:

gradient = change in y ÷ change in x
= (0.46 – 0.22) ÷ (5.0 – 1.4)
= 0.24 mol dm^{-3} ÷ 3.6 mins = 0.067 mol dm^{-3} min^{-1}

So, the rate of reaction at 3 mins is **0.067 mol dm^{-3} min^{-1}**.

Don't forget the units — you've divided mol dm^{-3} by mins, so it's mol dm^{-3} min^{-1}.

Rate Equations

*The **Rate Equation** links **Reaction Rate** to **Reactant Concentrations***

Rate equations look scary, but all they're telling you is how the **rate** is affected by the **concentrations of reactants**. For a general reaction **A + B → C + D**, the **rate equation** is:

The units of rate are mol dm^{-3} s^{-1}. → **Rate = k[A]^m[B]^n**

Remember — square brackets mean the concentration of whatever's inside them.

m and **n** are the orders of reaction and ***k*** is the rate constant. If you want to know what all that means then read on...

Orders of Reaction** Tell You How the **Reactant Concentrations** Affect the **Rate

1) **m** and **n** are the **orders of the reaction** with respect to reactant A and reactant B. An order of reaction is defined as the **power** to which the **concentration** of its reactant is raised in the rate equation.
2) **m** tells you how the **concentration of reactant A** affects the **rate** and **n** tells you the same for **reactant B**.

> If [A] changes and the rate **stays the same**, the order of reaction with respect to A is **0**.
> So if [A] doubles, the rate will stay the same. If [A] triples, the rate will stay the same.
>
> If the rate is **proportional to [A]**, then the order of reaction with respect to A is **1**.
> So if [A] doubles, the rate will double. If [A] triples, the rate will triple.
>
> If the rate is **proportional to [A]^2**, then the order of reaction with respect to A is **2**.
> So if [A] doubles, the rate will be 2^2 = 4 times faster. If [A] triples, the rate will be 3^2 = 9 times faster.

3) The **overall order of the reaction** is **m + n**.
4) You can only find **orders of reaction** from **experiments**. You **can't** work them out from chemical equations.

k** Relates the **Reactant Concentrations** to the **Rate** at a **Particular Temperature

1) ***k*** is the **rate constant** — a number that links the rate of the reaction to the concentration of the reactants. The bigger the value of ***k*** is, the **faster** the reaction.
2) The rate constant is **always the same** for a certain reaction at a **particular temperature** — but if you **increase** the temperature, the rate constant rises too.

> When you **increase** the **temperature** of a reaction, the **rate of reaction increases** — you're increasing the **number of collisions** between reactant molecules, and also the **energy** of each collision. But the **concentrations** of the reactants and the **orders of reaction** stay the same. So the value of ***k*** must **increase** for the rate equation to balance.

Pages 72-73 deal with the equation that links *k* with temperature.

*If You Know the **Rate Constant** and **Orders**, You can Calculate the **Rate***

Example: Propanone will react with iodine in the presence of an acid-catalyst: $CH_3COCH_{3(aq)} + I_{2(aq)} \xrightarrow{H^+_{(aq)}} CH_3COCH_2I_{(aq)} + H^+_{(aq)} + I^-_{(aq)}$

This reaction is first order with respect to propanone and $H^+_{(aq)}$ and zero order with respect to iodine. At a certain temperature, k was found to be 520 mol^{-1} dm^3 s^{-1} when $[CH_3COCH_3] = [I_2] = [H^+]$ = 1.50 × 10^{-3} mol dm^{-3}. Calculate the rate at this temperature.

1) The **rate equation** is: rate = $k[CH_3COCH_3]^1[H^+]^1[I_2]^0$
   [X]^1 is usually written as **[X]**. [X]^0 equals **1**.
   So you can **simplify** it to: **rate = $k[CH_3COCH_3][H^+]$**

   Even though H$^+$ is a catalyst, rather than a reactant, it can still be in the rate equation.

2) Now you have the rate equation, calculating the **rate** is as simple as putting in the values:
   rate = $k[CH_3COCH_3][H^+]$ = 520 × (1.50 × 10^{-3}) × (1.50 × 10^{-3}) = **1.17 × 10^{-3}**
3) You can find the **units** for the rate by putting the other units into the rate equation.
   The units for the rate are ~~mol^{-1}~~ ~~dm^3~~ s^{-1} × ~~mol~~ ~~dm^{-3}~~ × mol dm^{-3} = **mol dm^{-3} s^{-1}**.

   mol and mol^{-1} cancel each other out, as do dm^3 and dm^{-3}.

   So the answer is: rate = **1.17 × 10^{-3} mol dm^{-3} s^{-1}**

Rate Equations

You can Calculate the Rate Constant from the Orders and Rate of Reaction

Once the rate and the orders of the reaction have been found by experiment, you can work out the **rate constant**, ***k***. The **units of *k* vary**, so you'll need to work them out too.

Example: The reaction below is second order with respect to NO and zero order with respect to CO and O_2.

$$NO_{(g)} + CO_{(g)} + O_{2(g)} \rightarrow NO_{2(g)} + CO_{2(g)}$$

At a certain temperature, when $[NO_{(g)}] = [CO_{(g)}] = [O_{2(g)}] = 2.00 \times 10^{-3}$ mol dm^{-3} the rate is 1.76×10^{-3} mol dm^{-3} s^{-1}. Find the value of the rate constant, *k*, at this temperature.

1) First write out the **rate equation:** $\text{Rate} = k[NO]^2[CO]^0[O_2]^0 = k[NO]^2$

2) Next insert the **concentration** and the **rate**. **Rearrange** the equation and calculate the value of *k*:

$$\text{Rate} = k[NO]^2\text{, so } 1.76 \times 10^{-3} = k \times (2.00 \times 10^{-3})^2 \Rightarrow k = \frac{1.76 \times 10^{-3}}{(2.00 \times 10^{-3})^2} = 440$$

3) Find the **units of *k*** by putting the other units in the rate equation:

$$\text{Rate} = k[NO]^2\text{, so mol dm}^{-3}\text{ s}^{-1} = k \times (\text{mol dm}^{-3})^2 \Rightarrow k = \frac{\cancel{\text{mol dm}^{-3}}\text{ s}^{-1}}{(\cancel{\text{mol dm}^{-3}})(\text{mol dm}^{-3})} = \frac{\text{s}^{-1}}{(\text{mol dm}^{-3})} = \text{dm}^3\text{ mol}^{-1}\text{ s}^{-1}$$

So the answer is: $k = 440$ **dm^3 mol^{-1} s^{-1}**

Practice Questions

Q1 Write the general rate equation for the reaction A + B → C + D.

Q2 Explain what is meant by the term 'order of reaction'.

Q3 How would the value of *k* change if you decreased the temperature of a chemical reaction?

Exam Questions

Q1 Compounds X and Y react as in the equation below.

$$X + Y \rightarrow Z$$

a) From the graph on the right, work out the rate of reaction at 3 minutes. [3 marks]

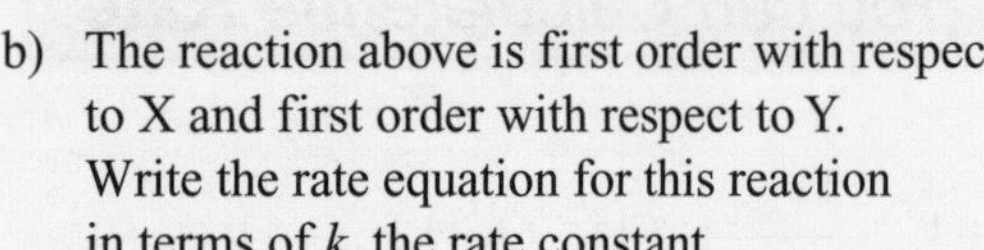

b) The reaction above is first order with respect to X and first order with respect to Y. Write the rate equation for this reaction in terms of *k*, the rate constant. [1 mark]

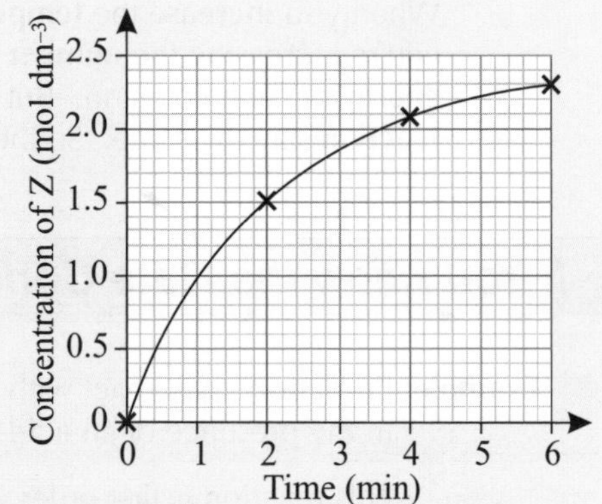

Q2 The following reaction is second order with respect to NO and first order with respect to H_2.

$$2NO_{(g)} + 2H_{2(g)} \rightarrow 2H_2O_{(g)} + N_{2(g)}$$

a) Write a rate equation for the reaction in terms of *k*, the rate constant. [1 mark]

b) The rate of the reaction at 800 °C was determined to be 0.00267 mol dm^{-3} s^{-1} when $[H_2] = 0.00200$ mol dm^{-3} and $[NO] = 0.00400$ mol dm^{-3}. Calculate a value for the rate constant at 800 °C, including units. [2 marks]

Rate of Chemistry Revision = k [Student] [Tea] ...

These rate equations might look a bit odd, but knowing how to use them can be really handy. Learn what all the different bits of the equation mean, and how to use it to find the overall order of reaction. Practise drawing some tangents to curved graphs and working out their gradients too — we're not through with all that graph stuff yet...

Rate Experiments

Told you, didn't I? Here are some more pages on rates and graphs for you to get your head around...

The Initial Rates Method can be used to work out Rate Equations

The **initial rate of a reaction** is the rate right at the **start** of the reaction. You can find this from a **concentration-time** graph by calculating the **gradient** of the **tangent** at **time = 0**.

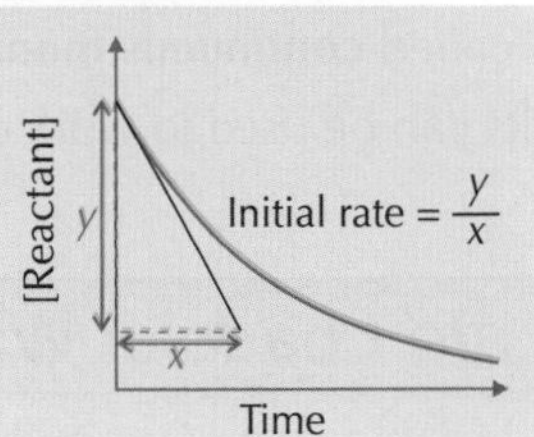

Here's a quick explanation of how to use the **initial rates method**:

1) Repeat an experiment several times using **different initial concentrations** of the reactants. You should usually only change **one** of the concentrations at a time, keeping the rest constant.

You might get exam questions where they change more than one concentration though — handily, there's an example of this below.

2) Calculate the **initial rate** for each experiment using the method above.
3) Finally, see how the **initial concentrations** affect the **initial rates** and figure out the **order** for each reactant. The example below shows you how to do this. Once you know the **orders**, you can work out the rate equation.

Example:
The table on the right shows the results of a series of initial rate experiments for the reaction:

$$NO_{(g)} + CO_{(g)} + O_{2(g)} \rightarrow NO_{2(g)} + CO_{2(g)}$$

The experiments were carried out at a constant temperature.

Write down the rate equation for the reaction.

Experiment number	[NO] ($mol\ dm^{-3}$)	[CO] ($mol\ dm^{-3}$)	[O_2] ($mol\ dm^{-3}$)	Initial rate ($mol\ dm^{-3}\ s^{-1}$)
1	2.0×10^{-2}	1.0×10^{-2}	1.0×10^{-2}	0.17
2	6.0×10^{-2}	1.0×10^{-2}	1.0×10^{-2}	1.53
3	2.0×10^{-2}	2.0×10^{-2}	1.0×10^{-2}	0.17
4	4.0×10^{-2}	1.0×10^{-2}	2.0×10^{-2}	0.68

1) Look at experiments 1 and 2 — when **[NO] triples** (and all the other concentrations stay constant) the rate is **nine times** faster, and $9 = 3^2$. So the reaction is **second order** with respect to NO.
2) Look at experiments 1 and 3 — when **[CO]** doubles (but all the other concentrations stay constant), the rate **stays the same**. So the reaction is **zero order** with respect to CO.
3) Look at experiments 1 and 4 — the rate of experiment 4 is **four times faster** than experiment 1. The reaction is **second order** with respect to **[NO]**, so the rate will **quadruple** when you **double** [NO]. But in experiment 4, **[O_2]** has also been **doubled**. As doubling [O_2] hasn't had any additional effect on the rate, the reaction must be **zero order** with respect to O_2.
4) Now that you know the order with respect to each reactant you can write the rate equation: **rate = k[NO]^2**.

You can calculate k at this temperature by putting the concentrations and the initial rate from one of the experiments into the rate equation (see page 65).

You Can Also Measure the Initial Reaction Rate

For some reactions there is a **sudden colour change** when a product reaches a certain concentration. The **rate of reaction** can be worked out from measuring the **time** it takes for the colour change to happen — the **shorter** the time, the **faster** the rate.

1) In an **iodine clock reaction**, the reaction you're monitoring is: $H_2O_{2(aq)} + 2I^-_{(aq)} + 2H^+_{(aq)} \rightarrow 2H_2O_{(l)} + I_{2(aq)}$

   A small amount of **sodium thiosulfate solution** and **starch** are added to the reaction mixture. The sodium thiosulfate reacts instantly with the iodine that's being formed. But once all of the sodium thiosulfate has been used up, any more iodine that's formed stays in solution. This turns the starch indicator blue-black.

   Varying the **concentration** of iodide or hydrogen peroxide while keeping everything else **constant** will give **different times** for the colour change. These can be used to work out the reaction order.
2) For reactions that produce a **precipitate** that clouds the solution, you can measure the **time** it takes for a **mark** underneath the reaction vessel to **disappear** from view.
3) For other reactions, you can work out the initial reaction rate by measuring the **time** taken for a **small amount of product** to be formed. E.g. the catalytic decomposition of hydrogen peroxide produces O_2, so you could measure the time it takes for 5 cm^3 of O_2 to be released.

Rate Experiments

You Can Measure the Rate of Reaction by Continuous Monitoring

Instead of working out the initial rate of several reactions. you can follow a reaction all the way though to its end by recording the **amount of product** (or **reactant**) you have at **regular time intervals** — this is called **continuous monitoring**.

The results can be used to work out how the **rate changes** over **time**.

There are Loads of Ways to Follow the Rate of a Reaction

Although there are a lot of ways to follow reactions, not every method works for every reaction. You've got to **pick a property** that **changes** as the reaction goes on. You can then use your results to calculate the **concentrations** of reactants at different time points.

Gas volume

If a **gas** is given off, you could **collect it** in a gas syringe and record how much you've got at **regular time intervals** (e.g. every 15 seconds). For example, this would work for the reaction between an **acid** and a **carbonate** in which **carbon dioxide gas** is given off.

To find the concentration of a reactant at each time point, use the **ideal gas equation** (p.15) to work out how many moles of gas you've got, then use the **molar ratio** to work out the concentration of the reactant.

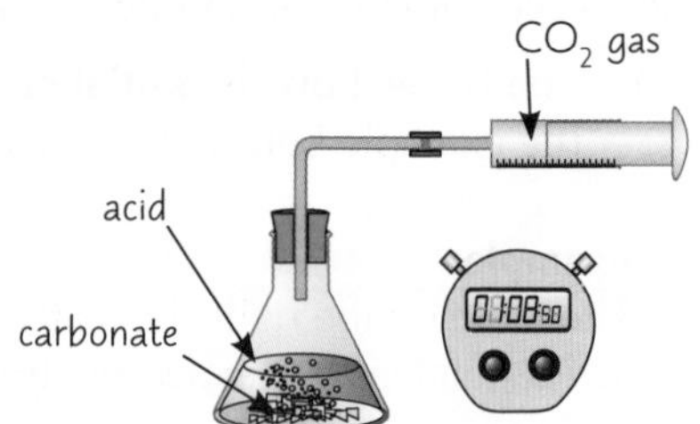

Loss of mass

If a **gas** is given off, the system will **lose mass**. You can measure this at regular intervals with a **balance**.

Use mole calculations to work out how much gas you've lost, and therefore how many moles of reactants are left.

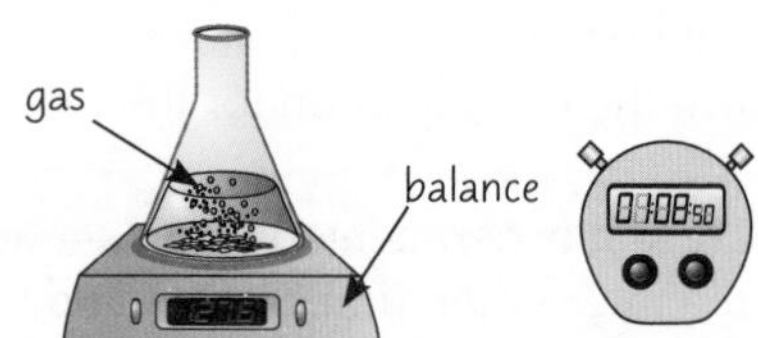

Colour change

You can sometimes track the colour change of a reaction using a gadget called a **colorimeter**. A colorimeter measures **absorbance** (the amount of light absorbed by the solution). The **more concentrated** the **colour** of the solution, the **higher** the **absorbance** is.

For example, in the reaction between propanone and iodine, the **brown** colour fades. So the absorbance of the solution will **decrease**.

$$CH_3COCH_{3(aq)} + I_{2(aq)} \rightarrow CH_3COCH_2I_{(aq)} + H^+_{(aq)} + I^-_{(aq)}$$

colourless (propanone), brown (iodine), colourless (products)

You measure the change in absorbance like this:

1) Plot a **calibration curve** — a graph of **known concentrations** of the coloured solution (in this case I_2) plotted against absorbance. (There's more about calibration curves on page 115.)
2) During the experiment, take a **small sample** from your reaction solution at **regular intervals** and read the **absorbance**.
3) Use your calibration curve to **convert** the absorbance at each time point into a **concentration**.

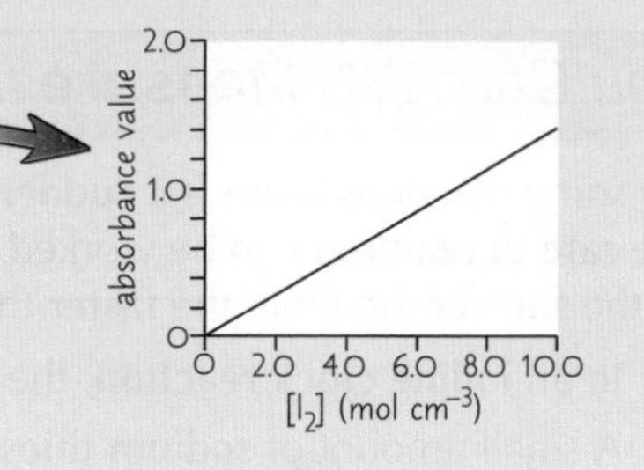

Change in pH

If the reaction produces or uses up H^+ ions, the pH of the solution will change. So you could measure the **pH** of the solution at **regular intervals** and calculate the **concentration of H^+**.

See page 84 for more about working out $[H^+]$ using pH.

You can use the data you've gathered from your rate of reaction experiment to draw a **concentration–time graph**. This'll help you work out the orders of reaction — more about that on the next page...

Rate Experiments

The Shape of a Rate-Concentration Graph Tells You the Order

You can use data from a **concentration–time graph** to construct a **rate-concentration graph**, which can tell you the **reaction order**. Here's how...

1) Find the **gradient** at various points on the graph. This will give you the **rate** at that particular **concentration**. With a **straight-line graph**, this is easy, but if it's a **curve**, you need to draw **tangents** and find their gradients.
2) Now plot each point on a new graph with the axes **rate** and **concentration**. Then draw a smooth line or curve through the points. The shape of the line will tell you the order of the reaction with respect to that reactant.

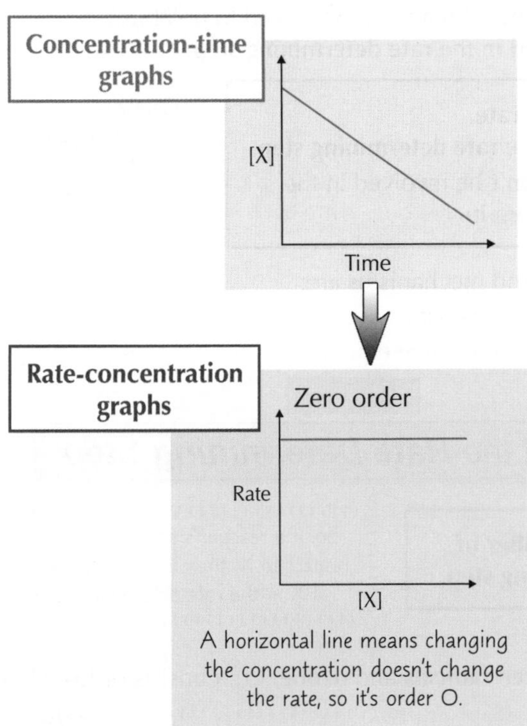

A horizontal line means changing the concentration doesn't change the rate, so it's order 0.

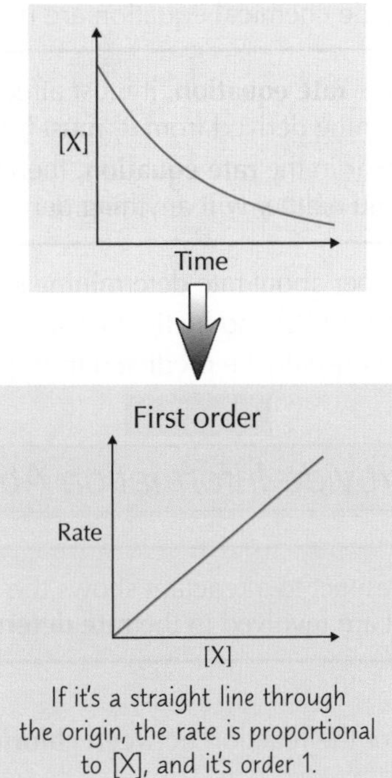

If it's a straight line through the origin, the rate is proportional to [X], and it's order 1.

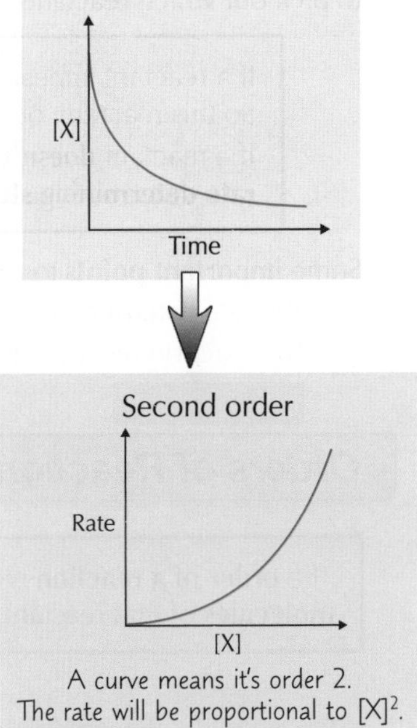

A curve means it's order 2. The rate will be proportional to $[X]^2$. (In theory, a curve could mean a higher order than 2 but you won't be asked about them.)

Practice Questions

Q1 How could you find the initial reaction rate from a concentration-time graph?

Q2 A solid carbonate is added to an acid to produce carbon dioxide gas. Describe how you could follow the rate of reaction for this reaction.

Q3 Sketch rate-concentration graphs for zero, first and second order reactions.

Exam Question

Q1 This table shows the results of a series of initial rate experiments for the reaction between substances D and E at a constant temperature:

Experiment	[D] ($mol\,dm^{-3}$)	[E] ($mol\,dm^{-3}$)	Initial rate × 10^{-3} ($mol\,dm^{-3}\,s^{-1}$)
1	0.2	0.2	1.30
2	0.4	0.2	2.60
3	0.1	0.1	0.65
4	0.6	0.1	

a) Find the order of reaction with respect to reactant D. [1 mark]

b) Find the order of reaction with respect to reactant E. [1 mark]

c) Work out the initial rate for experiment 4. [1 mark]

Describe the link between concentration and rate, soldier — that's an order...

Picking a way to measure the rate of reaction is easy enough once you figure out what physical changes are going to happen, like a colour changing or a gas being produced. But if you need to draw a graph, don't forget that to work out the reaction order you need to plot concentrations, not volumes or pH — so you may have to do some maths first .

The Rate Determining Step

If you want to work out the mechanism of a reaction then you're going to have to know about the rate determining step...

The *Rate Determining Step* is the *Slowest Step* in a Multi-Step Reaction

Mechanisms can have **one step** or a **series of steps**. In a series of steps, each step can have a **different rate**. The **overall rate** is decided by the step with the **slowest** rate — the **rate determining step**.

Otherwise known as the rate-limiting step.

Reactants in the *Rate Equation* Affect the *Rate*

The rate equation is handy for working out the **mechanism** of a chemical reaction. You need to be able to pick out which reactants from the chemical equation are involved in the **rate determining step**:

> If a reactant appears in the **rate equation**, it must affect the **rate**.
> So this reactant, or something derived from it, must be in the **rate determining step**.
> If a reactant **doesn't** appear in the **rate equation**, then it **won't** be involved in the **rate determining step** (and neither will anything derived from it).

Catalysts can appear in rate equations, so they can be in rate-determining steps too.

Some **important points** to remember about rate determining steps and mechanisms are:
1) The rate determining step **doesn't** have to be the first step in a mechanism.
2) The reaction mechanism **can't** usually be predicted from **just** the chemical equation.

Orders of Reaction Provide Information About the *Rate Determining Step*

> The **order of a reaction** with respect to a reactant shows the **number of molecules** of that reactant that are involved in the **rate determining step**.

So, if a reaction's second order with respect to X, there'll be two molecules of X in the rate determining step.

For example, the mechanism for the reaction between **chlorine free radicals** and **ozone**, O_3, consists of **two steps:**

$Cl\bullet_{(g)} + O_{3(g)} \rightarrow ClO\bullet_{(g)} + O_{2(g)}$ — **slow (rate determining step)**

$ClO\bullet_{(g)} + O\bullet_{(g)} \rightarrow Cl\bullet_{(g)} + O_{2(g)}$ — **fast**

The O• radical is formed when O_2 is broken down by UV light. (There's more about this reaction on page 139.)

Cl• and O_3 must both be in the rate equation, so the rate equation will be: **rate = $k[Cl\bullet]^m[O_3]^n$**.
There's only **one** Cl• radical and **one** O_3 molecule in the rate determining step, so the **orders**, m and n, are both **1**.

So the rate equation is **rate = $k[Cl\bullet][O_3]$**.

The rate determining step can sometimes involve an **intermediate** that isn't in the full equation.

Example: The reaction $2NO + O_2 \rightarrow 2NO_2$ has a two-step mechanism:

Step 1 — $NO + NO \rightarrow N_2O_2$

Step 2 — $N_2O_2 + O_2 \rightarrow 2NO_2$

The rate equation for this reaction is: **rate = $k[NO]^2[O_2]$**.
What is the rate determining step of this reaction?

1) From the rate equation, you know that the **rate determining step** must involve **2 molecules of NO** and **1 molecule of O_2**.
2) So the rate determining step can't be step 1, as step 1 does not involve O_2.
3) Step 2 doesn't contain all the molecules you'd expect from the rate equation, but it does contain N_2O_2. N_2O_2 is an **intermediate** molecule **derived from** 2 molecules of NO.
4) So **step 2** must be the rate determining step.

Geoffrey's rate-determining steps were all out of sync with his partner's.

The Rate Determining Step

The **Rate Determining Step** Can Help You Work Out the **Reaction Mechanism**

Knowing which reactants are in the **rate determining step** can help you to work out the **reaction mechanism**.

Example:
There are two possible mechanisms for the nucleophile **OH^-** substituting for **Br** in 2-bromo-2-methylpropane:

$$(CH_3)_3C{-}Br + OH^- \rightarrow (CH_3)_3C{-}OH + Br^-$$

or

$$(CH_3)_3C{-}Br \rightarrow (CH_3)_3C^+ + Br^- \quad \text{— slow (rate determining step)}$$

This involves breaking strong carbon-bromine bonds, so this will be a slow change.

$$(CH_3)_3C^+ + OH^- \rightarrow (CH_3)_3C{-}OH \quad \text{— fast}$$

If the OH^- concentration is high, the positive ion has a good chance of colliding with one. So this step will be fast.

The actual **rate equation** was worked out using rate experiments. It was found to be: **rate = $k[(CH_3)_3CBr]$**
OH^- isn't in the **rate equation**, so it **can't** be involved in the rate determining step.
The **second mechanism** is correct because OH^- **isn't** in the rate determining step.

Practice Questions

Q1 What is meant by the 'rate determining step' of a reaction?

Q2 Explain how knowing the rate equation for a reaction can help you to predict a reaction mechanism.

Q3 For a reaction between three reactants, X, Y and Z, the rate determining step is $X + 2Y \rightarrow M + N$.
What will the order of reaction be with respect to: a) reactant X? b) reactant Y? c) reactant Z?

Exam Questions

Q1 For the reaction $CH_3COOH_{(aq)} + C_2H_5OH_{(aq)} \rightarrow CH_3COOC_2H_{5(aq)} + H_2O_{(l)}$, the rate equation is:

$$\text{rate} = k[CH_3COOH][H^+]$$

What can you deduce about the role that H^+ plays in the reaction? Explain your answer. [2 marks]

Q2 Hydrogen reacts with iodine monochloride as in the equation $H_{2(g)} + 2ICl_{(g)} \rightarrow I_{2(g)} + 2HCl_{(g)}$.
The rate equation for this reaction is: rate = $k[H_2][ICl]$.

a) The mechanism for the reaction consists of two steps.
Identify the molecules that are in the rate determining step. Justify your answer. [3 marks]

b) A chemist suggested the following mechanism for the reaction:

$2ICl_{(g)} \rightarrow I_{2(g)} + Cl_{2(g)}$ slow

$H_{2(g)} + Cl_{2(g)} \rightarrow 2HCl_{(g)}$ fast

Suggest, with reasons, whether this mechanism is likely to be correct. [2 marks]

I found rate determining step aerobics a bit on the slow side...

These pages show you how rate equations, orders of reaction, and reaction mechanisms all tie together and how each actually means something in the grand scheme of A-Level Chemistry. It's all very profound. So get it all learnt, answer the questions and then you'll have plenty of time to practise the quickstep for your Strictly Come Dancing routine.

The Arrhenius Equation

The Arrhenius Equation. As the name suggests, it's a bit heinous to learn I'm afraid, but super useful. It links together reaction constants, activation energies and temperatures — all pretty important in the world of reaction rates.

The Arrhenius Equation Links k with Temperature and Activation Energy

The **Arrhenius equation** (nasty-looking thing in the green box) shows how the **rate constant** (k) varies with **temperature** (T) and **activation energy** (E_a, the minimum amount of kinetic energy particles need to react).

This is probably the **worst** equation you're going to meet. Luckily, it'll be given to you in your exams if you need it, so you don't have to learn it off by heart. But you do need to know what all the different bits **mean**, and how it works. Here it is:

What be a pirate's favourite part of chemistry? The ARR-henius equation!

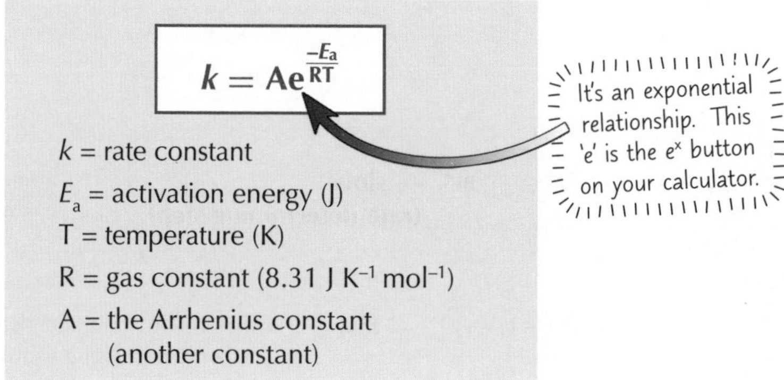

$$k = Ae^{\frac{-E_a}{RT}}$$

k = rate constant

E_a = activation energy (J)

T = temperature (K)

R = gas constant (8.31 J K^{-1} mol^{-1})

A = the Arrhenius constant (another constant)

It's an exponential relationship. This 'e' is the e^x button on your calculator.

1) As the activation energy, E_a, gets **bigger**, k gets **smaller**. You can **test** this out by trying **different numbers** for E_a in the equation... ahh go on, have a go.
2) So, a **large E_a** will mean a **slow rate**. This **makes sense** when you think about it — if a reaction has a **high activation energy**, then not many of the reactant particles will have enough energy to react. So only a **few** of the collisions will result in the reaction actually happening, and the rate will be **slow**.
3) The equation also shows that as the temperature **rises, k increases**. (You can **test** this out by trying **different numbers** for T as well. Will the fun never cease?)
4) The **temperature dependence** makes sense too. **Higher temperatures** mean reactant particles move around **faster** and with **more energy** so they're more likely to **collide** and more likely to collide with at least the **activation energy**, so the **reaction rate increases**.

Use the Arrhenius Equation to Calculate the Rate Constant or E_a

You might be given four of the five values from the Arrhenius equation and asked to use the equation to find the value of the fifth. Here's how you'd go about getting the answer...

Example: The decomposition of N_2O_5 at 308 K has a rate constant of 1.35×10^{-4} s^{-1}. The Arrhenius constant for this reaction is 4.79×10^{13} s^{-1}. Calculate the activation energy of this reaction. (R = 8.31 J K^{-1} mol^{-1})

First it's a good idea to get the Arrhenius equation into a **simpler** form so it's easier to use. That means getting rid of the nasty exponential bit — so you need to take the **natural log (ln)** of everything in the equation:

$$k = Ae^{\frac{-E_a}{RT}} \longrightarrow \ln k = \ln A - \frac{E_a}{RT}$$

Thankfully there should be a handy 'ln' button on your calculator for taking natural logs.

Now **rearrange** the equation to get E_a on the **left hand side**: $\frac{E_a}{RT} = \ln A - \ln k$

And another quick rearrangement to get E_a on its **own**: $E_a = (\ln A - \ln k) \times RT$

Now you can just pop the numbers from the question into this formula:

$E_a = (\ln (4.79 \times 10^{13}) - \ln (1.35 \times 10^{-4})) \times (8.31 \times 308)$

$= (31.5... - (-8.91...)) \times (8.31 \times 308) = 103\ 429.54$ J mol^{-1} = **103 kJ mol^{-1}** (3 s.f.)

The Arrhenius Equation

Use an *Arrhenius Plot* to Find the E_a and the *Arrhenius Constant*

1) As you saw on the previous page, putting the **Arrhenius equation** into **logarithmic form** generally makes it a bit easier to use. → $\ln k = -\frac{E_a}{RT} + \ln A$
2) You can use the equation in this form to create an **Arrhenius plot** by plotting **ln *k*** against $\frac{1}{T}$.
3) The line of best fit for the points on the Arrhenius plot will be a straight line graph with a **gradient** of $\frac{-E_a}{R}$.
4) Once you know the gradient, you can use it to find both the **activation energy** and the **Arrhenius constant**.

Example: The graph below shows an Arrhenius plot for the decomposition of hydrogen iodide. Calculate the activation energy and the Arrhenius constant for this reaction. R = 8.31 J K^{-1} mol^{-1}.

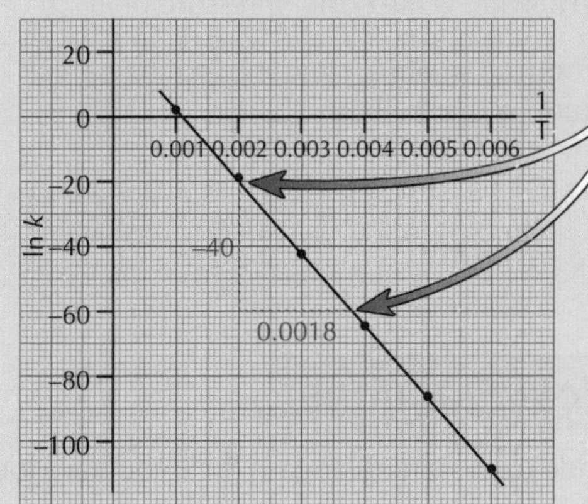

The values of k and T used to draw the Arrhenius plot come from experiments. To gather the data you repeat the same experiment at several different temperatures.

To find the gradient, pick two points on the line with 'nice' coordinates, e.g. (0.002, –20) and (0.0038, –60).

$$\text{gradient} = \frac{-40}{0.0018} = -22\,222. \text{ So } \frac{-E_a}{R} = -22\,222$$

E_a = –(–22 222 × 8.31) = 184 666 J mol^{-1}
= **185 kJ mol^{-1}** (3 s.f.)

1 kJ = 1000 J

To find ln A, substitute the gradient and the coordinates of any point on the line into $\ln k = -\frac{E_a}{RT} + \ln A$

$\frac{-E_a}{R} \times \frac{1}{T} = -\frac{E_a}{RT}$

At (0.002, –20): $-20 = (\frac{-40}{0.0018} \times 0.002) + \ln A$

–20 = –44.4 + ln A

ln A = 24.4, so A = e$^{24.4}$ = **4 × 10^{10} dm^3 mol^{-1} s^{-1}**

The value of ln A is given to one decimal place here, so you should give this answer to one significant figure.

Practice Questions

Q1 In the Arrhenius equation, what do the terms *k*, T and R represent?

Q2 How does increasing the temperature of a reaction affect the value of *k*?

Q3 The Arrhenius equation is $k = Ae^{-E_a/RT}$. Which one of the following answers is true as E_a increases?

A *k* increases and rate of reaction increases. B *k* increases and rate of reaction decreases.

C *k* decreases and rate of reaction increases. D *k* decreases and rate of reaction decreases.

Q4 Describe how you would find activation energy from a graph of ln *k* against 1/T.

Exam Question

Q1 The table on the right gives values for the rate constant of the reaction between hydroxide ions and bromoethane at different temperatures.

a) Complete the table. [2 marks]

b) Use the table to plot a graph that would allow you to calculate the activation energy of the reaction. [3 marks]

c) Calculate the activation energy of the reaction. (R = 8.31 J K^{-1} mol^{-1}) [2 marks]

d) Calculate the value of the Arrhenius constant, A. [1 mark]

T (K)	*k*	1/T (K^{1})	ln *k*
305	0.181	0.00328	–1.709
313	0.468		
323	1.34		
333	3.29	0.00300	1.191
344	10.1		
353	22.7	0.00283	3.122

Who knew rates of reaction could be such a pain in the ar...

...rhenius? You don't need to learn the Arrhenius equation, but you do need to know how to use it. There's some vicious-looking maths here, but the best way to get your head around it is to do loads of practice questions. Then, if any of the different rearrangements of the equation pop up in the exam, you can look them in the eye without panicking.

Gas Equilibria and K_p

It's easier to talk about gases in terms of their pressures rather than their molar concentrations. If you're dealing with an equilibrium where all the substances are gases, you use a slightly different equilibrium constant to K_c — it's called K_p.

The **Total Pressure** of a Gas is **Equal** to the **Sum** of the **Partial Pressures**

In a mixture of gases, each individual gas exerts its own pressure — this is called its **partial pressure**.

The **total pressure** of a gas mixture is the **sum** of all the **partial pressures** of the individual gases.

You might have to put this fact to use in pressure calculations:

Example: When 3.00 moles of the gas PCl_5 is heated, it decomposes into PCl_3 and Cl_2: $PCl_{5(g)} \rightleftharpoons PCl_{3(g)} + Cl_{2(g)}$
In a sealed vessel at 500 K, the equilibrium mixture contains chlorine with a partial pressure of 263 kPa. If the total pressure of the mixture is 714 kPa, what is the partial pressure of PCl_5?

From the equation you know that PCl_3 and Cl_2 are produced in equal amounts, so the partial pressures of these two gases are the **same** at equilibrium — they're both 263 kPa.

Total pressure = $p_{PCl_5} + p_{PCl_3} + p_{Cl_2}$

$714 = p_{PCl_5} + 263 + 263$

So the partial pressure of PCl_5 = 714 – 263 – 263 = **188 kPa**

p_X just means 'partial pressure of X'.

Partial Pressures can be Worked Out from **Mole Fractions**

A '**mole fraction**' is just the **proportion** of a gas mixture that is a made up of particular gas. So if you have four moles of gas in total and two of them are gas A, the mole fraction of gas A is ½. There are **two formulas** you need to know:

1) Mole fraction of a gas in a mixture = $\dfrac{\textbf{number of moles of gas}}{\textbf{total number of moles of gas in the mixture}}$
2) Partial pressure of a gas = **mole fraction of gas × total pressure of the mixture**

Example: When 3.00 mol of PCl_5 is heated in a sealed vessel, the equilibrium mixture contains 1.75 mol of chlorine. If the total pressure of the mixture is 714 kPa, what is the partial pressure of PCl_5?

From the equation, PCl_3 and Cl_2 are produced in equal amounts, so there'll be **1.75 moles** of PCl_3 too.
1.75 moles of PCl_5 must have decomposed so (3.00 – 1.75) = **1.25 moles** of PCl_5 must be left at equilibrium.
This means that the total number of moles of gas at equilibrium = 1.75 + 1.75 + 1.25 = **4.75**.
So the mole fraction of $PCl_5 = \frac{1.25}{4.75} = \mathbf{0.263}$
The partial pressure of PCl_5 = mole fraction × total pressure = 0.263 × 714 = **188 kPa**

The **Equilibrium Constant K_p** is Calculated from **Partial Pressures**

1) $\boldsymbol{K_p}$ is the equilibrium constant for a reversible reaction where all the reactants and products are **gases**.
2) The **expression** for K_p is just like the one for K_c (see p.50), except you use **partial pressures** instead of concentrations.
3) To calculate K_p you put the partial pressures in the expression. Then you work out the **units** like you did for K_c.

For the equilibrium
$aA_{(g)} + bB_{(g)} \rightleftharpoons dD_{(g)} + eE_{(g)}$ $\quad K_p = \dfrac{(p_D)^d (p_E)^e}{(p_A)^a (p_B)^b}$

Example: Find the value of K_p for the decomposition of PCl_5 gas at 500 K (as described above).
The partial pressures of each gas are: p_{PCl_5} = 188 kPa, p_{PCl_3} = 263 kPa, p_{Cl_2} = 263 kPa

$$K_p = \frac{p_{Cl_2} \times p_{PCl_3}}{p_{PCl_5}} = \frac{263 \times 263}{188} = 368$$

Then find the units by putting the units for the partial pressures in the expression instead of the numbers and cancelling: $K_p = \frac{kPa \times kPa}{kPa} = kPa$. So, $\boldsymbol{K_p}$ **= 368 kPa**

Gas Equilibria and K_p

Temperature Changes Alter K_p

1) Back in Unit 1: Section 5, you saw how changes in pressure and temperature affect the equilibrium position. The effects of Le Chatelier's principle (see pages 47-49) still apply to gas equilibria. Remember:

If you **increase** the **pressure**, the equilibrium shifts to the side with **fewer moles of gas.**
If you **decrease** the **pressure**, the equilibrium shifts to the side with **more moles of gas.**

If you **increase** the **temperature**, the equilibrium shifts in the **endothermic (positive ΔH) direction.**
If you **decrease** the **temperature**, the equilibrium shifts in the **exothermic (negative ΔH) direction.**

2) Just like K_c (see pages 50-51), the value of K_p is affected by **temperature**. A particular value of K_p is only valid for a **given temperature**. Changing the **temperature** changes how much product is formed at equilibrium. This changes the **mole fractions** of the gases present, which changes their **partial pressures**. For example:

The reaction on the right is exothermic in the forward direction. If the temperature is **increased**, the equilibrium shifts to the **left** to counteract the change. This means that **less product** is formed.

Exothermic ⟶

$$2SO_{2(g)} + O_{2(g)} \rightleftharpoons 2SO_{3(g)} \quad \Delta H = -197 \text{ kJ mol}^{-1}$$

⟵ Endothermic

$$K_p = \frac{(p_{SO_3})^2}{(p_{SO_2})^2 \times p_{O_2}}$$

⟵ The partial pressure of SO_3 will be reduced.
⟵ The partial pressures of SO_2 and O_2 will increase.

As the temperature **increases**, p_{SO_3} will **decrease** and p_{SO_2} and p_{O_2} will **increase**, so K_p will **decrease.**

3) Just as changing concentration doesn't change K_c, **changing pressure doesn't affect K_p** — the equilibrium will **shift** to keep it the same.
4) And, like K_c, adding a **catalyst won't affect K_p** — it just gets the system to equilibrium more quickly.

Practice Questions

Q1 If you knew the partial pressures of all the gases in a mixture, how would you find the total pressure?

Q2 How do you work out the mole fraction of a gas?

Q3 Write the expression for K_p for this equilibrium: $N_{2(g)} + 3H_{2(g)} \rightleftharpoons 2NH_{3(g)}$*

*Answer on page 218.

Exam Questions

Q1 At high temperatures, SO_2Cl_2 dissociates according to the equation $SO_2Cl_{2(g)} \rightleftharpoons SO_{2(g)} + Cl_{2(g)}$ ($\Delta H = +67 \text{ kJ mol}^{-1}$)
When 1.50 moles of SO_2Cl_2 dissociates at 700 K, the equilibrium mixture contains SO_2 with a partial pressure of 60.2 kPa. The mixture has a total pressure of 141 kPa.

a) Write an expression for K_p for this reaction. [1 mark]

b) Calculate the partial pressure of Cl_2 and the partial pressure of SO_2Cl_2 in the equilibrium mixture. [2 marks]

c) Calculate a value for K_p for this reaction and give its units. [2 marks]

d) Describe and explain the effect that increasing the temperature would have on the value of K_p. [3 marks]

Q2 When nitric oxide and oxygen were mixed in a 2:1 mole ratio at a constant temperature in a sealed flask, an equilibrium was set up according to the equation $2NO_{(g)} + O_{2(g)} \rightleftharpoons 2NO_{2(g)}$.
The partial pressure of the nitric oxide (NO) at equilibrium was 36 kPa and the total pressure in the flask was 99 kPa.

a) Deduce the partial pressure of oxygen in the equilibrium mixture. [1 mark]

b) Calculate the partial pressure of nitrogen dioxide in the equilibrium mixture. [1 mark]

c) Write an expression for the equilibrium constant, K_p, for this reaction and calculate its value at this temperature. State its units. [3 marks]

Pressure pushing down on me, pressing down on you... Under pressure...

Partial pressures are just like concentrations, but for gases. The more of a substance you've got in a solution, the higher its concentration — and the more of a gas that you've got squashed into a container, the higher its partial pressure.

Electrode Potentials

There are electrons toing and froing in redox reactions. And when electrons move, you get electricity.

Electrochemical Cells Make Electricity

An electrochemical cell can be made from **two different metals** dipped in salt solutions of their **own ions** and connected by a wire (the **external circuit**).

There are always **two** reactions within an electrochemical cell — one's an oxidation and one's a reduction — so it's a **redox process** (see page 52).

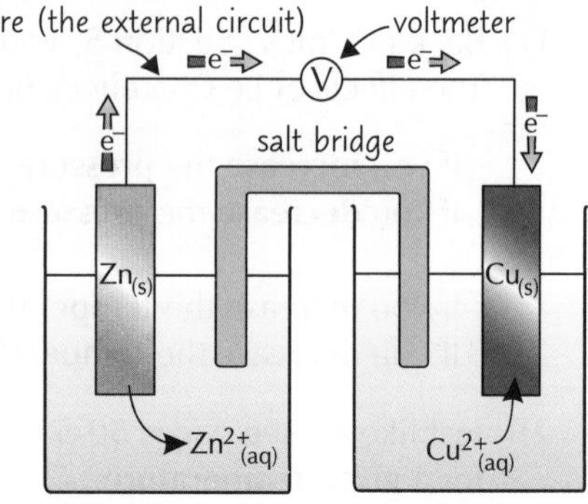

The solutions are connected by a **salt bridge** made from filter paper soaked in $KNO_{3(aq)}$. This allows ions to flow through and balance out the charges.

Here's what happens in the **zinc/copper** electrochemical cell on the right:

1) Zinc **loses electrons** more easily than copper. So in the half-cell on the left, zinc (from the zinc electrode) is **oxidised** to form $\mathbf{Zn^{2+}_{(aq)}}$ ions. This releases electrons into the external circuit.
2) In the other half-cell, the **same number of electrons** are taken from the external circuit, **reducing** the Cu^{2+} ions to copper atoms.

Electrons flow through the wire from the more reactive metal to the less reactive one.

A voltmeter in the external circuit shows the **voltage** between the two half-cells. This is the **cell potential** or **EMF** (which stands for electromotive force), also known as $\boldsymbol{E_{cell}}$.

You can also have half-cells involving **solutions of two aqueous ions of the same element**, such as $\mathbf{Fe^{2+}_{(aq)}/Fe^{3+}_{(aq)}}$. The conversion from Fe^{2+} to Fe^{3+} or vice versa happens on the surface of a platinum **electrode**. The electrode is made of **platinum** because it is an **inert metal**, so it won't react with the ions.

The Reactions Happening at Each Electrode are Reversible

1) The reactions that occur at each of the electrodes in an electrochemical cell are **reversible**.

Example: The reactions that are taking place at the electrodes in the zinc/copper cell above are:

$$Zn^{2+}_{(aq)} + 2e^- \rightleftharpoons Zn_{(s)}$$
$$Cu^{2+}_{(aq)} + 2e^- \rightleftharpoons Cu_{(s)}$$

The '**reversible**' arrows show that both reactions can go in **either direction**.

2) These equations are **half-equations**. When you write half-equations for electrochemical cells, they're always written with the **reduction reaction** going in the **forward** direction (with the **electrons** on the **left-hand side**).
3) When two half-cells are joined to make a cell, which direction each reaction will actually go in depends on how easily each metal **loses electrons** (i.e. how easily it's **oxidised**).
4) How easily a metal is oxidised is measured using **electrode potentials**. A metal that's easy to oxidise has a **very negative** electrode potential. One that's harder to oxidise has a **less negative** (or **positive**) electrode potential.

Example: The table below shows the electrode potentials for the copper and zinc half-cells:

Half-cell	Electrode potential $E^{\ominus}$(V)
$Zn^{2+}_{(aq)}/Zn_{(s)}$	–0.76
$Cu^{2+}_{(aq)}/Cu_{(s)}$	+0.34

There's more about how these values are worked out coming up on the next page.

The zinc half-cell has a **more negative** electrode potential than the copper half-cell.

So in a zinc/copper cell:
- zinc is **oxidised** (the half-equation shown above goes **backwards**).
- copper is **reduced** (the half-equation shown above goes **forwards**).

5) Once you know which direction each of the half-equations will go in, you can write the equation for the **overall reaction** that will happen in the cell:

$$Cu^{2+}_{(aq)} + Zn_{(s)} \rightleftharpoons Cu_{(s)} + Zn^{2+}_{(aq)}$$

Electrode Potentials

Electrode Potentials are Measured Against Standard Hydrogen Electrodes

You measure the electrode potential of a half-cell against a **standard hydrogen electrode**.
The cell potential is **affected** by the conditions (**temperature**, **pressure** and **concentration**).
To get around this, **standard conditions** are used to measure electrode potentials.

The **standard electrode potential**, $E^{\ominus}$, of a half-cell is the **voltage measured** under **standard conditions** when the **half-cell** is connected to a **standard hydrogen electrode**.

Standard conditions are:

1) Any solutions must have a concentration of **1.00 mol dm^{-3}**.
2) The temperature must be **298 K (25 °C)**.
3) The pressure must be **100 kPa**.

For example, this cell would allow you to find the standard electrode potential of the Zn^{2+}/Zn half-cell:

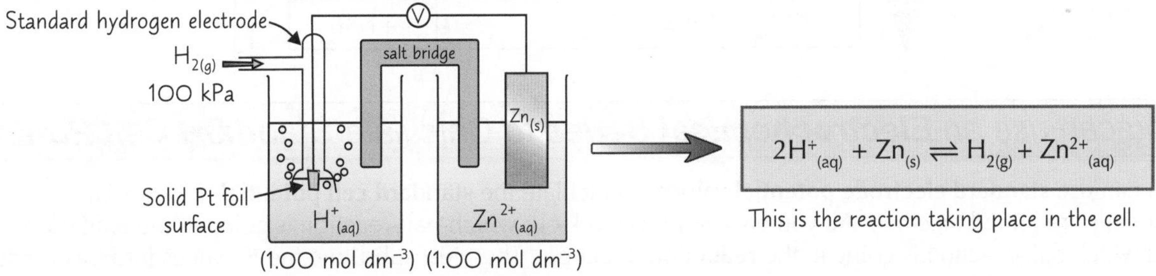

This is the reaction taking place in the cell.

If **standard conditions** are maintained, the **reading on the voltmeter** when a half-cell is connected to the **standard hydrogen electrode** will be the **standard electrode potential** of that half-cell.

Practice Questions

Q1 Sketch a diagram showing how you would set up an electrochemical cell using zinc and copper electrodes and solutions containing zinc and copper ions.

Q2 Here are the half-equations for the reactions taking place at the electrodes in a zinc/copper cell:
$Zn^{2+}_{(aq)} + 2e^- \rightleftharpoons Zn_{(s)}$, $E^{\ominus} = -0.76$ V $Cu^{2+}_{(aq)} + 2e^- \rightleftharpoons Cu_{(s)}$, $E^{\ominus} = +0.34$ V
When the cell is operating, which of these reactions will be moving backwards (in the oxidation direction)?

Q3 Why are standard conditions used to measure standard electrode potentials?

Exam Questions

Q1 A cell is made up of a lead and an iron plate, dipped in solutions of lead(II) nitrate and iron(II) nitrate, respectively. The two half-cells are connected using wires and a salt bridge. The electrode potentials for the two electrodes are:

$Fe^{2+}_{(aq)} + 2e^- \rightleftharpoons Fe_{(s)}$, $E^{\ominus} = -0.44$ V $Pb^{2+}_{(aq)} + 2e^- \rightleftharpoons Pb_{(s)}$, $E^{\ominus} = -0.13$ V

Which metal is oxidised in this cell? Explain your answer. [2 marks]

Q2 An electrochemical cell containing a zinc half-cell and a silver half-cell was set up.
The two half-cells were connected using wires and a salt bridge..

$Zn^{2+}_{(aq)} + 2e^- \rightleftharpoons Zn_{(s)}$, $E^{\ominus} = -0.76$ V $Ag^+_{(aq)} + e^- \rightleftharpoons Ag_{(s)}$, $E^{\ominus} = +0.80$ V

a) Write an equation for the overall cell reaction. [1 mark]

b) Which half-cell accepts electrons from the circuit? Explain your answer. [1 mark]

My mam always said I had potential...

OK, so this stuff is tricky, but don't panic. The most important thing to remember here is that in the half-cell with the more negative electrode potential the half-equation will go in the oxidation direction (and vice versa). And don't forget to learn what the standard hydrogen electrode is (and what it's used for) too. Then take a deep breath and relax...

The Electrochemical Series

Elements have different standard electrode potentials. So what do chemists do — they write a list of them all in order.

The Electrochemical Series Shows You What's Reactive and What's Not

An electrochemical series is basically a **list** of **standard electrode potentials** for different electrochemical **half-cells**. They look something like this:

Half-reaction	$E^\ominus$/V
$Mg^{2+}_{(aq)} + 2e^- \rightleftharpoons Mg_{(s)}$	–2.37
$Zn^{2+}_{(aq)} + 2e^- \rightleftharpoons Zn_{(s)}$	–0.76
$2H^+_{(aq)} + 2e^- \rightleftharpoons H_{2(g)}$	0.00
$Cu^{2+}_{(aq)} + 2e^- \rightleftharpoons Cu_{(s)}$	+0.34
$Fe^{3+}_{(aq)} + e^- \rightleftharpoons Fe^{2+}_{(aq)}$	+0.77
$Br_{2(aq)} + 2e^- \rightleftharpoons 2Br^-_{(aq)}$	+1.07
$Cl_{2(aq)} + 2e^- \rightleftharpoons 2Cl^-_{(aq)}$	+1.36

This isn't the whole series. You might see different lists. Every list works the same way though.

More positive electrode potentials mean that:
1. The left-hand substances are more easily reduced.
2. The right-hand substances are more stable.

More negative electrode potentials mean that:
1. The right-hand substances are more easily oxidised.
2. The left-hand substances are more stable.

You can use an Electrochemical Series to Calculate Standard Cell Potentials

You can use **standard electrode potential** values to calculate the **standard cell potential**, E°_{cell} (or EMF), when two half-cells are joined together. All you have to do is work out which half-reaction is going in the **oxidation** direction and which half-reaction is going in the **reduction** direction. Then just substitute the E° values into this formula:

$$E^\circ_{cell} = E^\circ_{reduced} - E^\circ_{oxidised}$$

Example: Calculate the standard cell potential of a Mg/Fe electrochemical cell using the two redox reaction equations shown in the series above.

First write the two **half-equations** down as reduction reactions:

$$Mg^{2+}_{(aq)} + 2e^- \rightleftharpoons Mg_{(s)} \qquad E^\circ = -2.37\text{ V}$$
$$Fe^{3+}_{(aq)} + e^- \rightleftharpoons Fe^{2+}_{(aq)} \qquad E^\circ = +0.77\text{ V}$$

The **Mg/Mg^{2+}** half-cell has the **more negative electrode potential**, so this half reaction will go in the direction of **oxidation**. The **Fe^{2+}/Fe^{3+}** half-cell has the **more positive electrode potential**, so this half reaction will go in the direction of **reduction**. The overall reaction is:

$$2Fe^{3+}_{(aq)} + Mg_{(s)} \rightleftharpoons 2Fe^{2+}_{(aq)} + Mg^{2+}_{(aq)}$$

So, $E^\circ_{cell} = E^\circ_{reduced} - E^\circ_{oxidised} = +0.77 - (-2.37) = $ **+3.14 V**

There's a Convention for Drawing Electrochemical Cells

It's a bit of a faff drawing pictures of electrochemical cells. There's a **shorthand** way of representing them though:

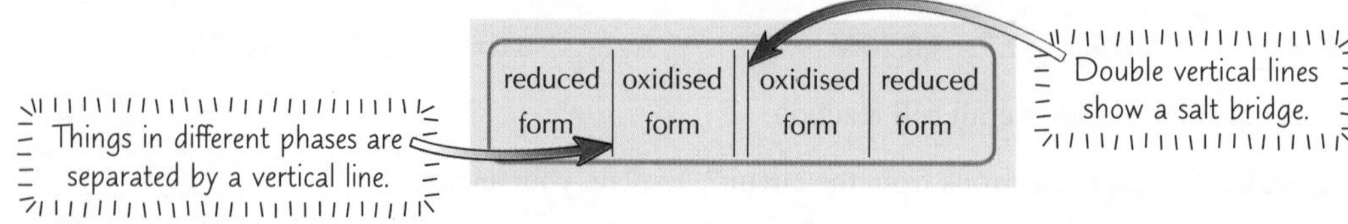

Things in different phases are separated by a vertical line.

Double vertical lines show a salt bridge.

There are a couple of important **conventions** when drawing cells:

1) The **half-cell** with the **more negative** potential goes on the **left**.
2) The **oxidised forms** go in the **centre** of the cell diagram.

So for the Zn/Cu cell (see page 76), the diagram would look like this:

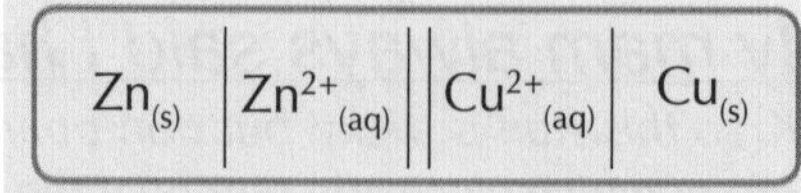

The Electrochemical Series

Use Electrode Potentials to Predict Whether a Reaction Will Happen

To figure out if a metal will react with the aqueous ions of another metal, you can use their E° values.

Example: Predict whether zinc metal reacts with aqueous copper(II) ions.

First write down the two **half-equations** as reduction reactions:

$$Zn^{2+}_{(aq)} + 2e^- \rightleftharpoons Zn_{(s)} \qquad E^\circ = -0.76\text{ V}$$
$$Cu^{2+}_{(aq)} + 2e^- \rightleftharpoons Cu_{(s)} \qquad E^\circ = +0.34\text{ V}$$

The half-reactions are written as reduction reactions but one will always have to move in the direction of oxidation. Zn^{2+}/Zn has a more negative electrode potential than Cu^{2+}/Cu, so the zinc will be oxidised.

Then look at the standard electrode potentials.

The half-equation with the **more negative** electrode potential will move to the **left**: $Zn_{(s)} \rightarrow Zn^{2+}_{(aq)} + 2e^-$

The half-equation with the **more positive** electrode potential will move to the **right**: $Cu^{2+}_{(aq)} + 2e^- \rightarrow Cu_{(s)}$

The two half-equations combine to give: $Zn_{(s)} + Cu^{2+}_{(aq)} \rightarrow Zn^{2+}_{(aq)} + Cu_{(s)}$

This is the **feasible** direction of the overall reaction. It matches the reaction described in the question, so zinc **will** react with aqueous copper ions.

For any **feasible reaction** E°_{cell} is **positive** — E°_{cell} for this reaction is +0.34 – (–0.76) = **+1.10 V**.

Practice Questions

Q1 Copper is less reactive than magnesium. Predict which of these half-reactions has the more negative standard electrode potential: A $Mg^{2+}_{(aq)} + 2e^- \rightleftharpoons Mg_{(s)}$ B $Cu^{2+}_{(aq)} + 2e^- \rightleftharpoons Cu_{(s)}$

Q2 Use electrode potentials to show that magnesium will reduce Zn^{2+} ions.

Q3 Use the table on the previous page to predict whether or not Zn^{2+} ions can oxidise Fe^{2+} ions to Fe^{3+} ions.

Exam Questions

Q1 Use the E° values in the table on the right and on the previous page to determine the outcome of mixing the following solutions. If there is a reaction, determine the overall E° value and write the equation. Explain how you know whether or not the reaction takes place.

a) Zinc metal and Ni^{2+} ions. [2 marks]

b) Acidified MnO_4^- ions and Sn^{2+} ions. [2 marks]

c) $Br_{2(aq)}$ and acidified $Cr_2O_7^{2-}$ ions. [2 marks]

Half-reaction	E°/V
$Ni^{2+}_{(aq)} + 2e^- \rightleftharpoons Ni_{(s)}$	–0.25
$Sn^{4+}_{(aq)} + 2e^- \rightleftharpoons Sn^{2+}_{(aq)}$	+0.14
$Cr_2O_7^{2-}{}_{(aq)} + 14H^+_{(aq)} + 6e^- \rightleftharpoons 2Cr^{3+}_{(aq)} + 7H_2O_{(l)}$	+1.33
$MnO_4^-{}_{(aq)} + 8H^+_{(aq)} + 5e^- \rightleftharpoons Mn^{2+}_{(aq)} + 4H_2O_{(l)}$	+1.51

Q2 Potassium manganate(VII), $KMnO_4$, and potassium dichromate $K_2Cr_2O_7$, are both used as oxidising agents. From their electrode potentials (given in the table above), which would you predict is the stronger oxidising agent? Explain why. [2 marks]

Q3 Use the following data to answer the questions below.

$$O_{2(g)} + 2H_2O_{(l)} + 4e^- \rightleftharpoons 4OH^-_{(aq)} \qquad E^\circ = +0.40\text{ V}$$
$$Fe^{2+}_{(aq)} + 2e^- \rightleftharpoons Fe_{(s)} \qquad E^\circ = -0.44\text{ V}$$

a) i) Draw a cell diagram for the reaction using the conventional representation. [1 mark]

ii) Calculate the EMF for the cell. [1 mark]

b) Use electrode potentials to explain why iron is oxidised in the presence of oxygen and water. [1 mark]

Why so series? Let's put a smile on that face...

To see if a reaction will happen, you basically find the two half-equations in the electrochemical series and check that the one you are predicting will go backwards is the one with the more negative electrode potential. Alternatively, you could calculate the cell potential for the reaction — if it's negative, it's never going to happen.

Batteries and Fuel Cells

It turns out that electrochemical cells aren't just found in the lab — they're all over your house too... whoa.

Electrochemical Cells Are Used as *Batteries*

1) **Batteries** are types of **electrochemical cell** which provide the **electricity** we use to power things like watches and mobile phones. Some types of cell are **rechargeable** while others can only be used until they **run out**.
2) **Non-rechargeable** batteries are **cheaper** than rechargeable ones. But since you can just **recharge** and **re-use** rechargeable batteries, they **last longer** and work out **cheaper** in the **long run**.

Lithium Batteries are *Rechargeable*

Rechargeable batteries are found in loads of devices, such as **mobile phones**, **laptops** and **cars**. For example:

Lithium cells are used in mobile phones and laptops. One type of lithium cell is made up of a **lithium cobalt oxide ($LiCoO_2$) electrode** and a **graphite electrode**. The **electrolyte** is a **lithium salt** in an **organic solvent**.

The **half-equations** are: $Li^+ + e^- \rightleftharpoons Li$ $\quad E^\circ = -3.04\text{ V}$

$Li^+ + CoO_2 + e^- \rightleftharpoons Li^+[CoO_2]^-$ $\quad E^\circ = +0.56\text{ V}$

The **Li^+/ Li** half-cell has the **more negative** E° value so goes in the direction of **oxidation** (backwards).

So the reactions which happen when the battery supplies power are:

At the **negative** electrode: $\mathbf{Li \rightarrow Li^+ + e^-}$

At the **positive** electrode: $\mathbf{Li^+ + CoO_2 + e^- \rightarrow Li^+[CoO_2]^-}$

Learn these two half-reactions — you could be asked to give them in the exams.

The **EMF** of this type of cell is: $E^\circ_{cell} = E^\circ_{reduced} - E^\circ_{oxidised} = +0.56 - (-3.04) = \mathbf{+3.60\text{ V}}$

To recharge these batteries, a **current** is supplied to force **electrons** to flow in the **opposite direction** around the circuit and **reverse the reactions**. The reactions that take place in **non-rechargeable** batteries are **difficult** or **impossible to reverse** in this way.

Fuel Cells can *Generate Electricity* From *Hydrogen* and *Oxygen*

In most cells the **chemicals** that generate the electricity are contained in the **electrodes** and the **electrolyte** that form the cell. In a **fuel cell** the chemicals are **stored separately** outside the cell and fed in when electricity is required. One example of this is the **alkaline hydrogen-oxygen fuel cell**, which can be used to **power electric vehicles**. **Hydrogen** and **oxygen gases** are fed into two separate platinum-containing electrodes. The electrodes are separated by an **anion-exchange membrane** that **allows anions** (OH^-) and water to pass through it, but **not hydrogen and oxygen gas**. The **electrolyte** is an aqueous alkaline (KOH) solution.

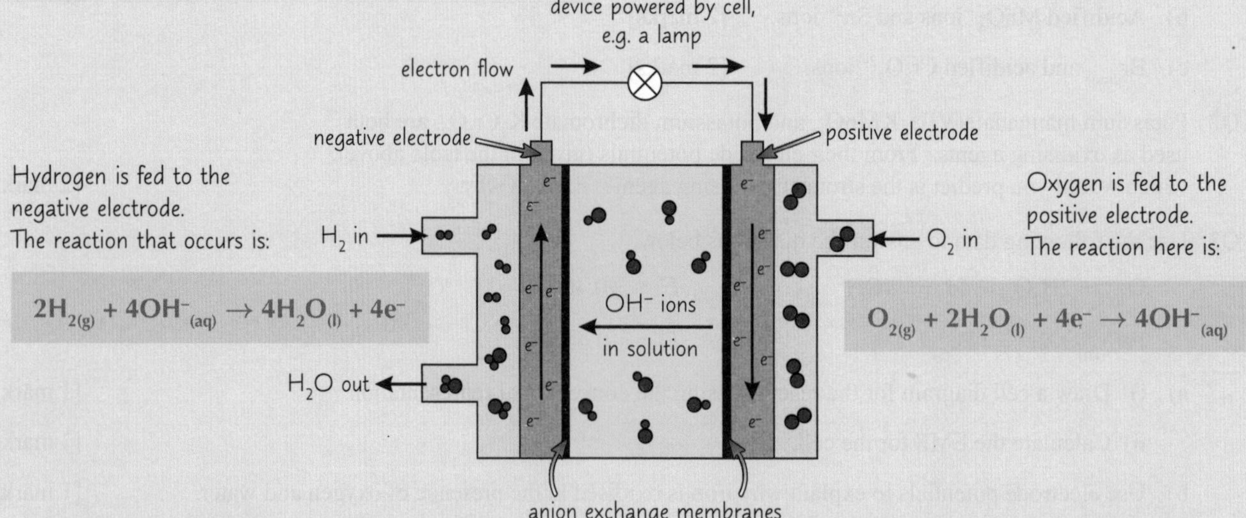

The **electrons** flow from the **negative electrode** through an **external circuit** to the **positive electrode**.
The **OH^- ions** pass through the **anion-exchange membrane** towards the negative electrode.

The **overall effect** is that H_2 and O_2 react to make **water**: $2H_{2(g)} + O_{2(g)} \rightarrow 2H_2O_{(g)}$

Batteries and Fuel Cells

Fuel Cells Have some Big Advantages

The **major advantage** of using **fuel cells** in cars, rather than the **internal combustion engine**, is that fuel cells are **more efficient** — they **convert more** of their **available energy** into **kinetic** energy to get the car moving. Internal combustion engines **waste** a lot of their **energy** producing **heat**. Other benefits are

- The only **waste product** is **water**, so there are **no nasty toxic chemicals** to dispose of and **no CO_2 emissions** from the cell itself.
- Fuel cells don't need to be recharged like batteries. As long as **hydrogen** and **oxygen** are supplied, the cell will continue to **produce electricity**.

The **downside** is that you **need energy** to produce a supply of **hydrogen** and **oxygen**. They can be produced from the **electrolysis of water**, i.e. by **reusing the waste product** from the fuel cell, but this requires **electricity** — and this **electricity** is normally generated by **burning fossil fuels**. So the whole process isn't usually carbon neutral. **Hydrogen** is also **highly flammable** so it needs to be handled carefully when it is **stored** or **transported**.

Practice Questions

Q1 Give one advantage and one disadvantage of non-rechargeable batteries.

Q2 Name a metal that is used in the electrodes in an alkaline hydrogen-oxygen fuel cell.

Q3 What electrolyte is used in an alkaline hydrogen-oxygen fuel cell?

Exam Questions

Q1 The diagram below shows the structure of an alkaline hydrogen-oxygen fuel cell.

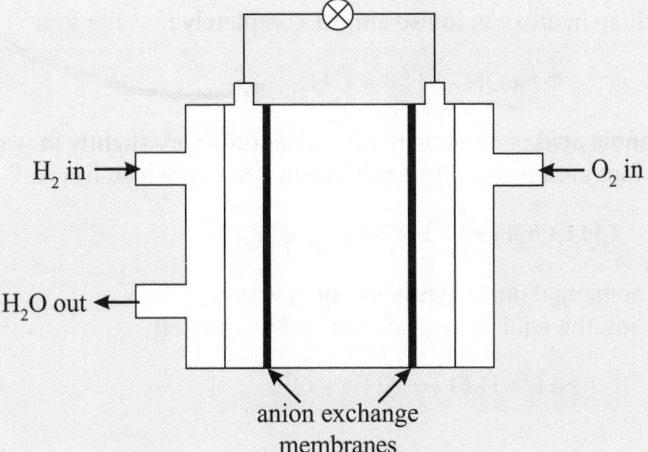

a) i) Label the site of oxidation and the site of reduction on the diagram. [1 mark]

ii) Draw an arrow to show the direction of the flow of electrons. [1 mark]

b) Write a half-equation for the reaction at each electrode. [2 marks]

c) Explain the purpose of the anion exchange membrane in the fuel cell. [1 mark]

Q2 a) Write the half-equations that take place in a lithium battery cell. [2 marks]

b) How is a lithium battery recharged? [1 mark]

c) What happens to the half-equations from part a) when the cell is being recharged? [1 mark]

Been charged with a salt in battery? Don't worry, it's reversible...

You've got to love batteries — they sit there in their shiny metal cases all ready to release the energy stored inside them just when you need it most. So, have some respect for batteries and don't just throw them in the bin when they go flat, recharge them if you can, and if not recycle them. Oh, and you'd probably best learn those half-equations too.

Acids, Bases and K_w

Remember this stuff? Well, it's all down to Brønsted and Lowry — they've got a lot to answer for.

*An Acid **Releases** Protons — a Base **Accepts** Protons*

You might see the term 'alkali' being used instead of 'base'. Don't panic though — an alkali is just a soluble base.

Brønsted-Lowry acids are **proton donors** — they release **hydrogen ions** (H^+) when they're mixed with water. You never get H^+ ions by themselves in water though — they're always combined with H_2O to form **hydroxonium ions, H_3O^+**.

HA is any old acid. ⟹ $HA_{(aq)} + H_2O_{(l)} \rightarrow H_3O^+_{(aq)} + A^-_{(aq)}$

Brønsted-Lowry bases are **proton acceptors**.
When they're in solution, they grab **hydrogen ions** from water molecules.

B is just a random base. ⟹ $B_{(aq)} + H_2O_{(l)} \rightarrow BH^+_{(aq)} + OH^-_{(aq)}$

Brønsted laid down the base.
Lowry rocked the flow.
That's just how chemists roll.

*Acids and Bases can be **Strong** or **Weak***

1) **Strong acids dissociate** (or ionise) **almost completely** in water — **nearly all** the H^+ ions will be released. **Hydrochloric acid** is a strong acid:

$$HCl \rightarrow H^+ + Cl^-$$

Strong bases (like sodium hydroxide) **ionise almost completely** in water too:

$$NaOH \rightarrow Na^+ + OH^-$$

These are really both reversible reactions, but the equilibrium lies extremely far to the right.

2) **Weak acids** (e.g. **ethanoic acid** or citric acid) dissociate only very **slightly** in water — so only small numbers of H^+ ions are formed. An **equilibrium** is set up which lies well over to the **left**:

$$CH_3COOH \rightleftharpoons CH_3COO^-_{(aq)} + H^+_{(aq)}$$

Weak bases (such as ammonia) **only slightly ionise** in water.
Just like with weak acids, the equilibrium lies well over to the **left**:

$$NH_3 + H_2O \rightleftharpoons NH_4^+ + OH^-$$

***Protons** are **Transferred** when **Acids** and **Bases** React*

Acids can't just throw away their protons — they can only get rid of them if there's a **base** to accept them.
In this reaction the **acid**, HA, **transfers** a proton to the **base**, B:

$$HA_{(aq)} + B_{(aq)} \rightleftharpoons BH^+_{(aq)} + A^-_{(aq)}$$

It's an **equilibrium**, so if you add more **HA** or **B**, the position of equilibrium moves to the **right**.
But if you add more **BH^+** or **A^-**, the equilibrium will move to the **left**. (See pages 47-49 for more on equilibria.)
When an acid is added to **water**, water acts as the **base** and accepts the proton:

$$HA_{(aq)} + H_2O_{(l)} \rightleftharpoons H_3O^+_{(aq)} + A^-_{(aq)}$$

The equilibrium's far to the left for weak acids, and far to the right for strong acids.

Acids, Bases and K_w

Water Dissociates Slightly

Water dissociates into **hydroxonium ions** and **hydroxide ions**.

So this **equilibrium** exists in water:

$H_2O + H_2O \rightleftharpoons H_3O^+ + OH^-$ or more simply $H_2O \rightleftharpoons H^+ + OH^-$

And, just as for any other equilibrium reaction, you can apply the equilibrium law and write an expression for the **equilibrium constant**:

$$K_c = \frac{[H^+][OH^-]}{[H_2O]}$$

Water only dissociates a **tiny amount**, so the equilibrium lies well over to the **left**. There's so much water compared to the amounts of H^+ and OH^- ions that the concentration of water is considered to have a **constant** value. So if you multiply the expression you wrote for K_c (which is a constant) by $[H_2O]$ (another constant), you get a **constant**. This new constant is called the **ionic product of water** and it's given the symbol K_w.

$K_w = K_c \times [H_2O] = [H^+][OH^-]$ ⟹ $\mathbf{K_w = [H^+][OH^-]}$

The units of K_w are always $mol^2\ dm^{-6}$.

K_w always has the **same value** for an aqueous solution at a **given temperature**. For example, at 298 K (25 °C), K_w has a value of 1.00×10^{-14} $mol^2\ dm^{-6}$. The value of K_w changes as temperature changes.

In **pure water**, there is always **one H^+ ion** for **each OH^- ion**. So $[H^+] = [OH^-]$. That means if you are dealing with **pure water**, then you can say that $\mathbf{K_w = [H^+]^2}$.

Practice Questions

Q1 Explain what is meant by the term 'strong acid' and give an example of one.

Q2 Which substance is acting as a Brønsted-Lowry base in the equation on the right? $HA + H_2O \rightleftharpoons H_3O^+ + A^-$

Q3 Define K_w and give its value at 298 K.

Exam Questions

Q1 Show, by writing appropriate equations, how HSO_4^- can behave as:

a) a Brønsted-Lowry acid, b) a Brønsted-Lowry base. [2 marks]

Q2 Hydrocyanic acid (HCN) is a weak acid.

a) Define the term 'weak acid'. [1 mark]

b) Write a balanced equation for the equilibrium that occurs when HCN dissolves in water. [1 mark]

Q3 A solution contains 2.50 g dm^{-3} of sodium hydroxide.
What is the molar concentration of the hydroxide ions in this solution? [2 marks]

Alright, this is a stick-up — hand over your protons and nobody gets hurt...

Don't confuse strong acids with concentrated acids, or weak acids with dilute acids. Strong and weak are to do with how much an acid ionises, whereas concentrated and dilute are to do with the concentration of the acid. You can have a strong dilute acid, or a weak concentrated acid. And it works just the same way with bases too.

pH Calculations

Just when you thought it was safe to turn the page, there's even more about acids and bases. This page is positively swarming with calculations and constants...

The pH Scale is a Measure of the Hydrogen Ion Concentration

The **concentration of hydrogen ions** in a solution can vary enormously, so those wise chemists of old decided to express the concentration on a **logarithmic scale**.

$$\text{pH} = -\log_{10} [\text{H}^+]$$

$[H^+]$ is the concentration of hydrogen ions in a solution, measured in mol dm^{-3}.

The pH scale normally goes from **0** (very acidic) to **14** (very basic). **pH 7** is regarded as being **neutral**.

You Can Calculate pH from Hydrogen Ion Concentration...

If you know the **hydrogen ion concentration** of a solution, you can calculate its **pH** by sticking the numbers into the **formula**.

Pippa's an expert at finding logs.

Example: A solution of hydrochloric acid has a hydrogen ion concentration of 0.01 mol dm^{-3}. What is the pH of this solution?

$\text{pH} = -\log_{10} [\text{H}^+] = -\log_{10} (0.01) = \mathbf{2.0}$

Use the 'log' button on your calculator for this.

...Or Hydrogen Ion Concentration From pH

If you've got the **pH** of a solution, and you want to know its **hydrogen ion concentration**, then you need the **inverse** of the pH formula:

$$[\text{H}^+] = 10^{-\text{pH}}$$

Now you can use this formula to find $[H^+]$.

Example: A solution of sulfuric acid has a pH of 1.5.
What is the hydrogen ion concentration of this solution?

$[\text{H}^+] = 10^{-\text{pH}} = 10^{-1.5} = 0.03 \text{ mol dm}^{-3} = \mathbf{3 \times 10^{-2} \text{ mol dm}^{-3}}$

For Strong Monoprotic Acids, [H⁺] = [Acid]

1) **Strong acids** such as hydrochloric acid and nitric acid **ionise fully** in solution.
2) Hydrochloric acid (HCl) and nitric acid (HNO_3) are also **monoprotic**, which means that **each molecule** of acid will release **one proton** when it dissociates. This means **one mole of acid** produces **one mole of hydrogen ions**. So the H^+ concentration is the **same** as the acid concentration.

E.g. For **0.10 mol dm^{-3} HCl**, $[H^+]$ is also 0.10 mol dm^{-3}. So the **pH** $= -\log_{10} [\text{H}^+] = -\log_{10} (0.10) = \mathbf{1.00}$.
Or for **0.050 mol dm^{-3} HNO_3**, $[H^+]$ is also 0.050 mol dm^{-3}, giving **pH** $= -\log_{10} (0.050) = \mathbf{1.30}$.

For Strong Diprotic Acids, [H⁺] = 2[Acid]

There's more about diprotic acids on page 90.

1) Each molecule of a **strong diprotic acid** releases **2 protons** when it dissociates. So, diprotic acids produce **2 mol of hydrogen ions** for **each mole of acid**.
2) Sulfuric acid is an example of a **strong diprotic acid**: $H_2SO_{4(l)} + \text{water} \rightarrow 2H^+_{(aq)} + SO_4{}^{2-}{}_{(aq)}$

E.g. For **0.10 mol dm^{-3} H_2SO_4**, $[H^+]$ is 0.20 mol dm^{-3}. So the pH $= -\log_{10} [\text{H}^+] = -\log_{10} (0.20) = \mathbf{0.70}$.

pH Calculations

Use K_w to Find the pH of a Strong Base

1) Sodium hydroxide (NaOH) and potassium hydroxide (KOH) are **strong bases** that **fully ionise** in water:

$NaOH \rightarrow Na^+ + OH^-$ $KOH \rightarrow K^+ + OH^-$

2) They donate **one mole of OH^- ions** per mole of base.
This means that the concentration of OH^- ions is the **same** as the **concentration of the base**.
So for 0.02 mol dm^{-3} sodium hydroxide solution, $[OH^-]$ is also **0.02 mol dm^{-3}**.
3) But to work out the **pH** you need to know **$[H^+]$**
— luckily this is linked to **$[OH^-]$** through the **ionic product of water, K_w**:

$K_w = [H^+][OH^-]$

4) So if you know K_w and $[OH^-]$ for a **strong aqueous base** at a certain temperature, you can work out **$[H^+]$** and then the **pH**.

Example: Find the pH of 0.10 mol dm^{-3} NaOH at 298 K, given that K_w at 298 K is 1.0×10^{-14} mol^2 dm^{-6}.

1) First put all the values you know into the expression for the ionic product of water, K_w:

$$1.0 \times 10^{-14} = [H^+] \times 0.10$$

2) Now rearrange the expression to find $[H^+]$:

$$[H^+] = \frac{1.0 \times 10^{-14}}{0.10} = 1.0 \times 10^{-13}\ \text{mol dm}^{-3}$$

3) Use your value of $[H^+]$ to find the pH of the solution:

$$\text{pH} = -\log_{10}[H^+] = -\log_{10}(1.0 \times 10^{-13}) = \mathbf{13.00}$$

Practice Questions

Q1 Write the formula for calculating the pH of a solution.

Q2 What can you assume about $[H^+]$ for a strong monoprotic acid?

Q3 Explain how you'd find the pH of a strong base.

Exam Questions

Q1 a) What's the pH of a solution of the strong acid, hydrobromic acid (HBr), if it has a concentration of 0.32 mol dm^{-3}? [1 mark]

b) Hydrobromic acid is a stronger acid than ethanoic acid.
Explain what that means in terms of hydrogen ions and pH. [1 mark]

Q2 Nitric acid, HNO_3, is a strong monoprotic acid.

a) Explain what is meant by a monoprotic acid. [1 mark]

b) Find the concentration of a solution of nitric acid that has a pH value of 0.55. [1 mark]

Q3 A solution contains 11.22 g dm^{-3} of potassium hydroxide. K_w at 298 K is 1.0×10^{-14} mol^2 dm^{-6}.

a) What is the molar concentration of the hydroxide ions in this solution? [2 marks]

b) Calculate the pH of this solution. [2 marks]

James and the Giant pH — a bit too basic for me if I'm being honest...

You know things are getting serious when maths stuff like logs start appearing. It's fine really though, just practise a few questions and make sure you know how to use the log button on your calculator. And make sure you've learned the equation for K_w and both pH equations. And while you're up, go and make me a nice cup of tea, lots of milk, no sugar.

More pH Calculations

More acid calculations to come, so you'll need to get that calculator warmed up... Either hold it for a couple of minutes in your armpit, or even better, sit on it for a while. OK, done that? Good stuff...

K_a is the *Acid Dissociation Constant*

1) Weak acids (like CH_3COOH) and weak bases **dissociate only slightly** in aqueous solution, so the $[H^+]$ **isn't** the same as the acid concentration. This makes it a **bit trickier** to find their pH. You have to use yet another **equilibrium constant**, K_a (the acid dissociation constant).

- For a weak aqueous acid, HA, you get the following equilibrium: $HA_{(aq)} \rightleftharpoons H^+_{(aq)} + A^-_{(aq)}$
- As only a **tiny amount** of HA dissociates, you can assume that $[HA_{(aq)}]_{equilibrium} \approx [HA_{(aq)}]_{start}$.
- So if you apply the equilibrium law, you get: $K_a = \frac{[H^+][A^-]}{[HA]_{start}}$
- You can also assume that dissociation of the **acid** is much greater than dissociation of **water**. This means you can assume that all the H^+ ions in solution come from the **acid**, so $[H^+_{(aq)}] \approx [A^-_{(aq)}]$.

So, for a weak acid: $K_a = \frac{[H^+]^2}{[HA]}$ ← The units of K_a are mol dm^{-3}.

2) The assumptions made above to find K_a only work for **weak acids**, because strong acids dissociate almost completely in solution. (So for a strong acid, $[HA_{(aq)}]_{equilibrium}$ is not equal to $[HA_{(aq)}]_{start}$.)

To Find the *pH* of a *Weak Acid*, You Use K_a

K_a is an **equilibrium constant** just like K_c or K_w. It applies to a particular acid at a **specific temperature**, regardless of the **concentration**. You can use this fact to find the **pH** of a known concentration of a weak acid.

Example: Calculate the hydrogen ion concentration and the pH of a 0.0200 mol dm^{-3} solution of propanoic acid (CH_3CH_2COOH) at 298 K. K_a for propanoic acid at 298 K is 1.30×10^{-5} mol dm^{-3}.

First, write down your expression for K_a and rearrange to find $[H^+]$.

$$K_a = \frac{[H^+]^2}{[CH_3CH_2COOH]} \Longrightarrow [H^+]^2 = K_a[CH_3CH_2COOH] = 1.30 \times 10^{-5} \times 0.0200 = 2.60 \times 10^{-7}$$

$$\Longrightarrow [H^+] = \sqrt{(2.60 \times 10^{-7})} = \mathbf{5.10 \times 10^{-4}\ mol\ dm^{-3}}$$

You can now use your value for $[H^+]$ to find pH: $pH = -\log_{10}(5.10 \times 10^{-4}) = \mathbf{3.292}$

You Might Have to Find the *Concentration* or K_a of a *Weak Acid*

You don't need to know anything new for this type of calculation. You usually just have to find $[H^+]$ from the pH, then fiddle around with the **K_a expression** to find the missing bit of information.

This bunny may look cute, but he can't help Horace with his revision.

Example: The pH of an ethanoic acid (CH_3COOH) solution was 3.02 at 298 K. Calculate the molar concentration of this solution. K_a of ethanoic acid is 1.75×10^{-5} mol dm^{-3} at 298 K.

First, use the pH to find $[H^+]$: $[H^+] = 10^{-pH} = 10^{-3.02} = \mathbf{9.55 \times 10^{-4}\ mol\ dm^{-3}}$

Then rearrange the expression for K_a and plug in your values to find $[CH_3COOH]$:

$$K_a = \frac{[H^+]^2}{[CH_3COOH]} \Longrightarrow [CH_3COOH] = \frac{[H^+]^2}{K_a} = \frac{(9.55 \times 10^{-4})^2}{1.75 \times 10^{-5}} = \mathbf{0.0521\ mol\ dm^{-3}}$$

More pH Calculations

$pK_a = -\log_{10} K_a$ and $K_a = 10^{-pK_a}$

pK_a is calculated from K_a in exactly the same way as pH is calculated from $[H^+]$ — and vice versa.

Example: i) If an acid has a K_a value of 1.50×10^{-7} mol dm^3, what is its pK_a?

$pK_a = -\log_{10}(1.50 \times 10^{-7}) = \mathbf{6.824}$

ii) What is the K_a value of an acid if its pK_a is 4.32?

$K_a = 10^{-4.32} = \mathbf{4.8 \times 10^{-5}}$ **mol dm^{-3}**

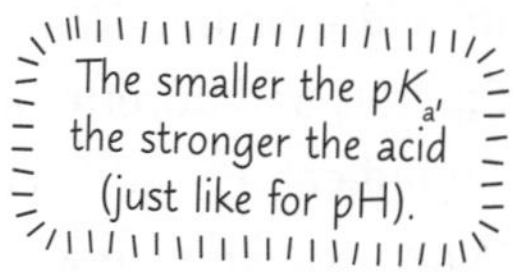

The smaller the pK_a, the stronger the acid (just like for pH).

Just to make things that bit more complicated, you might be given a **pK_a** value in a question to work out concentrations or pH. If so, you just need to convert it to K_a so that you can use the **K_a expression**.

Example: Calculate the pH of 0.0500 mol dm^{-3} methanoic acid (HCOOH). Methanoic acid has a pK_a of 3.75 at this temperature.

$K_a = 10^{-pK_a} = 10^{-3.75} = 1.8 \times 10^{-4}$ mol dm^{-3} ⟸ First you have to convert the pK_a to K_a.

$K_a = \frac{[H^+]^2}{[HCOOH]}$ ⟹ $[H^+]^2 = K_a \times [HCOOH] = 1.78 \times 10^{-4} \times 0.0500 = 8.90 \times 10^{-6}$

$[H^+] = \sqrt{(8.90 \times 10^{-6})} = 2.98 \times 10^{-3}$ mol dm^{-3}

pH $= -\log_{10}(2.98 \times 10^{-3}) = \mathbf{2.526}$

Bob the revision goat.

You might also be asked to work out a **pK_a** value from concentrations or pH. In this case, you just work out the K_a value as usual and then convert it to **pK_a** — and Bob's your revision goat.

Practice Questions

Q1 What are the units of K_a?

Q2 What assumptions do you have to make when calculating K_a for a weak acid? Why aren't these assumptions true for a strong acid?

Q3 Would you expect strong acids to have higher or lower pK_a values than weak acids?

Exam Questions

Q1 The value of K_a for the weak acid HA, at 298 K, is 5.60×10^{-4} mol dm^{-3}.

a) Write an expression for K_a for HA. [1 mark]

b) Calculate the pH of a 0.280 mol dm^{-3} solution of HA at 298 K. [2 marks]

Q2 The pH of a 0.150 mol dm^{-3} solution of a weak monoprotic acid, HX, is 2.64 at 298 K.

a) Calculate the value of K_a for the acid HX at 298 K. [2 marks]

b) Using your answer from part a), calculate pK_a for this acid. [1 mark]

Q3 Benzoic acid is a weak acid that is used as a food preservative. It has a pK_a of 4.2 at 298 K. Find the pH of a 1.6×10^{-4} mol dm^{-3} solution of benzoic acid at 298 K. [3 marks]

Fluffy revision animals... aaawwwww...

Strong acids have high K_a values and weak acids have low K_a values. For pK_a values, it's the other way round — the stronger the acid, the lower the pK_a. If something's got p in front of it, like pH, pK_w or pK_a, it'll mean $-\log_{10}$ of whatever. Oh and did you like the cute animals? Did they really make your day? Good, I'm really pleased about that.

pH Curves and Indicators

If you add a base to an acid, the pH changes in a squiggly sort of way.

Use **Titration** to Find the **Concentration** of an **Acid** or **Base**

Titration is covered, in detail, on pages 18-20 so if it's a little hazy in your mind, best to go back and brush up. This section will just go over **the basics**:

1) In **acid-base** titrations, you add a **standard solution** of acid to a **measured quantity** of base (or vice versa).
2) **Pipettes** and **burettes** are used so that you know **precisely** how much acid and base is used.
3) An **indicator** is added to the base to show you **exactly** when it's **neutralised** by the acid.
4) When you know how much acid it takes to neutralise the base, you can work out the **concentration of the base**. (Remember, you could swap it round and titrate an acid with a standard solution of a base instead — then you'd be working out the **concentration of the acid**.)

pH Curves Plot **pH** Against **Volume** of **Acid** or **Base** Added

The graphs below show the pH curves for the different combinations of **strong and weak** monoprotic acids and bases.

strong acid/strong base

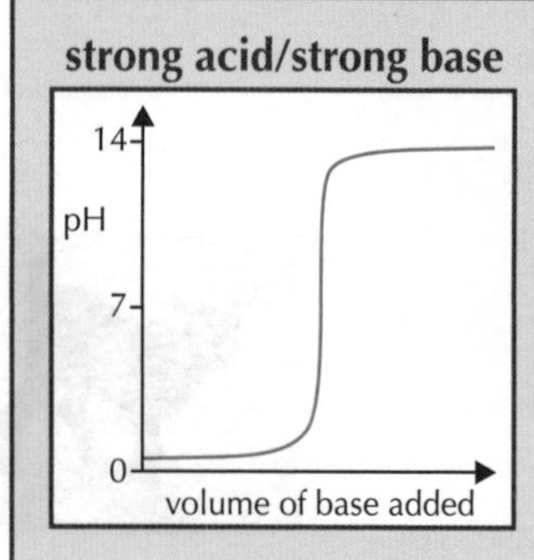

The pH starts around **1**, as there's an excess of **strong acid**.

It finishes up around pH **13**, when you have an excess of **strong base**.

strong acid/weak base

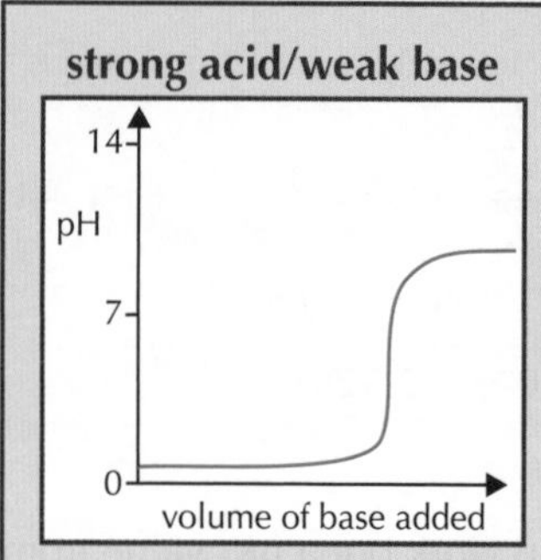

The pH starts around **1**, as there's an excess of **strong acid**.

It finishes up around pH **9**, when you have an excess of **weak base**.

weak acid/strong base

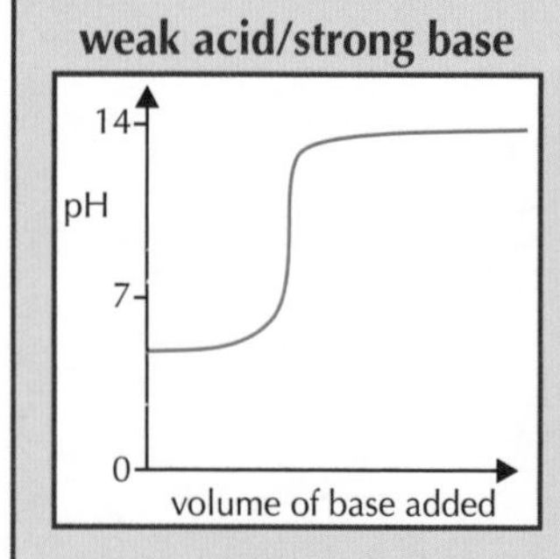

The pH starts around **5**, as there's an excess of **weak acid**.

It finishes up around pH **13**, when you have an excess of **strong base**.

weak acid/weak base

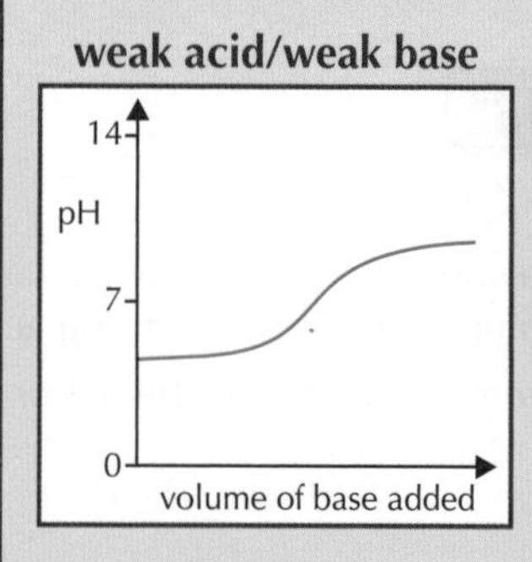

The pH starts around **5**, as there's an excess of **weak acid**.

It finishes up around pH **9**, when you have an excess of **weak base**.

All the graphs, apart from the weak acid/weak base graph, have a bit that's vertical — this is the **equivalence point** or **end point**. At this point, a tiny amount of base causes a sudden, big change in pH — it's here that all the acid is just **neutralised**.

You don't get such a sharp change in a **weak acid/weak base** titration.
If you used an indicator for this type of titration, its colour would change very **gradually**, and it would be very tricky to see the exact end point. So you're usually better off using a **pH meter** for this type of titration.

If you titrate a **base** with an **acid** instead, the **shapes** of the curves **stay the same**, but they **flip** over:

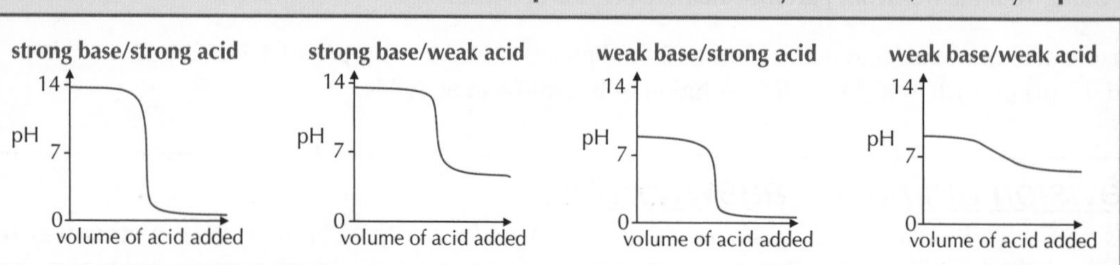

pH Curves and Indicators

pH Curves can Help you Decide which Indicator to Use

When you use an **indicator**, you need it to change colour exactly at the **end point** of your titration. So you need to pick one that changes colour over a **narrow pH range** that lies **entirely** on the **vertical part** of the **pH curve**.

E.g. For this titration, the curve is vertical between **pH 8** and **pH 11** — so a very small amount of base will cause the pH to **change** from 8 to 11.

So you want an indicator that changes **colour** somewhere between pH 8 and pH 11.

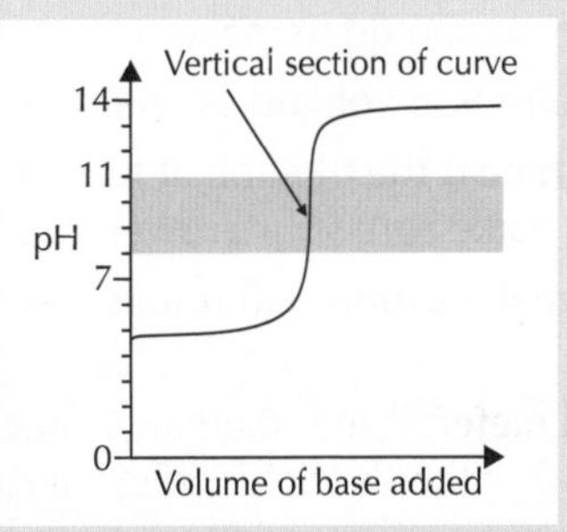

Methyl orange and **phenolphthalein** are **indicators** that are often used for acid-base titrations. They each change colour over a **different pH range**:

Name of indicator	Colour at low pH	Approx. pH of colour change	Colour at high pH
Methyl orange	red	3.1 – 4.4	yellow
Phenolphthalein	colourless	8.3 – 10	pink

- For a **strong acid/strong base** titration, you can use **either** of these indicators — there's a rapid pH change over the range for **both** indicators.
- For a **strong acid/weak base** only **methyl orange** will do. The pH changes rapidly across the range for methyl orange, but not for phenolphthalein.
- For a **weak acid/strong base**, **phenolphthalein** is the stuff to use. The pH changes rapidly over phenolphthalein's range, but not over methyl orange's.
- For **weak acid/weak base** titrations there's no sharp pH change, so **neither** of these indicators works. In fact, there aren't **any** indicators you can use in weak acid/weak base titrations, so you should just use a pH meter.

Practice Questions

Q1 Sketch the pH curve for a strong acid/weak base titration.

Q2 What indicator should you use for a strong acid/weak base titration — methyl orange or phenolphthalein?

Q3 What colour is methyl orange at low pH?

Exam Questions

Q1 NaOH (a strong base) is added separately to samples of nitric acid (a strong acid) and ethanoic acid (a weak acid). Sketch the pH curves for each of these titrations. [2 marks]

Q2 A sample of ethanoic acid (a weak acid) was added to a solution of potassium hydroxide (a strong base).

From the table on the right, select the best indicator for this titration, and explain your choice. [2 marks]

Name of indicator	pH range
bromophenol blue	3.0 – 4.6
methyl red	4.2 – 6.3
bromothymol blue	6.0 – 7.6
thymol blue	8.0 – 9.6

Q3 Ethanoic acid (a weak acid) was added to ammonia (a weak base).

a) Sketch the pH curve for this titration. [1 mark]

b) Explain why an indicator can't be used to accurately determine the end point of this titration. [2 marks]

My face turns luminous red when I reach my end point with revision...

Titrations involve playing with big bits of glassware that you're told not to break as they're really expensive — so you instantly become really clumsy. If you manage not to smash the burette, you'll find it easier to get accurate results if you use a dilute acid or base — drops of dilute acid and base contain fewer particles so you're less likely to overshoot.

Titration Calculations

OK, so you can carry out titrations like a boss, but here's what you can do with the results...

You Can Use **Titration Results** to Calculate **Concentrations**

When you've done a titration you can use your results to calculate the **concentration** of your acid or base.

There are a few things you can do to make sure your titration **results** are as **accurate** as possible:

1) Measure the neutralisation volume as precisely as you possibly can (this will usually be to the **nearest 0.05 cm^3**).
2) It's a good idea to **repeat** the titration at least three times and take a **mean** titre value. That'll help you to make sure your answer is **reliable**.
3) Don't use any **anomalous** (unusual) results — all your results should be within 0.1 cm^3 of each other.

If you use a **pH meter**, rather than an indicator, you can draw a pH curve of the titration and use it to work out how much acid or base is needed for neutralisation.

You do this by finding the **equivalence point** (the mid-point of the line of rapid pH change) and drawing a **vertical line downwards** until it meets the x-axis. The value at this point on the x-axis is the volume of acid or base needed.

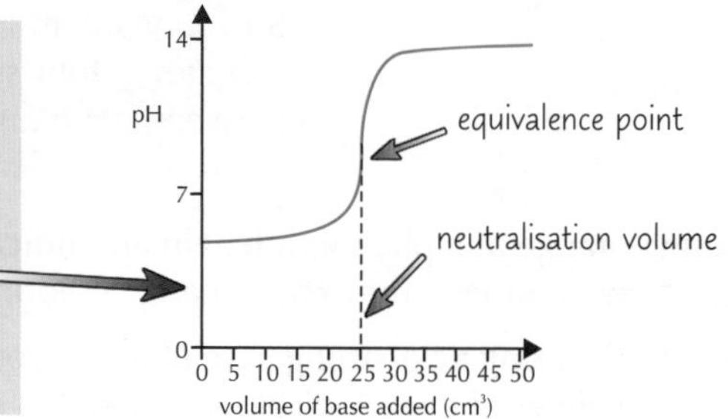
14
pH
7
0
0 5 10 15 20 25 30 35 40 45 50
volume of base added (cm^3)
equivalence point
neutralisation volume

Here's an **Example Calculation...**

You should have seen these types of calculations before but it never hurts to refresh the important stuff.

Example: 40 cm^3 of 0.75 mol dm^{-3} HNO_3 was needed to neutralise 60 cm^3 of KOH solution. Calculate the concentration of the potassium hydroxide solution.

$$HNO_3 + KOH \rightarrow KNO_3 + H_2O$$

Work out how many **moles of HNO_3** you have:

$$\text{Number of moles of } HNO_3 = \frac{\text{conc.} \times \text{volume (cm}^3)}{1000} = \frac{0.75 \times 40}{1000} = 0.030 \text{ mol}$$

You should remember this formula — you divide by 1000 to get the volume from cm^3 to dm^3.

From the equation, you know 1 mol of HNO_3 neutralises 1 mol of KOH.

So 0.030 mol of HNO_3 must neutralise **0.030** mol of KOH.

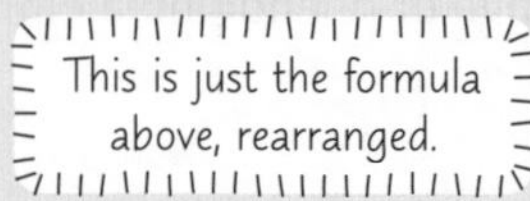
This is just the formula above, rearranged.

$$\text{Concentration of KOH} = \frac{\text{moles of } HNO_3 \times 1000}{\text{volume (cm}^3)} = \frac{0.030 \times 1000}{60} = \mathbf{0.50 \text{ mol dm}^{-3}}$$

A **Diprotic Acid** Releases **Two Protons** When it Dissociates

A **diprotic acid** is one that can release **two protons** when it's in solution. **Ethanedioic acid** (HOOC-COOH) is diprotic. When ethanedioic acid reacts with a **base** like sodium hydroxide, it's **neutralised**.

But the reaction happens in **two stages**, because the **two protons** are removed from the acid **separately**.

This means that when you titrate **ethanedioic acid** with a **strong base** you get a pH curve with two **equivalence points**:

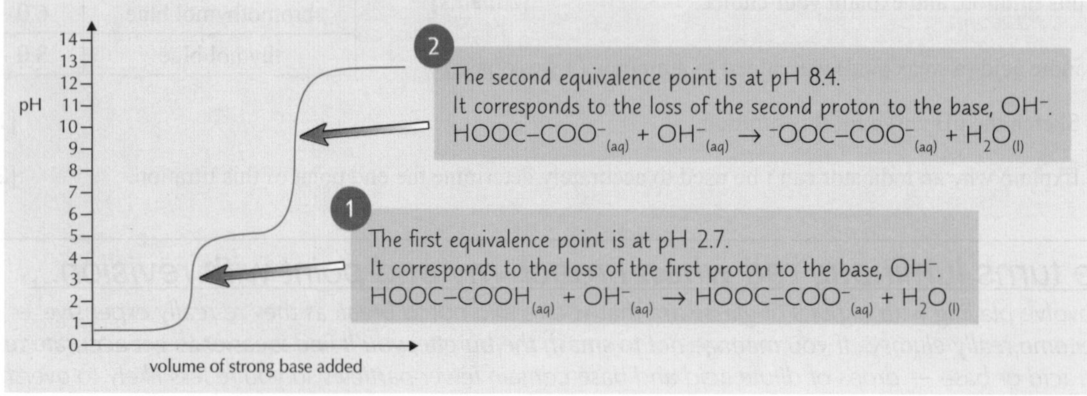
14
13
12
pH 11
10
9
8
7
6
5
4
3
2
1
0
volume of strong base added
2
The second equivalence point is at pH 8.4.
It corresponds to the loss of the second proton to the base, OH^-.
$HOOC\text{-}COO^-_{(aq)} + OH^-_{(aq)} \rightarrow {}^-OOC\text{-}COO^-_{(aq)} + H_2O_{(l)}$
1
The first equivalence point is at pH 2.7.
It corresponds to the loss of the first proton to the base, OH^-.
$HOOC\text{-}COOH_{(aq)} + OH^-_{(aq)} \rightarrow HOOC\text{-}COO^-_{(aq)} + H_2O_{(l)}$

Titration Calculations

You Can Find the Concentration of a Diprotic Acid From Titration Results Too

You can calculate the concentration of a **diprotic** acid from titration data in the same way as you did for a monoprotic acid.

Example: 25 cm³ of ethanedioic acid, $C_2H_2O_4$, was completely neutralised by 20 cm³ of 0.10 mol dm⁻³ NaOH solution. Calculate the concentration of the ethanedioic acid solution.

Write a **balanced equation** and decide **what you know** and what you **need to know:**

$$C_2H_2O_4 + 2NaOH \rightarrow Na_2C_2O_4 + 2H_2O$$

$C_2H_2O_4$	2NaOH
25 cm³	**20 cm³**
?	**0.10 mol dm⁻³**

Because it's a diprotic acid, you need twice as many moles of base as moles of acid.

Now work out how many **moles of NaOH** you have:

$$\text{Number of moles of NaOH} = \frac{\text{conc.} \times \text{volume (cm}^3\text{)}}{1000} = \frac{0.10 \times 20}{1000} = 0.0020 \text{ mol}$$

You know from the equation that you need 2 mol of NaOH to neutralise 1 mol of $C_2H_2O_4$.
So 0.0020 mol of NaOH must neutralise (0.002 ÷ 2) = **0.0010 mol of $C_2H_2O_4$.**
Now find the **concentration of $C_2H_2O_4$.**

$$\text{Concentration of } C_2H_2O_4 = \frac{\text{moles of } C_2H_2O_4 \times 1000}{\text{volume (cm}^3\text{)}} = \frac{0.0010 \times 1000}{25} = \mathbf{0.040 \text{ mol dm}^3}$$

Practice Questions

Q1 What is a diprotic acid?

Q2 How many moles of NaOH would you need to neutralise one mole of a diprotic acid?

Exam Questions

Q1 A student performed a titration with 25 cm³ of hydrochloric acid, adding 0.10 mol dm⁻³ sodium hydroxide from a burette. The student's results are shown in the table below.

	Titration 1	Titration 2	Titration 3
Titre volume (cm³ of NaOH)	25.60	25.65	25.55

a) Write a balanced equation for the reaction. [1 mark]

b) i) Calculate the average titre of sodium hydroxide needed to neutralise the hydrochloric acid. [1 mark]

ii) Use this to find the number of moles of sodium hydroxide that were needed to neutralise the acid. [1 mark]

c) Find the concentration of the hydrochloric acid. [1 mark]

Q2 Sulfuric acid is a diprotic acid.
25.0 cm³ of this acid is needed to neutralise 35.6 cm³ of 0.100 mol dm⁻³ sodium hydroxide.

a) Write a balanced equation for the reaction. [1 mark]

b) Calculate:

i) the number of moles of sodium hydroxide present in the 35.6 cm³ sample. [1 mark]

ii) the number of moles of sulfuric acid needed to neutralise the sodium hydroxide. [1 mark]

iii) the concentration of the sulfuric acid used. [1 mark]

Diprotic acids — double the protons, double the fun...

Don't forget, if it's a diprotic acid that you're using, you need twice as many moles of NaOH to neutralise it as you would for a monoprotic acid. But when it comes down to it, it's just the same story as any other chemistry calculation — write out the equation, compare the number of moles, and put the values into the right formula.

Buffer Action

Some solutions resist becoming more acidic if you add acid to them. Why would they want to? Read on to find out...

Buffers Resist Changes in **pH**

A **buffer** is a solution that **resists** changes in pH when **small** amounts of acid or base are added, or when it's **diluted**.

A buffer **doesn't** stop the pH from changing completely — it does make the changes **very slight** though. Buffers only work for small amounts of acid or base. You get **acidic buffers** and **basic buffers**.

Acidic Buffers are Made from a **Weak Acid** and one of its **Salts**

Acidic buffers have a pH of less than 7 — they're made by mixing a **weak acid** with one of its **salts**. **Ethanoic acid** and **sodium ethanoate** ($CH_3COO^- Na^+$) is a good example:

The ethanoic acid is a **weak acid**, so it only **slightly** dissociates: $CH_3COOH_{(aq)} \rightleftharpoons H^+_{(aq)} + CH_3COO^-_{(aq)}$.

But the salt **fully** dissociates into its ions when it dissolves: $CH_3COONa_{(s)} + \text{water} \rightarrow CH_3COO^-_{(aq)} + Na^+_{(aq)}$.

So in the solution you've got lots of **undissociated ethanoic acid molecules**, and lots of **ethanoate ions** from the salt.

When you alter the **concentration** of **H^+** or **OH^- ions** in the buffer solution, the **equilibrium position** moves to **counteract** the change (this is down to **Le Chatelier's principle** — see page 47). Here's how it all works:

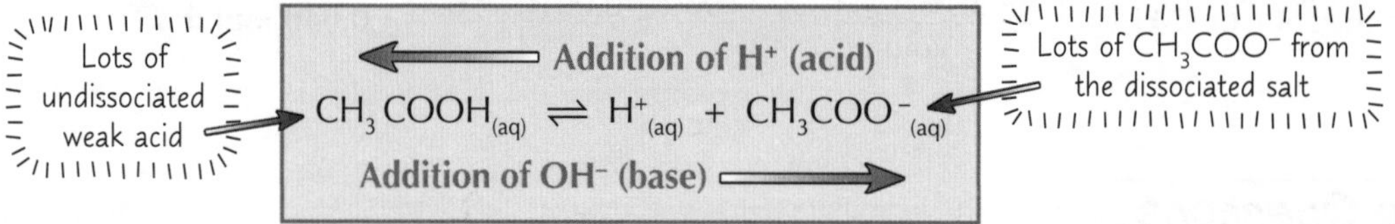

1) If you add a **small** amount of **acid**, the **H^+ concentration** increases. Most of the extra H^+ ions combine with CH_3COO^- ions to form CH_3COOH. This shifts the equilibrium to the **left**, reducing the H^+ concentration to close to its original value. So the **pH** doesn't change.
2) If a **small** amount of **base** (e.g. NaOH) is added, the **OH^- concentration** increases. Most of the extra OH^- ions react with H^+ ions to form water — removing H^+ ions from the solution. This causes more CH_3COOH to **dissociate** to form H^+ ions — shifting the equilibrium to the **right**. The H^+ concentration increases until it's close to its original value, so the **pH** doesn't change.

Basic Buffers are Made from a **Weak Base** and one of its **Salts**

Basic buffers have a pH greater than 7 — and they're made by mixing a **weak base** with one of its **salts**.

A solution of **ammonia** (NH_3, a weak base) and **ammonium chloride** (NH_4Cl, a salt of ammonia) acts as a **basic** buffer.

The **salt** fully dissociates in solution: $NH_4Cl_{(aq)} \rightarrow NH_4^+{}_{(aq)} + Cl^-_{(aq)}$.

Some of the NH_3 molecules will also react with water molecules: $NH_{3\,(aq)} + H_2O_{(aq)} \rightleftharpoons NH_4^+{}_{(aq)} + OH^-_{(aq)}$.

So the solution will contain loads of **ammonium ions** (NH_4^+), and lots of **ammonia** molecules too.

The **equilibrium position** of this reaction can move to **counteract** changes in pH:

Addition of H^+ (acid) ⟶

$$NH_{3(aq)} + H_2O_{(l)} \rightleftharpoons NH_4^+{}_{(aq)} + OH^-_{(aq)}$$

⟵ **Addition of OH^- (base)**

1) If a small amount of **base** is added, the OH^- concentration **increases**, making the solution more **basic**. Most of the extra OH^- ions will react with the NH_4^+ ions, to form NH_3 and H_2O. So the equilibrium will shift to the **left**, removing OH^- ions from the solution, and stopping the pH from changing much.
2) If a small amount of **acid** is added, the H^+ concentration **increases**, making the solution more **acidic**. Some of the H^+ ions react with **OH^-** ions to make **H_2O**. When this happens the equilibrium position **moves to the right** to replace the OH^- ions that have been used up. Some of the H^+ ions react with **NH_3** molecules to form NH_4^+: $NH_3 + H^+ \rightleftharpoons NH_4^+$. These reactions will **remove** most of the extra H^+ ions that were added — so the pH **won't** change much.

Buffer Action

Here's How to Calculate the pH of a Buffer Solution

Calculating the **pH** of an acidic buffer isn't too tricky. You just need to know the K_a of the weak acid and the **concentrations** of the weak acid and its salt. Here's how to go about it:

Example: A buffer solution contains 0.40 mol dm^{-3} methanoic acid, HCOOH, and 0.60 mol dm^{-3} sodium methanoate, $HCOO^- Na^+$. For methanoic acid, $K_a = 1.6 \times 10^{-4}$ mol dm^{-3}. What is the pH of this buffer?

First, write the expression for K_a of the weak acid:

$$HCOOH_{(aq)} \rightleftharpoons H^+_{(aq)} + HCOO^-_{(aq)} \longrightarrow K_a = \frac{[H^+_{(aq)}] \times [HCOO^-_{(aq)}]}{[HCOOH_{(aq)}]}$$

Remember — these are all equilibrium concentrations.

Then rearrange the expression and stick in the data to calculate $[H^+_{(aq)}]$:

$$[H^+_{(aq)}] = K_a \times \frac{[HCOOH_{(aq)}]}{[HCOO^-_{(aq)}]}$$

$$[H^+_{(aq)}] = 1.6 \times 10^{-4} \times (0.40 \div 0.60) = 1.07 \times 10^{-4} \text{ mol dm}^{-3}$$

You have to make a **few assumptions** here:

- $HCOO^- Na^+$ is fully dissociated, so assume that the equilibrium concentration of $HCOO^-$ is the same as the initial concentration of $HCOO^- Na^+$.
- HCOOH is only slightly dissociated, so assume that its equilibrium concentration is the same as its initial concentration.

Finally, convert $[H^+_{(aq)}]$ to pH:

$$pH = -\log_{10}[H^+_{(aq)}] = -\log_{10}(1.07 \times 10^{-4}) = \mathbf{3.971}$$

Buffers are Really Handy

Most **shampoos** contain a pH 5.5 buffer. Human hair becomes rougher if it's exposed to alkaline conditions — the buffer in the shampoo stops this from happening, keeping hair smooth and shiny.

Biological washing powders also contain buffers. They keep the pH at the right level for the enzymes to work most efficiently.

There are lots of **biological buffer** systems in our bodies too, making sure our tissues are kept at the **right pH**. For example, it's vital that **blood** stays at a pH close to 7.4, so it contains a buffer system.

Ed resolved to use a shampoo with a buffer in future.

Practice Questions

Q1 What's a buffer solution? Give two uses of buffer solutions.

Q2 How can a mixture of ethanoic acid and sodium ethanoate act as a buffer?

Q3 Describe how to make a basic buffer.

Exam Questions

Q1 A buffer solution contains 0.40 mol dm^{-3} propanoic acid, CH_3CH_2COOH, and 0.20 mol dm^{-3} sodium propanoate, $CH_3CH_2COO^- Na^+$. At 25 °C, K_a for propanoic acid is 1.3×10^{-5} mol dm^{-3}.

a) Calculate the pH of the buffer solution. [3 marks]

b) Explain the effect on the buffer of adding a small quantity of dilute sulfuric acid. [2 marks]

Q2 A buffer was prepared by mixing solutions of butanoic acid, $CH_3(CH_2)_2COOH$, and sodium butanoate, $CH_3(CH_2)_2COO^- Na^+$, so that they had the same concentration.

a) Write a chemical equation to show butanoic acid acting as a weak acid. [1 mark]

b) Given that K_a for butanoic acid is 1.5×10^{-5} mol dm^{-3}, calculate the pH of the buffer solution. [3 marks]

Old buffers are often resistant to change...

So that's how buffers work. There's a pleasing simplicity and neatness about it that I find rather elegant. Like a fine glass of red wine with a nose of berry and undertones of raspberries, oak and... OK, I'll shut up now.

Periodicity

Periodicity is one of those words you hear a lot in chemistry without ever really knowing what it means. Well it basically means trends that occur (in physical and chemical properties) as you move across the periods. E.g. metal to non-metal is a trend that occurs going left to right in each period... The trends repeat each period.

The **Periodic Table** arranges Elements by **Proton Number**

1) The periodic table is arranged into **periods** (rows) and **groups** (columns), by atomic (proton) number.
2) All the elements **within a period** have the same number of **electron shells** (if you don't worry about s and p sub-shells). E.g. the elements in Period 2 have 2 electron shells.
3) All the elements **within a group** have the **same number** of electrons in their **outer shell** — so they have **similar properties**.
4) The **group number** tells you the number of electrons in the outer shell, e.g. Group 1 elements have 1 electron in their outer shell, Group 4 elements have 4 electrons and so on. The **exception** is **Group 0**. Group 0 elements all have full outer shells — that's two electrons for Helium, and eight electrons for all the others.

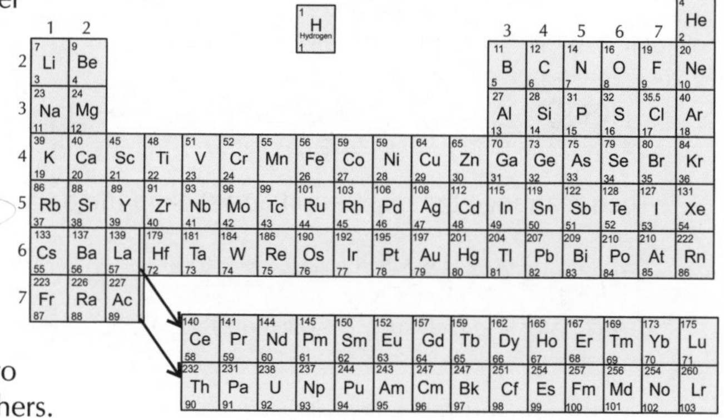

You can use the Periodic Table to work out **Electronic Configurations**

The periodic table can be split into an **s block**, **d block**, **p block** and **f block** like this: Doing this shows you which sub-shells all the electrons go into.

See page 8 if this sub-shell malarkey doesn't ring a bell.

When you've got the periodic table **labelled** with the **shells** and **sub-shells** like the one on the right, it's pretty easy to read off the **electronic structure** of any element. Just start at the top and work your way across and down until you get to your element.

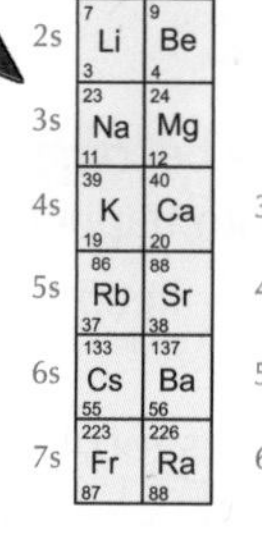

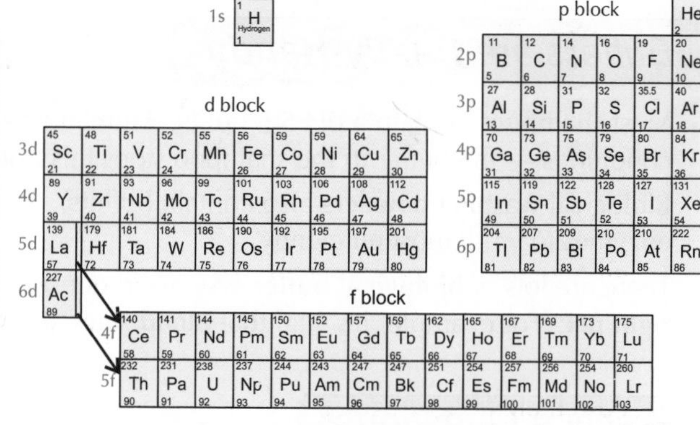

Example:

Electronic structure of phosphorus (P):
Period 1 — $1s^2$
Period 2 — $2s^2\ 2p^6$ ← Complete sub-shells
Period 3 — $3s^2\ 3p^3$ ← Incomplete outer sub-shell
So the full electronic structure of phosphorus is: $1s^2\ 2s^2\ 2p^6\ 3s^2\ 3p^3$

Example:

Electronic structure of cobalt (Co):
Period 1 — $1s^2$
Period 2 — $2s^2\ 2p^6$
Period 3 — $3s^2\ 3p^6$
Period 4 — $3d^7\ 4s^2$
So the full electronic structure of cobalt is: $1s^2\ 2s^2\ 2p^6\ 3s^2\ 3p^6\ 3d^7\ 4s^2$

See page 8 for more on electronic structure.

Atomic Radius **Decreases** across a Period

1) As the number of protons increases, the **positive charge** of the nucleus increases. This means electrons are **pulled closer** to the nucleus, making the atomic radius smaller.
2) The extra electrons that the elements gain across a period are added to the **outer energy level** so they don't really provide any extra shielding effect (shielding is mainly provided by the electrons in the inner shells).

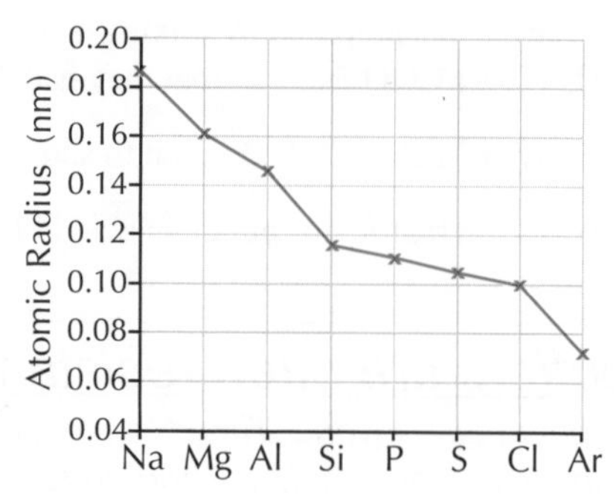

Periodicity

Melting Point is linked to Bond Strength and Structure

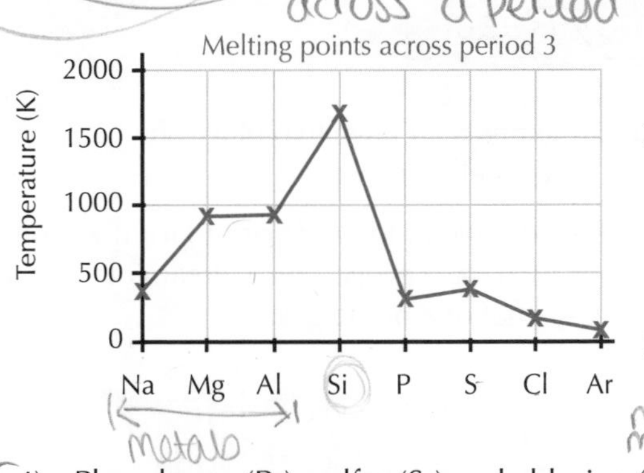

1) Melting points vary across a period as they depend on the **structure** of elements and the **bonding** within them. The graph on the left shows the melting points across **Period 3**.
2) Sodium, magnesium and aluminium are **metals**. Their melting and boiling points **increase** across the period because the **metal-metal bonds** get stronger. The bonds get stronger because the metal ions have an increasing positive charge, an increasing number of delocalised electrons and a decreasing radius.
3) Silicon is **macromolecular**, with a tetrahedral structure — **strong covalent bonds** link all its atoms together. **A lot** of energy is needed to break these bonds, so silicon has a **high** melting point.
4) Phosphorus (P_4), sulfur (S_8) and chlorine (Cl_2) are all **molecular substances**. Their melting points depend upon the strength of the **van der Waals forces** (see page 31) between the molecules. Van der Waals forces are weak and easily overcome so these elements have **low** melting points.
5) More atoms in a molecule mean stronger van der Waals forces. Sulfur is the **biggest molecule** (S_8), so it's got a higher melting point than phosphorus or chlorine.
6) Argon has a **very low** melting point because it exists as **individual atoms** (it's monatomic) resulting in **very weak** van der Waals forces.

Sam is looking hot in the latest periodic trends.

Ionisation Energy Generally Increases across a Period

This is because of the **increasing attraction** between the outer shell electrons and the nucleus, due to the number of **protons** increasing (there are a few blips in the trend however — check back to pages 12-13 for more details).

Practice Questions

Q1 Which elements of Period 3 are found in the s block of the periodic table?

Q2 Write down the electronic configuration of sodium.

Q3 Which element in Period 3 has the largest atomic radius?

Q4 Which element in Period 3 has the highest melting point?

Exam Questions

Q1 Explain why the melting point of magnesium is higher than that of sodium.

Q2 This table shows the melting points for the Period 3 elements.

Element	Na	Mg	Al	Si	P	S	Cl	Ar
Melting point / K	371	923	933	1687	317	388	172	84

In terms of structure and bonding explain why:

a) silicon has a high melting point. [2 marks]

b) the melting point of sulfur is higher than phosphorus. [2 marks]

Q3 State and explain the trend in atomic radius across Period 3. [3 marks]

Periodic trends — isn't that just another name for retro chic...

OK, I'll admit it, when it comes to fashion, I'm behind the times. I refuse to stop wearing my favourite pair of neon legwarmers just because they're a tiny bit out of date. The thing is though, every now and then eighties stuff comes back into fashion for a while and suddenly I'm bang on trend. Mind you, that doesn't stop me looking totally ridiculous.

Group 2 — The Alkaline Earth Metals

Group 2, AKA the alkaline earth metals, are in the "s block" of the periodic table. There are four pages about these jolly fellas and their compounds, so we've got a lot to do — best get on...

Group 2 Elements **Lose Two Electrons** when they React

Element	Atom	Ion
Be	$1s^2 2s^2$	$1s^2$
Mg	$1s^2 2s^2 2p^6 3s^2$	$1s^2 2s^2 2p^6$
Ca	$1s^2 2s^2 2p^6 3s^2 3p^6 4s^2$	$1s^2 2s^2 2p^6 3s^2 3p^6$

Group 2 elements all have two electrons in their outer shell (s^2).

They lose their two outer electrons to form **2+ ions**. Their ions then have every atom's dream electronic structure — that of a **noble gas**.

Atomic Radius **Increases** Down the Group

As you go **down** a group in the periodic table, the **atomic radius** gets **larger**. This is because extra **electron shells** are added as you go down the group.

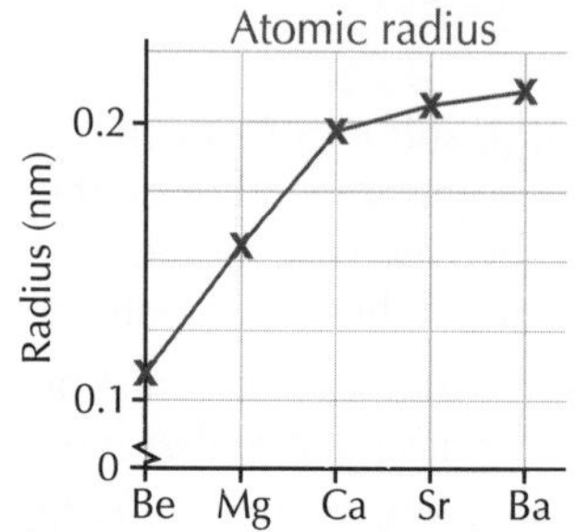

First Ionisation Energy **Decreases** Down the Group

1) Each element down Group 2 has an **extra electron shell** compared to the one above.
2) The extra inner shells **shield** the outer electrons from the attraction of the nucleus.
3) Also, the extra shell means that the outer electrons are **further away** from the nucleus, which greatly reduces the nucleus's attraction.

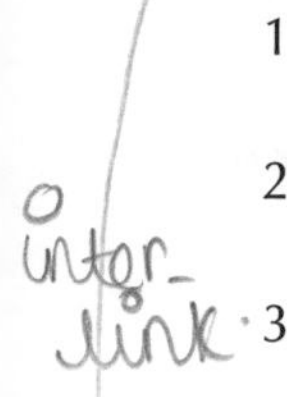

> Both of these factors make it **easier** to remove outer electrons, resulting in a **lower first ionisation energy**.

Mr Kelly has one final attempt at explaining electron shielding to his students...

The positive charge of the nucleus does increase as you go down a group (due to the extra protons), but this effect is overridden by the effect of the extra shells.

Reactivity **Increases** Down the Group

1) As you go down the group, the **first ionisation energy** decreases. This is due to the increasing atomic radius and shielding effect (see above).
2) When Group 2 elements react they **lose electrons**, forming positive ions. The easier it is to lose electrons (i.e. the lower the first and second ionisation energies), the more reactive the element, so **reactivity increases** down the group.

Melting Points Generally **Decrease** Down the Group

1) The Group 2 elements have typical **metallic structures**, with **positive ions** in a **crystal structure** surrounded by **delocalised electrons** from the outer electron shells.
2) Going **down** the group the metal ions get **bigger**. But the number of delocalised electrons per atom doesn't change (it's always 2) and neither does the charge on the ion (it's always +2).
3) The **larger** the ionic radius, the **further away** the delocalised electrons are from the positive nuclei and the **less attraction** they feel. So it takes **less energy** to break the bonds, which means the melting points generally decrease as you go down the group. However, there's a big 'blip' at magnesium, because the crystal structure (the arrangement of the metallic ions) changes.

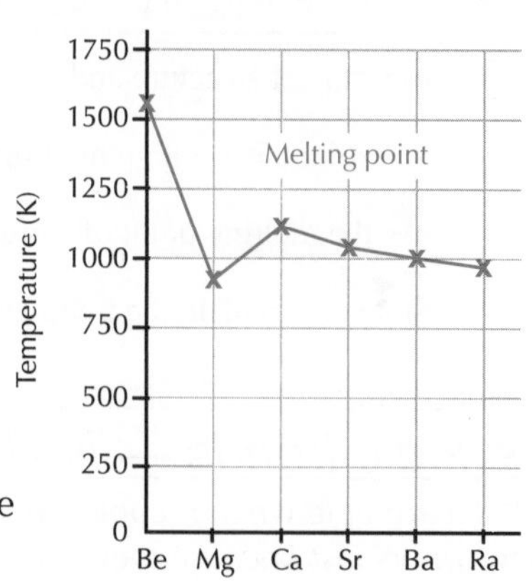

Group 2 — The Alkaline Earth Metals

Group 2 Elements React With Water

When Group 2 elements react, they are **oxidised** from a state of **0** to **+2**, forming M^{2+} ions.

	$M \rightarrow M^{2+} + 2e^-$	E.g.	$\mathbf{Ca \rightarrow Ca^{2+}} + 2e^-$
Oxidation state:	0 +2		**0** **+2**

The Group 2 metals react with water to give a **metal hydroxide and hydrogen.**

$$M_{(s)} + 2H_2O_{(l)} \rightarrow M(OH)_{2(aq)} + H_{2(g)}$$

Oxidation state: 0 → +2

e.g. $$\mathbf{Ca}_{(s)} + 2H_2O_{(l)} \rightarrow \mathbf{Ca(OH)}_{2(aq)} + H_{2(g)}$$

The water is the oxidising agent in this reaction.

They react **more readily** down the group because the **ionisation energies** decrease.

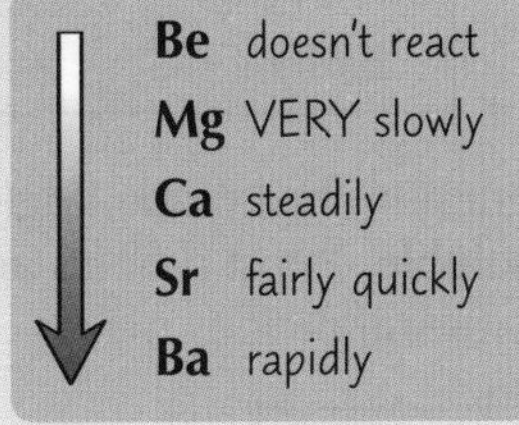

Be doesn't react
Mg VERY slowly
Ca steadily
Sr fairly quickly
Ba rapidly

Practice Questions

Q1 Do Group 2 elements lose or gain electrons when they react? How many electrons do they lose/gain?

Q2 Which is the least reactive metal in Group 2?

Q3 Which of the following increases in size down Group 2: atomic radius, first ionisation energy, melting point?

Q4 Why does reactivity with water increase down Group 2?

Exam Questions

Q1 a) Write out the electron configurations of magnesium and calcium. [2 marks]

b) Explain the difference between the first ionisation energies of magnesium and calcium. [3 marks]

Q2 The table shows the atomic radii of three elements from Group 2.

Element	Atomic Radius (nm)
X	0.112
Y	0.219
Z	0.194

a) Predict which element would react most rapidly with water. [1 mark]

b) Explain your answer. [2 marks]

Q3 Which element has been oxidised in the reaction below? State the change in its oxidation number.

$$Ba_{(s)} + 2H_2O_{(l)} \rightarrow Ba(OH)_{2(aq)} + H_{2(g)}$$

[2 marks]

Q4 The table shows the melting points of three elements from Group 2.

Element	Calcium	Strontium	Barium
Melting point (K)	1115	1050	1000

Explain the pattern in melting points that this table shows. [3 marks]

Bored of Group 2 trends? Me too. Let's play noughts and crosses...

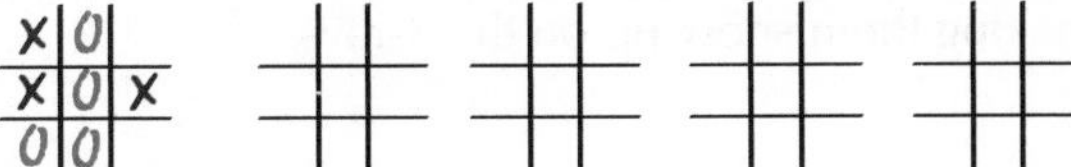

Noughts and crosses is pretty rubbish really, isn't it? It's always a draw. Ho hum. Back to Chemistry then, I guess...

Uses of the Group 2 Elements

Nice theories are all well and good but theories don't pay the bills. It's all about applications these days. So how can all this knowledge about Group 2 elements be put to good use? Read on to find out...

Solubility Trends in Group 2 Depend on the Compound Anion

Generally, compounds of Group 2 elements that contain **singly charged** negative ions (e.g. OH^-) **increase** in solubility down the group, whereas compounds that contain **doubly charged** negative ions (e.g. SO_4^{2-}) **decrease** in solubility down the group.

Group 2 element	hydroxide (OH^-)	sulfate (SO_4^{2-})
magnesium	least soluble	most soluble
calcium	↓	↑
strontium	↓	↑
barium	most soluble	least soluble

Compounds like magnesium hydroxide, $Mg(OH)_2$, which have **very low** solubilities are said to be **sparingly soluble**. Most sulfates are soluble in water, but **barium sulfate** ($BaSO_4$) is **insoluble**. The test for sulfate ions makes use of this property...

Test for sulfate ions

If acidified barium chloride ($BaCl_2$) is added to a solution containing sulfate ions then a white precipitate of barium sulfate is formed.

$$Ba^{2+}_{(aq)} + SO_4^{2-}{}_{(aq)} \rightarrow BaSO_{4\,(s)}$$

E.g. $$BaCl_{2(aq)} + FeSO_{4(aq)} \rightarrow BaSO_{4\,(s)} + FeCl_{2(aq)}$$

You need to acidify the solution with hydrochloric acid to get rid of any lurking sulfites or carbonates, which will also produce a white precipitate.

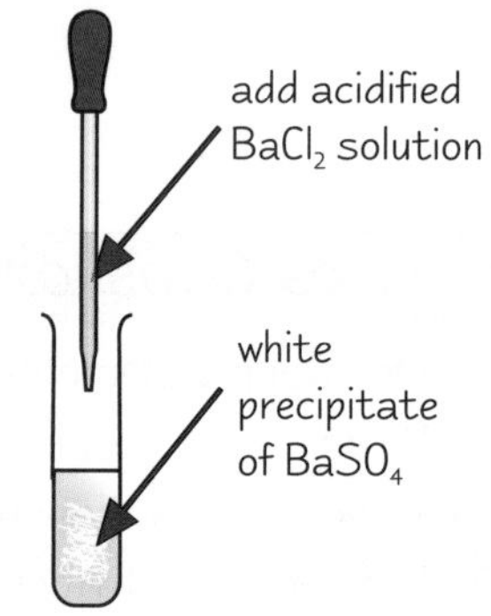

Group 2 Compounds are used to Neutralise Acidity

Group 2 elements are known as the **alkaline earth metals**, and many of their common compounds are used for neutralising acids. Here are a couple of common examples:

1) Calcium hydroxide (slaked lime, $Ca(OH)_2$) is used in **agriculture** to neutralise acid soils.

2) Magnesium hydroxide ($Mg(OH)_2$) is used in some indigestion tablets as an **antacid** — this is a substance which neutralises excess stomach acid.

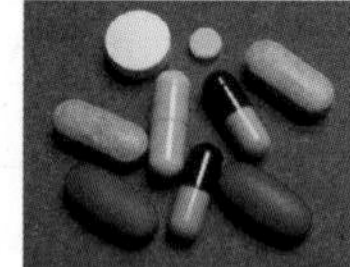

In both cases, the ionic equation for the neutralisation is

$$H^+_{(aq)} + OH^-_{(aq)} \rightarrow H_2O_{(l)}$$

Barium Sulfate is used in 'Barium Meals'

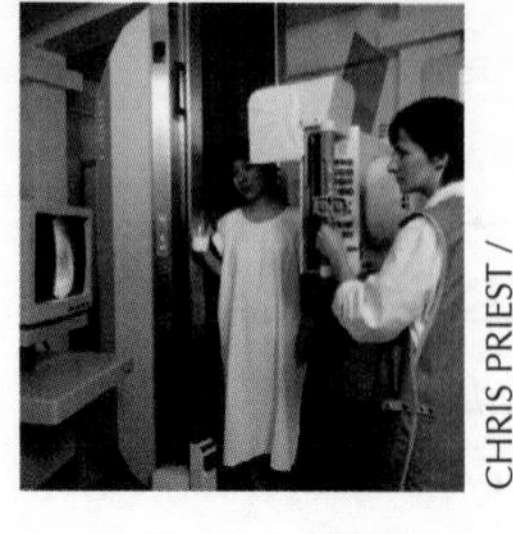

CHRIS PRIEST / SCIENCE PHOTO LIBRARY

X-rays are great for finding broken bones, but they pass straight through soft tissue — so soft tissues, like the digestive system, don't show up on conventional X-ray pictures.

1) Barium sulfate is **opaque** to X-rays — they won't pass through it. It's used in '**barium meals**' to help diagnose problems with the oesophagus, stomach or intestines.
2) A patient swallows the barium meal, which is a suspension of **barium sulfate**. The barium sulfate **coats** the tissues, making them show up on the X-rays, showing the structure of the organs.

Uses of the Group 2 Elements

Magnesium is used in the Extraction of Titanium

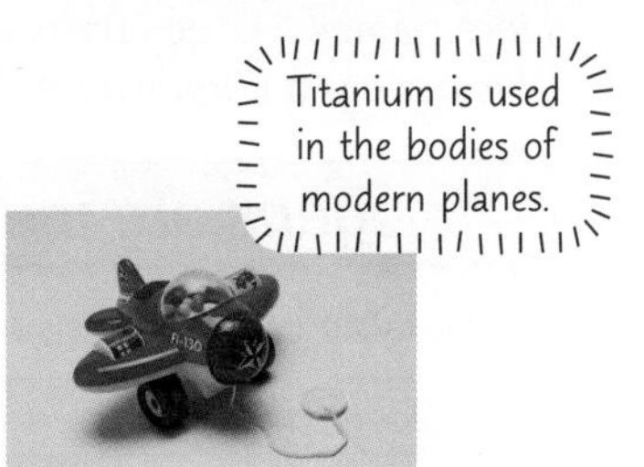

Titanium is used in the bodies of modern planes.

1) Magnesium is used as part of the process of **extracting titanium** from its ore.
2) The main titanium ore, titanium(IV) oxide (TiO_2) is first converted to **titanium(IV) chloride** ($TiCl_4$) by heating it with carbon in a stream of chlorine gas.
3) The titanium chloride is then purified by fractional distillation, before being **reduced by magnesium** in a furnace at almost 1000 °C.

$$TiCl_{4(g)} + 2Mg_{(l)} \rightarrow Ti_{(s)} + 2MgCl_{2(l)}$$

Mg is the reducing agent.

Calcium Oxide and Calcium Carbonate Remove Sulfur Dioxide

1) Burning fossil fuels to produce electricity also produces **sulfur dioxide**, which pollutes the atmosphere.
2) The acidic sulfur dioxide can be **removed** from **flue gases** by reacting with an alkali — this process is called **wet scrubbing**.
3) Powdered **calcium oxide** (lime, CaO) and **calcium carbonate** (limestone, $CaCO_3$) can both be used for this.
4) A **slurry** is made by **mixing** the calcium oxide or calcium carbonate with **water**. It's then sprayed onto the flue gases. The sulfur dioxide reacts with the alkaline slurry and produces a solid waste product, **calcium sulfite**.

Flue gases are the gases emitted from industrial exhausts and chimneys.

$$CaO_{(s)} + 2H_2O_{(l)} + SO_{2(g)} \rightarrow CaSO_{3(s)} + 2H_2O_{(l)}$$

$$CaCO_{3(s)} + 2H_2O_{(l)} + SO_{2(g)} \rightarrow CaSO_{3(s)} + 2H_2O_{(l)} + CO_{2(g)}$$

Practice Questions

Q1 Which is less soluble, barium sulfate or magnesium sulfate?

Q2 How is the solubility of magnesium hydroxide often described?

Q3 Give a use of calcium hydroxide.

Q4 Which Group 2 element can be used to extract titanium from titanium chloride?

Q5 Write the equation for the removal of sulfur dioxide from flue gases by calcium oxide.

Exam Questions

Q1 Describe how you could use acidified barium chloride solution to distinguish between solutions of zinc chloride and zinc sulfate. Give the expected observations and an appropriate balanced equation including state symbols. [2 marks]

Q2 Choose the Group 2 element, labelled A-D below, which best fits each of the following descriptions.

A magnesium B calcium C strontium D barium

a) Forms hydroxide and sulfate compounds, only one of which is soluble. [1 mark]

b) Forms a hydroxide used to neutralise acidic soils. [1 mark]

c) Forms a very soluble sulfate compound but sparingly soluble hydroxide compound. [1 mark]

Q3 Describe how barium sulfate can be used to diagnose problems with the digestive system. [2 marks]

Wet scrubbing — I thought that's what you did in the shower...

The Group 2 elements and compounds have lots of uses — they're used in agriculture, medicine, for reducing pollution, saving the world... Make sure you know which compound can be used for what though — you don't want to feed limestone to someone who has a problem with their digestive system, things could get horribly clogged up.

Group 7 — The Halogens

Here comes a page jam-packed with nuggets of halogen fun. Oh yes, I kid you not. This page is a chemistry roller coaster... white-knuckle excitement all the way...

Halogen is used to describe the atom (X) or molecule (X_2). Halide is used to describe the negative ion (X^-).

Halogens are the **Highly Reactive Non-Metals** of Group 7

The table below gives some of the main properties of the first 4 halogens.

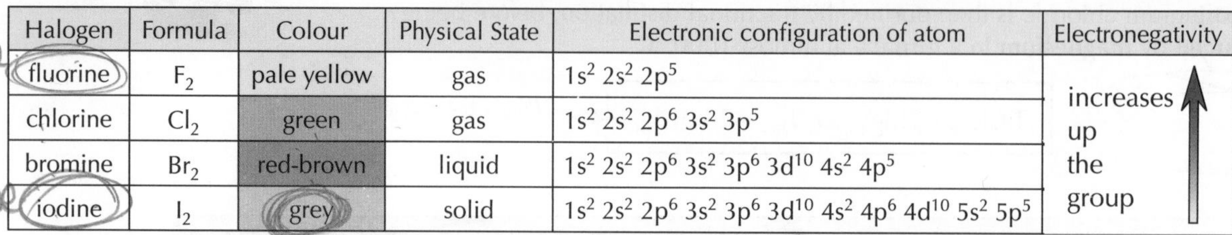

Halogen	Formula	Colour	Physical State	Electronic configuration of atom	Electronegativity
fluorine	F_2	pale yellow	gas	$1s^2\ 2s^2\ 2p^5$	increases up the group
chlorine	Cl_2	green	gas	$1s^2\ 2s^2\ 2p^6\ 3s^2\ 3p^5$	
bromine	Br_2	red-brown	liquid	$1s^2\ 2s^2\ 2p^6\ 3s^2\ 3p^6\ 3d^{10}\ 4s^2\ 4p^5$	
iodine	I_2	grey	solid	$1s^2\ 2s^2\ 2p^6\ 3s^2\ 3p^6\ 3d^{10}\ 4s^2\ 4p^6\ 4d^{10}\ 5s^2\ 5p^5$	

1) **Their boiling points increase down the group.**
This is due to the increasing strength of the **van der Waals forces** as the size and relative mass of the molecules increases. This trend is shown in the changes of **physical state** from fluorine (gas) to iodine (solid).

2) **Electronegativity decreases down the group.**
Electronegativity, remember, is the tendency of an atom to **attract** a bonding pair of **electrons**. The halogens are all highly electronegative elements. But larger atoms attract electrons **less** than smaller ones. This is because the electrons are **further** from the nucleus and are **shielded** by more electrons. (See page 10 for more on shielding).

Halogens **Displace** Less Reactive Halide Ions from Solution

1) When the halogens react, they **gain an electron**. They get **less reactive down the group**, because the atoms become larger. The outer shell is further from the nucleus, so electrons are less strongly attracted to it. So you can also say that the halogens become **less oxidising** down the group.

2) The **relative oxidising strengths** of the halogens can be seen in their **displacement reactions** with the halide ions:

Halogen	Displacement reaction	Ionic equation
Cl	chlorine (Cl_2) will displace bromide (Br^-) and iodide (I^-)	$Cl_{2(aq)} + 2Br^-_{(aq)} \rightarrow 2Cl^-_{(aq)} + Br_{2(aq)}$ $Cl_{2(aq)} + 2I^-_{(aq)} \rightarrow 2Cl^-_{(aq)} + I_{2(aq)}$
Br	bromine (Br_2) will displace iodide (I^-)	$Br_{2(aq)} + 2I^-_{(aq)} \rightarrow 2Br^-_{(aq)} + I_{2(aq)}$
I	no reaction with F^-, Cl^-, Br^-	

The basic rule is:

> A **halogen** will **displace a halide** from solution if the halide is **below it** in the periodic table.

3) These displacement reactions can be used to help **identify** which halogen (or halide) is present in a solution.

	Potassium chloride solution $KCl_{(aq)}$ — colourless	Potassium bromide solution $KBr_{(aq)}$ — colourless	Potassium iodide solution $KI_{(aq)}$ — colourless
Add chlorine water $Cl_{2(aq)}$ — colourless	no reaction	orange solution (Br_2) formed	brown solution (I_2) formed
Add bromine water $Br_{2(aq)}$ — orange	no reaction	no reaction	brown solution (I_2) formed
Add iodine solution $I_{2(aq)}$ — brown	no reaction	no reaction	no reaction

Chlorine and **Sodium Hydroxide** make **Bleach**

If you mix chlorine gas with cold, dilute, aqueous sodium hydroxide you get **sodium chlorate(I) solution**, $NaClO_{(aq)}$, which just happens to be common household **bleach** (which kills bacteria).

$$2NaOH_{(aq)} + Cl_{2(g)} \rightarrow NaClO_{(aq)} + NaCl_{(aq)} + H_2O_{(l)}$$

Ox. state: Cl_2: 0; NaClO: +1; NaCl: −1

ClO^- is the chlorate(I) ion. Chlorine's oxidation state is +1 in this ion.

In this reaction, the oxidation state of Cl goes up and down, meaning that chlorine is both oxidised and reduced. This is called disproportionation.

> The sodium chlorate(I) solution (bleach) has loads of uses — it's used in **water treatment**, to bleach **paper** and **textiles**... and it's good for **cleaning toilets**, too. Handy...

Group 7 — The Halogens

Chlorine is used to Kill Bacteria in Water

When you mix **chlorine** with **water**, it undergoes **disproportionation**.

You end up with a mixture of **chloride ions** and **chlorate(I) ions.**

$$Cl_{2(g)} + H_2O_{(l)} \rightleftharpoons 2H^+_{(aq)} + Cl^-_{(aq)} + ClO^-_{(aq)}$$

Ox. state: 0 (Cl), −1 (Cl^-), +1 (ClO^-)

In sunlight, chlorine can also decompose water to form chloride ions and **oxygen**.

$$2Cl_{2(g)} + 2H_2O_{(l)} \rightleftharpoons 4H^+_{(aq)} + 4Cl^-_{(aq)} + O_{2(g)}$$

Chlorate(I) ions **kill bacteria**. So, **adding chlorine** (or a compound containing chlorate(I) ions) to water can make it safe to **drink** or **swim** in. On the downside, chlorine is toxic.

1) In the UK our drinking water is **treated** to make it safe. **Chlorine** is an important part of water treatment:

- It **kills disease-causing microorganisms**.
- Some chlorine persists in the water and **prevents reinfection** further down the supply.
- It prevents the growth of **algae**, eliminating **bad tastes** and **smells**, and **removes discolouration** caused by organic compounds.

2) However, there are risks from using chlorine to treat water:

- **Chlorine gas** is **very harmful** if it's breathed in — it irritates the **respiratory system**. **Liquid chlorine** on the skin or eyes causes severe **chemical burns**. Accidents involving chlorine could be really serious, or fatal.
- Water contains a variety of organic compounds, e.g. from the decomposition of plants. Chlorine reacts with these compounds to form **chlorinated hydrocarbons**, e.g. chloromethane (CH_3Cl), and many of these chlorinated hydrocarbons are carcinogenic (cancer-causing). However, this increased cancer risk is small compared to the risks from untreated water — a cholera epidemic, say, could kill thousands of people.

3) We have to weigh up these risks and benefits when making decisions about whether we should add chemicals to drinking water supplies.

Practice Questions

Q1 Place the halogens F, Cl, Br and I in order of increasing: a) boiling point, b) electronegativity.

Q2 What would be seen when chlorine water is added to potassium iodide solution?

Q3 How is common household bleach formed?

Exam Questions

Q1 a) Write an ionic equation for the reaction between iodine solution and sodium astatide (NaAt). [1 mark]

b) For the equation in a), state which substance is oxidised. [1 mark]

Q2 a) Describe and explain the trends in:

i) the boiling points of Group 7 elements as you move down the Periodic Table. [3 marks]

ii) electronegativity of Group 7 elements as you move down the Periodic Table. [3 marks]

b) Which halogen is the most powerful oxidising agent? [1 mark]

Q3 Chlorine is added to the water in public swimming baths in carefully controlled quantities.

a) Write an equation for the reaction of chlorine with water to form chlorate(I) ions. [1 mark]

b) Why is chlorine added to the water in swimming baths, and why must the quantity added be carefully controlled? [2 marks]

Remain seated until the page comes to a halt. Please exit to the right...

Oooh, what a lovely page, if I do say so myself. There's nowt too taxing here — you just need to learn the colours of the solutions, all the equations, the advantages and disadvantages of adding chlorine to water supplies... it never ends.

Halide Ions

OK, a quick reminder of the basics first. Halides are compounds with the –1 halogen ion (e.g. Cl^-, Br^-, I^-) like KI, HCl and NaBr. They all end in "-ide" — chloride, bromide, iodide. Got that? Good. Now, you're ready to go in...

*The **Reducing Power** of Halides **Increases** Down the Group...*

To reduce something, the halide ion needs to lose an electron from its outer shell.
How easy this is depends on the **attraction** between the **nucleus** and the outer **electrons**.
As you go down the group, the attraction gets **weaker** because:

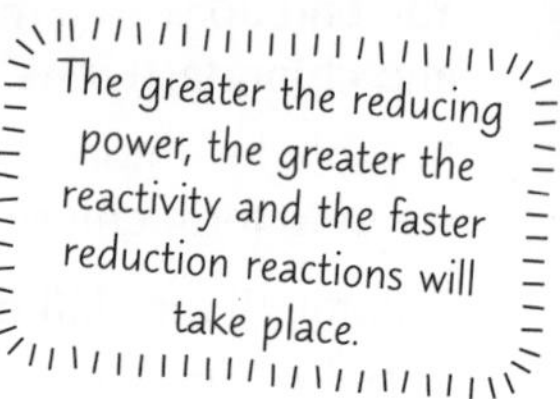

The greater the reducing power, the greater the reactivity and the faster reduction reactions will take place.

1) the ions get bigger, so the electrons are **further** away from the positive nucleus,
2) there are extra inner electron shells, so there's a greater **shielding** effect.

So, the further down the group the halide ion is, the easier it loses electrons and the greater its reducing power.

A good example of halogens doing a bit of reduction is the good old halogen / halide displacement reactions (the ones you learned on page 100... yes, those ones). And here come some more examples to learn...

*...which Explains their Reactions with **Sulfuric Acid***

All the halides react with concentrated sulfuric acid to give a **hydrogen halide** as a product to start with. But what happens next depends on which halide you've got...

Reaction of NaF or NaCl with H_2SO_4

$$NaF_{(s)} + H_2SO_{4(aq)} \rightarrow NaHSO_{4(s)} + HF_{(g)}$$

$$NaCl_{(s)} + H_2SO_{4(aq)} \rightarrow NaHSO_{4(s)} + HCl_{(g)}$$

1) Hydrogen fluoride (HF) or hydrogen chloride gas (HCl) is formed. You'll see misty fumes as the gas comes into contact with moisture in the air.
2) But HF and HCl aren't strong enough reducing agents to reduce the sulfuric acid, so the reaction stops there.
3) It's not a redox reaction — the oxidation states of the halide and sulfur stay the same (–1 and +6).

Reaction of NaBr with H_2SO_4

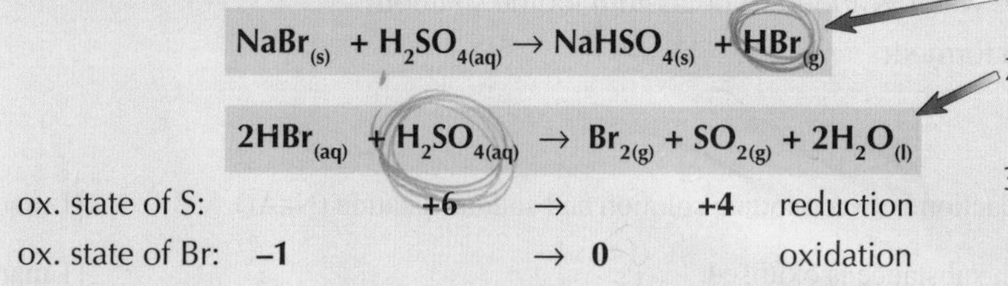

$$NaBr_{(s)} + H_2SO_{4(aq)} \rightarrow NaHSO_{4(s)} + HBr_{(g)}$$

$$2HBr_{(aq)} + H_2SO_{4(aq)} \rightarrow Br_{2(g)} + SO_{2(g)} + 2H_2O_{(l)}$$

ox. state of S: +6 → +4 reduction
ox. state of Br: –1 → 0 oxidation

1) The first reaction gives misty fumes of hydrogen bromide gas (HBr).
2) But the HBr is a stronger reducing agent than HCl and reacts with the H_2SO_4 in a redox reaction.
3) The reaction produces choking fumes of SO_2 and orange fumes of Br_2.

Reaction of NaI with H_2SO_4

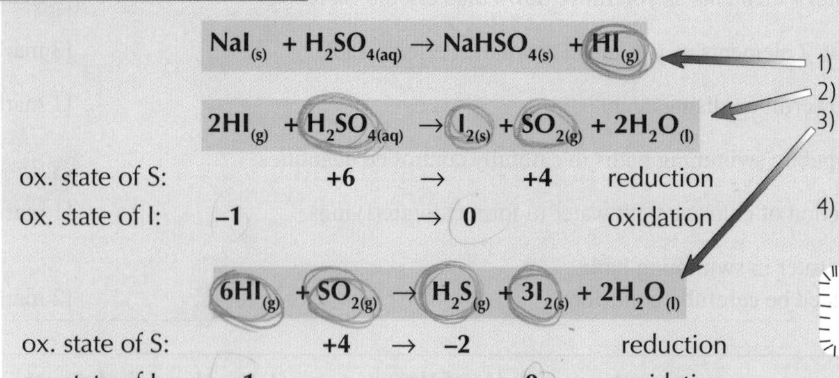

$$NaI_{(s)} + H_2SO_{4(aq)} \rightarrow NaHSO_{4(s)} + HI_{(g)}$$

$$2HI_{(g)} + H_2SO_{4(aq)} \rightarrow I_{2(s)} + SO_{2(g)} + 2H_2O_{(l)}$$

ox. state of S: +6 → +4 reduction
ox. state of I: –1 → 0 oxidation

$$6HI_{(g)} + SO_{2(g)} \rightarrow H_2S_{(g)} + 3I_{2(s)} + 2H_2O_{(l)}$$

ox. state of S: +4 → –2 reduction
ox. state of I: –1 → 0 oxidation

1) Same initial reaction giving HI gas.
2) The HI then reduces H_2SO_4 like above.
3) But HI (being very strong as far as reducing agents go) keeps going and reduces the SO_2 to H_2S.
4) Solid iodine is also formed by this reaction.

H_2S gas is toxic and smells of bad eggs.

Halide Ions

Silver Nitrate Solution is used to **Test for Halides**

The test for halides is dead easy. First you add **dilute nitric acid** to remove ions which might interfere with the test. Then you just add a few drops of **silver nitrate solution** ($AgNO_{3(aq)}$). A **precipitate** is formed (of the silver halide).

$$Ag^+_{(aq)} + X^-_{(aq)} \rightarrow AgX_{(s)}$$ ...where X is F, Cl, Br or I

1) The **colour** of the precipitate identifies the halide.

SILVER NITRATE TEST FOR HALIDE IONS...		
Fluoride F^-:	no precipitate	
Chloride Cl^-:	white precipitate	forms slowest
Bromide Br^-:	cream precipitate	↓
Iodide I^-:	yellow precipitate	forms fastest

2) Then to be extra sure, you can test your results by adding **ammonia solution**. Each silver halide has a different solubility in ammonia.

SOLUBILITY OF SILVER HALIDE PRECIPITATES IN AMMONIA...		
Chloride Cl^-:	white precipitate, dissolves in dilute $NH_{3(aq)}$	most soluble
Bromide Br^-:	cream precipitate, dissolves in conc. $NH_{3(aq)}$	↑
Iodide I^-:	yellow precipitate, insoluble in conc. $NH_{3(aq)}$	least soluble

Practice Questions

Q1 Give two reasons why a bromide ion is a more powerful reducing agent than a chloride ion.

Q2 Name the gaseous products formed when sodium bromide reacts with concentrated sulfuric acid.

Q3 What is produced when sodium iodide reacts with concentrated sulfuric acid?

Q4 How would you test whether an aqueous solution contained chloride ions?

Exam Questions

Q1 Describe the test you would carry out in order to distinguish between solid samples of sodium chloride and sodium bromide using silver nitrate solution and aqueous ammonia. State your observations. [6 marks]

Q2 The halogen below iodine in Group 7 is astatine (At).

a) Which of the following shows the products that would form when concentrated sulfuric acid is added to a solid sample of sodium astatide.

A $NaHSO_{4(aq)}$, $At_{2(g)}$, $SO_{2(g)}$, $2H_2O_{(l)}$

B $NaHSO_{4(aq)}$, $HAt_{(s)}$, $H_2S_{(g)}$, $2H_2O_{(l)}$

C $NaHSO_{4(aq)}$, $At_{2(s)}$, $H_2S_{(g)}$, $2H_2O_{(l)}$

D $NaHSO_{4(aq)}$, $At_{2(s)}$, $HAt_{(g)}$, $2H_2O_{(l)}$ [1 mark]

b) Predict what would be observed if silver astatide was added to a concentrated ammonia solution. Explain your answer. [2 marks]

Testing times — for the halides and for you...

Chemistry exams. What a bummer, eh... No one ever said it was going to be easy. Not even your teacher would be that cruel. There are plenty more equations on this page to learn. As well as that, make sure you really understand everything... the trend in the reducing power of halides... and the reactions with sulfuric acid...

Tests for Ions

It is a truth universally acknowledged that chemistry students need to know how to work out which ions are hanging around in a random ionic solution. Here are some tests that will help. (You have met a few of them before, thankfully...)*

**OK, I admit it, 'universally acknowledged' is probably a slight exaggeration...*

You Can Use Chemical Tests To Identify Positive Ions

Positive ions (or **cations**) include things like the ions of **Group 2 metals** and **ammonium ions**.
Here are the **chemical tests** that you need to know to help you identify them:

Use Flame Tests to Identify Group 2 Ions

Compounds of some **Group 2 metals** burn with characteristic **colours**.
You can identify them using a **flame test**. Here's how to do them:

1) Dip a **nichrome wire loop** in concentrated hydrochloric acid.
2) Then dip the wire loop into the **unknown compound**.
3) Hold the loop in the clear blue part of a Bunsen burner flame.
4) Observe the **colour change** in the flame.

Metal ion	Flame colour
Calcium, Ca^{2+}	**brick red**
Strontium, Sr^{2+}	**red**
Barium, Ba^{2+}	**pale green**

Use Red Litmus Paper and NaOH to Test for Ammonium Ions

1) **Ammonia gas** (NH_3) is alkaline — so you can test for it using a damp piece of **red litmus paper**. (The litmus paper needs to be damp so the ammonia gas can dissolve.) If there's ammonia present, the paper will turn **blue**.
2) If you add **hydroxide ions** to (OH^-) a solution containing **ammonium ions** (NH_4^+), they will react to produce **ammonia gas** and water, like this:

$$NH_4^+{}_{(aq)} + OH^-{}_{(aq)} \rightarrow NH_{3(g)} + H_2O_{(l)}$$

3) You can use this reaction to test whether a substance contains **ammonium ions** (NH_4^+). Add some dilute **sodium hydroxide** solution to your mystery substance in a test tube and **gently heat** the mixture. If there's ammonia given off, ammonium ions must be present.

You Can Use Chemical Tests To Identify Negative Ions Too

Negative ions (or **anions**) include things like **halide ions**, **hydroxide ions**, **sulfate ions** and **carbonate ions**.
Here are the **chemical tests** for these ions:

Test for Sulfates with HCl and Barium Chloride

You've already met this test back on page 98, but here's a quick reminder in case you've forgotten.

To identify a **sulfate** ion (SO_4^{2-}), you add a little dilute hydrochloric acid, followed by **barium chloride solution**, $BaCl_{2(aq)}$.

$$Ba^{2+}{}_{(aq)} + SO_4^{2-}{}_{(aq)} \rightarrow BaSO_{4\,(s)}$$

The hydrochloric acid is added to get rid of any traces of carbonate ions before you do the test. (These would also produce a precipitate, so they'd confuse the results.)

If a **white precipitate** of **barium sulfate** forms, it means the original compound contained a sulfate.

Use a pH Indicator to Test for Hydroxides

Hydroxide ions make solutions alkaline. So if you think a solution might contain hydroxide ions, you can use a **pH indicator** to test it. For example:

1) Dip a piece of **red litmus paper** into the solution.
2) If hydroxide ions are present, the paper will turn **blue**.

Tests for Ions

Test for **Halides** with **Silver Nitrate** Solution

To test for **chloride** (Cl^-), **bromide** (Br^-) or **iodide** (I^-) ions, you just add dilute **nitric acid** (HNO_3), followed by **silver nitrate** solution ($AgNO_3$).

A **chloride** gives a **white** precipitate of **silver chloride**.

A **bromide** gives a **cream** precipitate of **silver bromide**.

An **iodide** gives a **yellow** precipitate of **silver iodide**.

There's more info about this test on page 103.

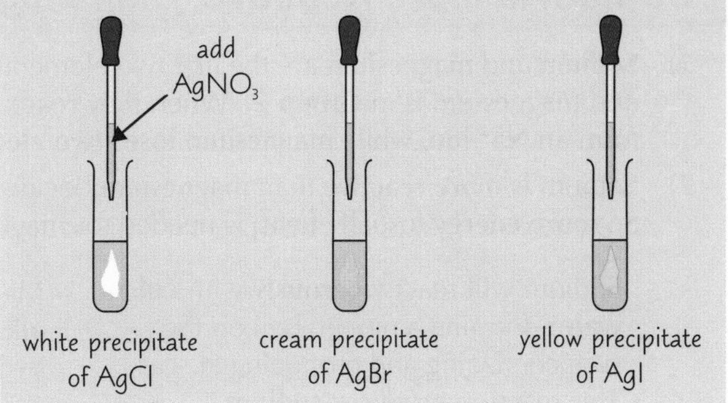

Hydrochloric Acid Can Help Detect **Carbonates**

You can test to see if a solution contains carbonate ions (CO_3^{2-}) by adding an acid. Here's how to do it:

When you add dilute **hydrochloric acid**, a solution containing **carbonate ions** will fizz.
This is because the carbonate ions react with the hydrogen ions in the acid to give **carbon dioxide**:

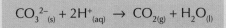

$$CO_3^{2-}{}_{(s)} + 2H^+_{(aq)} \rightarrow CO_{2(g)} + H_2O_{(l)}$$

You can test for carbon dioxide using **limewater**. Carbon dioxide **turns limewater cloudy** — just bubble the gas through a test tube of limewater and watch what happens. If the limewater goes cloudy, your solution contains **carbonate ions**.

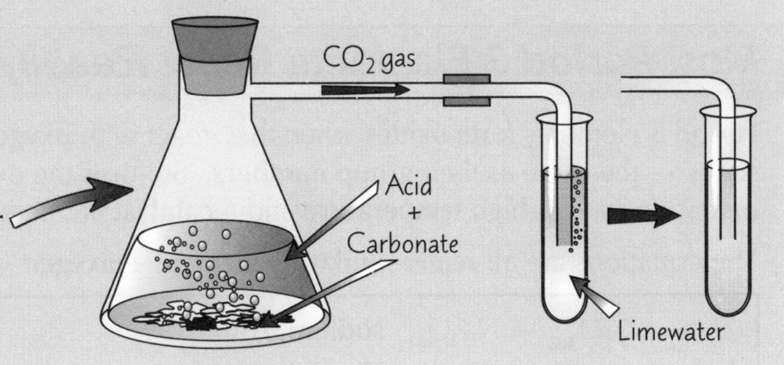

Practice Questions

Q1 Describe how to carry out a flame test.

Q2 What colour would damp, red litmus paper turn in the presence of ammonia gas?

Q3 Describe how you could test a solution to see if it contained hydroxide ions.

Exam Questions

Q1 A student adds dilute nitric acid and silver nitrate to a solution of an unknown ionic compound. A yellow precipitate forms. Which anion is present in the solution?

A Bromide B Carbonate C Sulfate D Iodide [1 mark]

Q2 A student has a solution of an unknown ionic compound. He performs a flame test on a sample of the solution and notes that the compound burns with a pale green flame. He then reacts another sample of the solution with dilute hydrochloric acid. The reactants fizz and the gas produced turns limewater cloudy. What is the formula of the ionic compound? [2 marks]

Q3 You are given a sample of an ionic compound in solution and asked to confirm that the compound is ammonium sulfate. Describe the tests that you could perform to confirm the identity of the compound. [4 marks]

I've got my ion you...

...and you better have your ion these pages. There are lots of tests to learn and you best not get them all muddled up. Go through each test and make sure you know all the reagents you'd need to do the test, how you'd actually do it, and what a positive result looks like. Then once you know them all, treat yourself with a biscuit — that's a positive result too.

Period 3 Elements and Oxides

Period 3's the third row down on the periodic table — the one that starts with sodium and ends with argon.

Sodium is More Reactive Than Magnesium

1) **Sodium** and **magnesium** are the first two elements in **Period 3**. Sodium is in **Group 1**, and magnesium is in **Group 2**. When they react, sodium **loses one electron** to form an **Na^+ ion**, while **magnesium loses two electrons** to form **Mg^{2+}**.
2) Sodium is **more reactive** than **magnesium** because it takes **less energy** to lose **one electron** than to lose two. So **more energy** (usually **heat**) is needed for magnesium to react. This is shown in their reactions with **water**.

Sodium will react **vigorously** with **cold water**, forming a molten ball on the surface, fizzing and producing H_2 gas. This reaction produces **sodium hydroxide**, so it creates a strongly **alkaline** solution (pH 12 – 14).

$$2Na_{(s)} + 2H_2O_{(l)} \rightarrow 2NaOH_{(aq)} + H_{2(g)}$$

Magnesium reacts **very slowly** with **cold water**. It forms a **weakly alkaline** solution (pH 9 – 10) and a thin coating of magnesium hydroxide forms on the surface of the metal. The solution is only weakly alkaline because magnesium hydroxide is **not very soluble** in water, so relatively **few hydroxide ions** are produced.

$$Mg_{(s)} + 2H_2O_{(l)} \rightarrow Mg(OH)_{2(aq)} + H_{2(g)}$$

Magnesium reacts much faster with **steam** (i.e. when there is **more energy**), to form **magnesium oxide**.

$$Mg_{(s)} + H_2O_{(g)} \rightarrow MgO_{(s)} + H_{2(g)}$$

Most Period 3 Elements React Readily With Oxygen

Period 3 elements form **oxides** when they react with **oxygen**. They're usually oxidised to their **highest** oxidation states — the same as their **group numbers**. Sulfur is the exception to this — it forms **SO_2**, in which it's only got a **+4** oxidation state (a **high temperature** and a **catalyst** are needed to make **SO_3**, where sulfur has an oxidation state of +6).

The equations are all **really similar** — **element + oxygen → oxide**:

$2Na_{(s)} + ½O_{2(g)} \rightarrow Na_2O_{(s)}$	**sodium oxide**	$Mg_{(s)} + ½O_{2(g)} \rightarrow MgO_{(s)}$	**magnesium oxide**
$2Al_{(s)} + 1½O_{2(g)} \rightarrow Al_2O_{3(s)}$	**aluminium oxide**	$Si_{(s)} + O_{2(g)} \rightarrow SiO_{2(s)}$	**silicon dioxide**
$P_{4(s)} + 5O_{2(g)} \rightarrow P_4O_{10(s)}$	**phosphorus(V) oxide**	$S_{(s)} + O_{2(g)} \rightarrow SO_{2(g)}$	**sulfur dioxide**

SO_2 reacts with oxygen and a vanadium catalyst to form SO_3 (see p.122).

The **more reactive metals** (Na, Mg) and the **non-metals** (P, S) react **readily** in air, while **Al** and **Si** react **slowly**.

Element	Na	Mg	Al	Si	P	S
Formula of oxide	Na_2O	MgO	Al_2O_3	SiO_2	P_4O_{10}	SO_2
Reaction of element in air	Vigorous	Vigorous	Slow (faster if powdered)	Slow	Spontaneously combusts	Burns steadily

Bonding and Structure Affect Melting Points

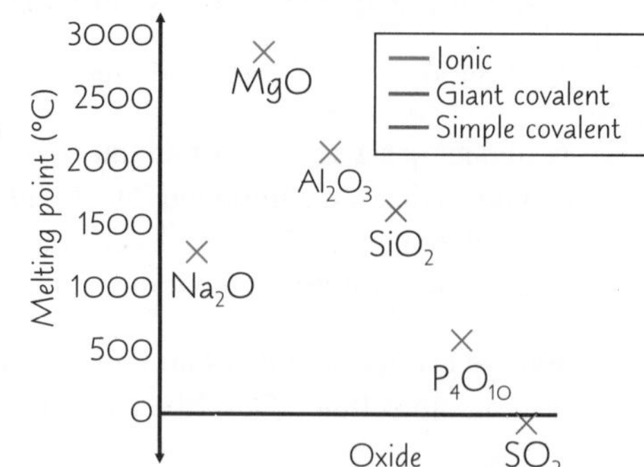

1) **Na_2O, MgO** and **Al_2O_3** are metal oxides. They have **high melting points** because they form **giant ionic lattices**. The **strong forces of attraction** between each ion mean it takes a lot of heat energy to **break the bonds** and melt them.
2) MgO has a **higher melting point** than Na_2O because Mg forms **2+ ions**, so bonds more strongly than the 1+ Na ions in Na_2O.
3) Al_2O_3 has a **lower melting point** than you might expect because the highly charged Al^{3+} ions distort the oxygen's electron cloud making the bonds **partially covalent**.
4) **SiO_2** has a **higher melting point** than the other non-metal oxides because it has a **giant macromolecular** structure. In order to melt, the strong covalent bonds between atoms need to be broken, and this requires a lot of energy.
5) **P_4O_{10}** and **SO_2** have relatively **low melting points** because they form **simple molecular** structures. The molecules are bound by **weak intermolecular forces** (dipole-dipole and van der Waals), which take little energy to break.

Period 3 Elements and Oxides

Ionic Oxides are *Alkaline, Covalent* Oxides are *Acidic*

1) The **ionic oxides** of the **metals** Na and Mg both contain oxide ions (O^{2-}). When they dissolve in water, the O^{2-} ions accept protons from the water molecules to form hydroxide ions. The solutions are both **alkaline**, but **sodium hydroxide** is more soluble in water, so it forms a **more alkaline** solution than magnesium hydroxide.

$Na_2O_{(s)} + H_2O_{(l)} \rightarrow 2NaOH_{(aq)}$ **pH 12 - 14**

$MgO_{(s)} + H_2O_{(l)} \rightarrow Mg(OH)_{2(aq)}$ **pH 9 - 10**

2) The **simple covalent oxides** of the **non-metals** phosphorus and sulfur form **acidic** solutions. All of the acids are **strong** and so the pH of their solutions is about **0 – 2** (for solutions with a concentration of at least 1 mol dm^{-3}). They will **dissociate** (split up into ions) in solution, forming hydrogen ions and a negative ion (sometimes called a **conjugate base**).

$P_4O_{10(s)} + 6H_2O_{(l)} \rightarrow 4H_3PO_{4(aq)}$	**phosphoric(V) acid**	$H_3PO_{4(aq)} \rightarrow 3H^+_{(aq)} + PO_4{}^{3-}{}_{(aq)}$
$SO_{2(g)} + H_2O_{(l)} \rightarrow H_2SO_{3(aq)}$	**sulfurous acid (or sulfuric(IV) acid)**	$H_2SO_{3(aq)} \rightarrow 2H^+_{(aq)} + SO_3{}^{2-}{}_{(aq)}$
$SO_{3(l)} + H_2O_{(l)} \rightarrow H_2SO_{4(aq)}$	**sulfuric(VI)acid**	$H_2SO_{4(aq)} \rightarrow 2H^+_{(aq)} + SO_4{}^{2-}{}_{(aq)}$

3) The **giant covalent structure** of **silicon dioxide** means that it is **insoluble** in water. However, it will **react with bases** to form salts so it is classed as **acidic**.
4) **Aluminium oxide**, which is partially **ionic** and partially **covalently** bonded, is also **insoluble** in water. But, it will react with **acids and bases** to form salts — i.e. it can act as an acid or a base, so it's classed as **amphoteric**.

Acid + Base → Salt + Water

The equation for **neutralising** an **acid** with a **base** is a classic (**acid + base → salt + water**) and it's no different for the Period 3 oxides. You may be asked to **write equations** for these reactions, so here are some examples:

1) Basic oxides neutralise acids:

$Na_2O_{(s)} + 2HCl_{(aq)} \rightarrow 2NaCl_{(aq)} + H_2O_{(l)}$

$MgO_{(s)} + H_2SO_{4(aq)} \rightarrow MgSO_{4(aq)} + H_2O_{(l)}$

An ironic ox-side?

2) Acidic oxides neutralise bases:

$SiO_{2(s)} + 2NaOH_{(aq)} \rightarrow Na_2SiO_{3(aq)} + H_2O_{(l)}$

$P_4O_{10(s)} + 12NaOH_{(aq)} \rightarrow 4Na_3PO_{4(aq)} + 6H_2O_{(l)}$

$SO_{2(g)} + 2NaOH_{(aq)} \rightarrow Na_2SO_{3(aq)} + H_2O_{(l)}$

$SO_{3(g)} + 2NaOH_{(aq)} \rightarrow Na_2SO_{4(aq)} + H_2O_{(l)}$

3) Amphoteric oxides neutralise acids and bases:

$Al_2O_{3(s)} + 3H_2SO_{4(aq)} \rightarrow Al_2(SO_4)_{3(aq)} + 3H_2O_{(l)}$

$Al_2O_{3(s)} + 2NaOH_{(aq)} + 3H_2O_{(l)} \rightarrow 2NaAl(OH)_{4(aq)}$

Practice Questions

Q1 Why is Na more reactive than Mg with water?

Q2 What type of bonding is in the following oxides: a) Na_2O, b) P_4O_{10}.

Q3 Write an equation for the reaction of Na_2O with water.

Q4 Explain why MgO forms a less alkaline solution than Na_2O.

Exam Question

Q1 X and Y are oxides of Period 3 elements. The element in X has an oxidation state of +6 and X forms an acidic solution in water. The element in Y has an oxidation state of +1 and Y has a high melting point.

a) Identify compound X and write an equation for its reaction with water. [2 marks]

b) i) Identify compound Y and write an equation for its reaction with water. [2 marks]

ii) Explain why compound Y has a high melting point. [2 marks]

These pages have got more trends than a school disco...

Hang on a minute, I hear you cry — what about chlorine and argon? Aren't they in Period 3 too? Well, yes, they are, but you don't need to know about them. Argon's a noble gas, anyway, so it doesn't really react with anything... yawn.

Transition Metals — The Basics

This section's all about transition metals — and there's a lot of it. It's obviously important stuff in the Chemistry world.

Transition Elements are Found in the *d-Block*

The **d-block** is the block of elements in the middle of the periodic table. Most of the elements in the d-block are **transition elements** (or transition metals).

You mainly need to know about the ones in the first row of the d-block. These are the elements from **titanium to copper**.

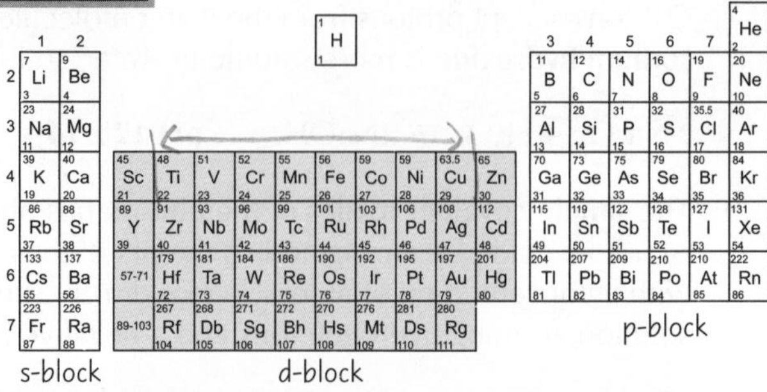

Transition Metals Have *Partially Filled d Sub-levels* in their *Atoms or Ions*

Here's the definition of a transition metal:

> A **transition metal** is a metal that can form **one or more stable ions** with a **partially filled d sub-level.**

A d-orbital can contain **up to 10** electrons. So transition metals must form **at least one ion** that has **between 1 and 9 electrons** in the d-orbital. All the Period 4 d-block elements are transition metals apart from **scandium** and **zinc**. Here are their electron configurations:

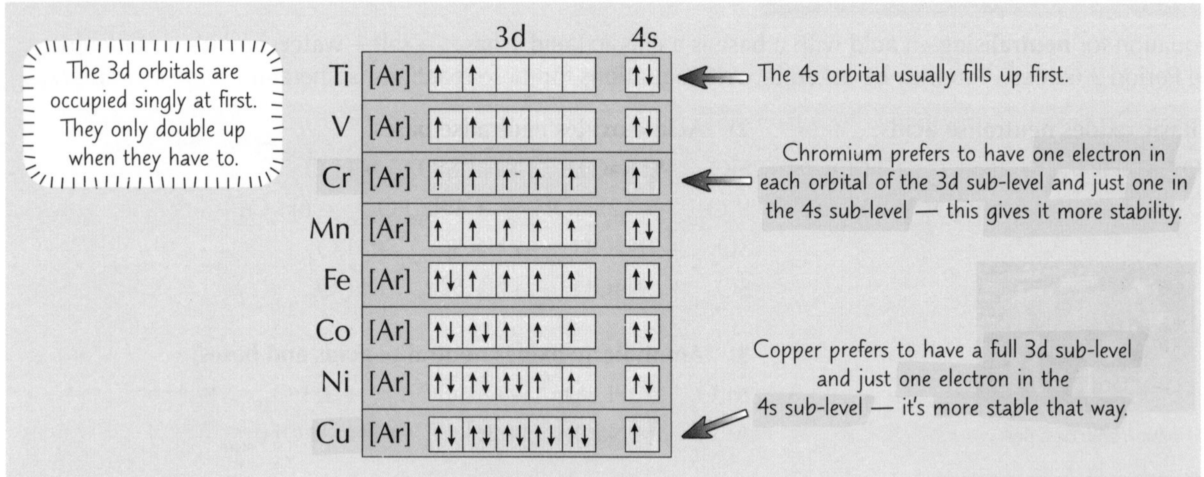

It's the **incomplete** d sub-level that causes the **special chemical properties** of transition metals on the next page.

Sc and Zn ***Aren't*** *Transition Metals*

1) **Scandium** only forms one ion, $\mathbf{Sc^{3+}}$, which has an **empty d sub-level**. Scandium has the electron configuration $\mathbf{[Ar]3d^1\,4s^2}$, so when it loses three electrons to form Sc^{3+}, it ends up with the electron configuration **[Ar]**.
2) **Zinc** only forms one ion, $\mathbf{Zn^{2+}}$, which has a **full d sub-level**. Zinc has the electron configuration $\mathbf{[Ar]3d^{10}\,4s^2}$. When it forms Zn^{2+} it loses 2 electrons, both from the 4s sub-level. This means it keeps its full 3d sub-level.

When *Ions* are Formed, the *s Electrons* are *Removed First*

Transition metal atoms form **positive** ions. When this happens, the **s electrons** are removed **first, then** the d electrons.

Example: Iron forms both Fe^{2+} and Fe^{3+} ions. What are the electron configurations of these ions?

When iron forms 2+ ions, it loses **both its 4s electrons.** $\mathbf{Fe = [Ar]3d^6\,4s^2 \rightarrow Fe^{2+} = [Ar]3d^6}$

Only once the 4s electrons are removed can a **3d electron** be removed. E.g. $\mathbf{Fe^{2+} = [Ar]3d^6 \rightarrow Fe^{3+} = [Ar]3d^5}$

Transition Metals — The Basics

The *Transition Metals* All Have Similar *Physical Properties*

The transition elements don't gradually change across the periodic table like you might expect. They're all typical metals and have **similar physical properties**:

1) They all have a **high density**.
2) They all have **high melting** and **high boiling points**.
3) Their **ionic radii** are more or less the same.

Transition Metals Have *Special Chemical Properties*

1) They can form **complex ions** — see pages 110-113. E.g. iron forms a **complex ion with water** — $[Fe(H_2O)_6]^{2+}$.
2) They form **coloured ions** — see pages 114-115. E.g. Fe^{2+} ions are **pale green** and Fe^{3+} ions are **yellow**.
3) They're **good catalysts** — see pages 122-123. E.g. iron is the catalyst used in the **Haber process**.
4) They can exist in **variable oxidation states** — see pages 118-119.
   E.g. iron can exist in the **+2** oxidation state as $\mathbf{Fe^{2+}}$ ions and in the **+3** oxidation state as $\mathbf{Fe^{3+}}$ ions.

Some common **coloured** ions and **oxidation states** are shown below. The colours refer to the **aqueous ions**.

oxidation state	+7	+6	+5	+4	+3	+2
			VO_2^+ (yellow)	VO^{2+} (blue)	V^{3+} (green)	V^{2+} (violet)
		$Cr_2O_7^{2-}$ (orange)			Cr^{3+} (green/violet)	
	MnO_4^- (purple)					Mn^{2+} (pale pink)
					Fe^{3+} (yellow)	Fe^{2+} (pale green)
						Co^{2+} (pink)
						Ni^{2+} (green)
						Cu^{2+} (blue)

When Cr^{3+} ions are surrounded by 6 water ligands (see page 124) they're violet. But the water ligands are often substituted (see page 116), so this solution usually looks green instead.

These elements show **variable** oxidation states because the **energy levels** of the 4s and the 3d sub-levels are **very close** to one another. So different numbers of electrons can be gained or lost using fairly **similar** amounts of energy.

Practice Questions

Q1 What is the definition of a transition metal?

Q2 Give the electron arrangement of: a) a vanadium atom, b) a V^{2+} ion.

Q3 State four chemical properties which are characteristic of transition elements.

Exam Question

Q1 When solid copper(I) sulfate is added to water, a blue solution forms with a red-brown precipitate of copper metal.

a) Give the electron configuration of copper(I) ions. [1 mark]

b) Does the formation of copper(I) ions show copper acting as a transition metal? Explain your answer. [2 marks]

c) Identify the blue solution. [1 mark]

s electrons — like rats leaving a sinking ship...

Definitely have a quick read of the electron configuration stuff in Unit 1 Section 1 if it's been pushed to a little corner of your mind labelled, "Well, I won't be needing that again in a hurry". It should come flooding back pretty quickly. This page is just an overview of transition metal properties. They're all looked at in lots more detail in the coming pages...

Complex Ions

Transition metals are always forming complex ions. These aren't as complicated as they sound, though. Honest.

Complex Ions are *Metal Ions* Surrounded by *Ligands*

A **complex** is a central **metal atom** or **ion** surrounded by **co-ordinately bonded ligands.**

1) A **co-ordinate bond** (or dative covalent bond) is a covalent bond in which **both electrons** in the shared pair come from the **same atom**. In a complex, they come from the **ligands**.
2) So, a **ligand** is an atom, ion or molecule that **donates a pair of electrons** to a central **transition metal ion** to form a **co-ordinate bond**.
3) The **co-ordination number** is the **number** of **co-ordinate bonds** that are formed with the central metal ion.
4) The usual co-ordination numbers are **6** and **4**. If the ligands are **small**, like H_2O or NH_3, **6** can fit around the central metal ion. But if the ligands are **larger**, like Cl^-, **only 4** can fit around the central metal ion.

6 CO-ORDINATE BONDS MEAN AN OCTAHEDRAL SHAPE

Here are a few examples.

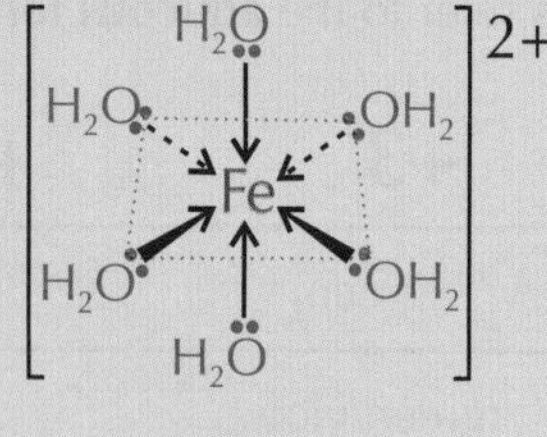

$[Fe(H_2O)_6]^{2+}_{(aq)}$

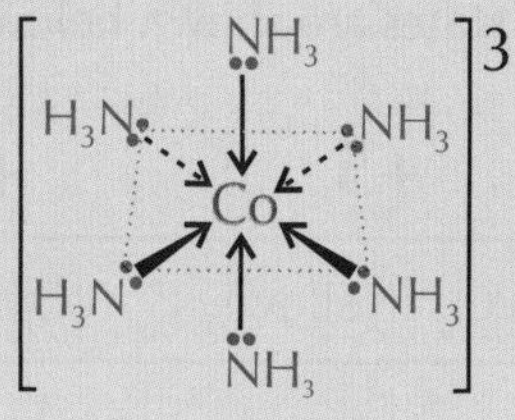

$[Co(NH_3)_6]^{3+}_{(aq)}$

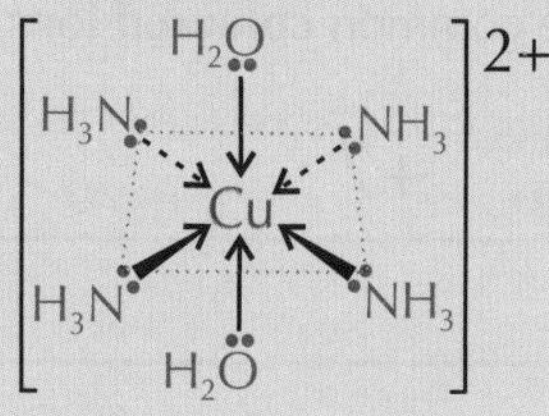

$[Cu(NH_3)_4(H_2O)_2]^{2+}_{(aq)}$

The **bond angles** are all **90°**.

The ligands don't have to be all the same.

4 CO-ORDINATE BONDS USUALLY MEAN A TETRAHEDRAL SHAPE...

E.g. the $[CuCl_4]^{2-}$ complex, which is yellow, and the $[CoCl_4]^{2-}$complex ion, which is blue.

The **bond angles** are **109.5°**.

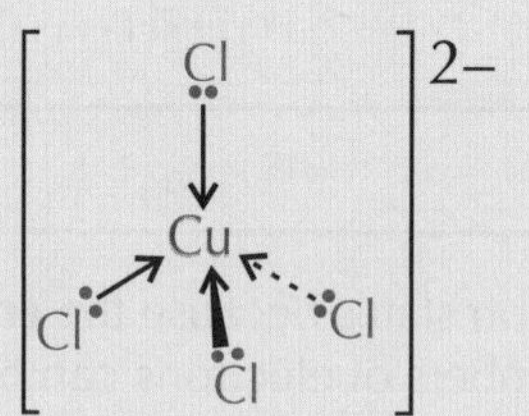

Make sure you learn the shapes of these complexes.

The Thomson family were proud of their colour co-ordination.

...BUT CAN FORM A SQUARE PLANAR SHAPE

In a **few** complexes, e.g. **cisplatin** (shown on the right), **4 co-ordinate bonds** form a **square planar** shape.

The **bond angles** are **90°**.

This compound is called cisplatin. It's used as an anti cancer drug (see page 189).

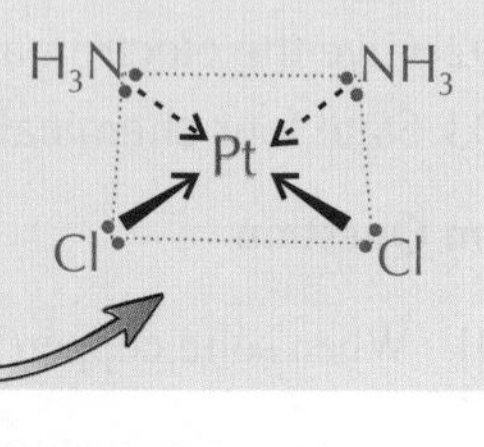

SOME SILVER COMPLEXES HAVE 2 CO-ORDINATE BONDS AND FORM A LINEAR SHAPE

$[Ag(NH_3)_2]^+$ forms a **linear shape**, as shown.

The bond angles are **180°**.

$[H_3N{:}\rightarrow Ag \leftarrow{:}NH_3]^+$

$[Ag(NH_3)_2]^+$ is also called Tollens' reagent — have a look at page 162 to see how it's used.

The different types of **bond arrow** used in the examples above show that the complexes are **3D**. The **wedge-shaped arrows** represent bonds coming **towards you** and the **dashed arrows** represent bonds **sticking out behind** the molecule.

Complex Ions

Complex Ions Have an **Overall Charge** or **Total Oxidation State**

The **overall charge** on the complex ion is its **total oxidation state.** It's put **outside** the **square** brackets. For example:

$[Cu(H_2O)_6]^{2+}_{(aq)}$ ← Overall charge is 2+.

You can work out the **oxidation state of the metal**:

The oxidation state of the metal ion = **the total oxidation state – the sum of the oxidation states of the ligands**

Example: Give the oxidation state of the metal ions in the following complexes:

a) The cobalt ion in $[CoCl_4]^{2-}$.

The total oxidation state is **–2** and each Cl^- ligand has an oxidation state of **–1**. So in this complex, cobalt's oxidation state = –2 – (4 × –1) = **+2**.

b) The chromium ion in $[CrCl_2(H_2O)_4]^+$.

The total oxidation state is **+1**. Each Cl^- ligand has an oxidation state of **–1**, and each H_2O ligand has an oxidation state of **0**. So in this complex, cobalt's oxidation state = +1 – (2 × –1) – (4 × 0) = **+3**.

Practice Questions

Q1 What is meant by the term 'complex ion'?

Q2 Describe how a ligand, such as ammonia, bonds to a central metal ion.

Q3 Draw the shape of the complex ion $[Co(NH_3)_6]^{3+}$. Name the shape.

Q4 What is the size of the bond angles in an octahedral complex?

Exam Questions

Q1 a) Using $[Ag(NH_3)_2]^+$ as an example, explain what is meant by the following terms:

i) ligand [2 marks]

ii) co-ordinate bond [2 marks]

iii) co-ordination number [2 marks]

b) Predict the shape of the complex $[Ag(S_2O_3)_2]^{3-}$. [1 mark]

Q2 When concentrated hydrochloric acid is added to an aqueous solution of $Cu^{2+}_{(aq)}$ a yellow solution is formed.

a) State the co-ordination number and shape of the $Cu^{2+}_{(aq)}$ complex ion in the initial solution. [2 marks]

b) State the co-ordination number, shape and formula of the complex ion responsible for the yellow solution. [3 marks]

c) What are the bond angles of the complex ion responsible for the yellow solution? [1 mark]

d) Explain why the co-ordination number of the Cu^{2+} complex ion in the yellow solution is different to the co-ordination number of the Cu^{2+} complex ion in the starting aqueous solution of Cu^{2+}. [2 marks]

Put your hands up — we've got you surrounded...

You never get transition metal ions floating round by themselves in a solution — they'll always be surrounded by other molecules. It's kind of like what'd happen if you put a dish of sweets in a room of eight (or eighteen) year-olds. When you're drawing complex ions, don't forget to include the dashed and wedge-shaped bonds to show that it's 3D.

More on Complex Ions

The complex ions on this page are a bit more complicated than the ones you've seen before, but not much. There's some thinking in 3D coming up, so if it helps, grab some modelling clay and matchsticks and try making some models.

A Ligand Must Have **at Least One Lone Pair of Electrons**

A ligand must have **at least one lone pair of electrons**, or it won't have anything to use to form a **co-ordinate bond**.

1) Ligands that can only form **one co-ordinate bond** are called **monodentate** — e.g. $H_2\ddot{O}$, $\ddot{N}H_3$, $\ddot{Cl}^-$.
2) Ligands that can form **more than one co-ordinate bond** are called **multidentate** — e.g. **$EDTA^{4-}$** has six lone pairs (it's **hexadentate** to be precise) so it can form **six co-ordinate bonds** with a metal ion (see below).
3) **Bidentate** ligands are multidentate ligands that can form **two co-ordinate bonds** e.g. ethane-1,2-diamine, $\ddot{N}H_2CH_2CH_2\ddot{N}H_2$, or ethanedioate, $[\ddot{O}OCCO\ddot{O}]^{2-}$. These compounds both have two lone pairs, so can each form **two co-ordinate bonds** with a metal ion.

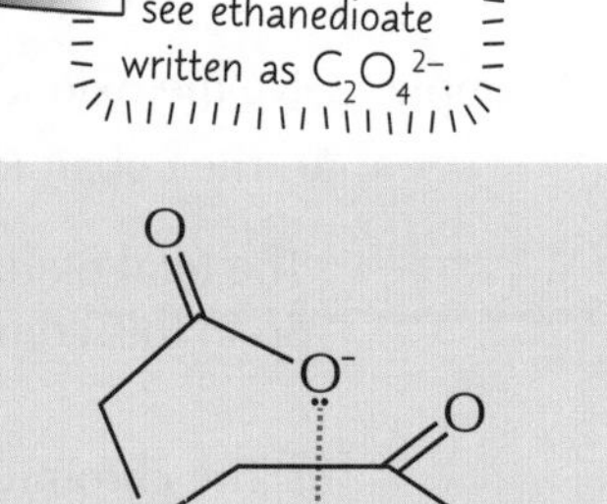

You'll normally see ethanedioate written as $C_2O_4^{2-}$.

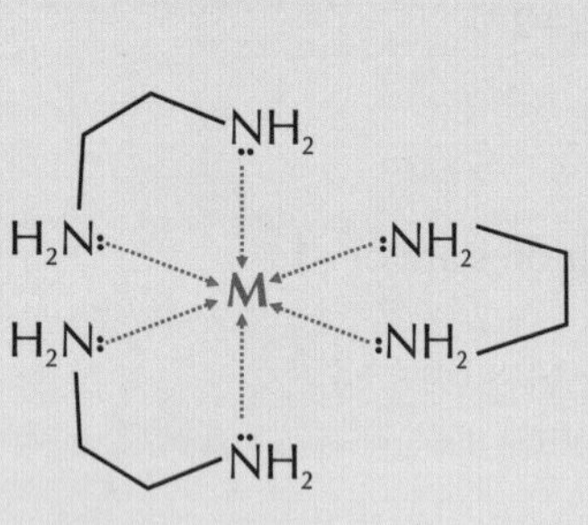

Each **ethane-1,2-diamine molecule** has 2 lone pairs and forms 2 **co-ordinate bonds** with the **metal ion.**

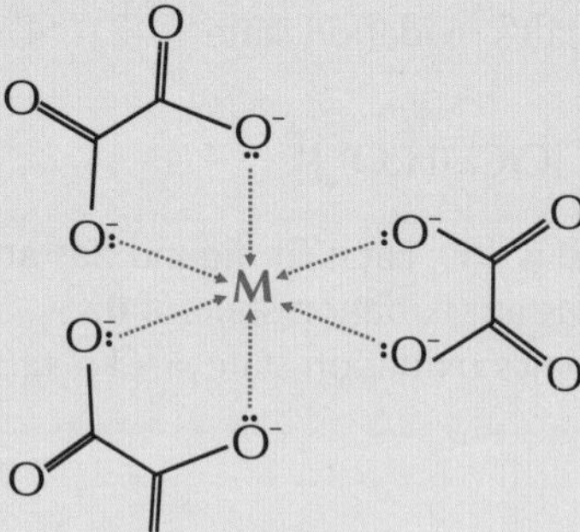

Each **ethanedioate** molecule forms 2 **co-ordinate bonds** with the **metal ion.**

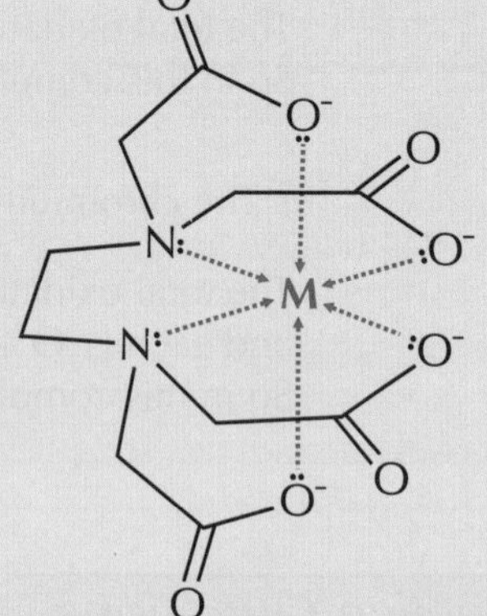

The **$EDTA^{4-}$ ion** has 6 lone pairs, so it forms 6 **co-ordinate bonds** with the **metal ion.**

Haem in **Haemoglobin** Contains a **Multidentate Ligand**

1) **Haemoglobin** is a protein found in **blood** that helps to **transport oxygen** around the body.
2) **Haemoglobin** contains Fe^{2+} ions, which are **hexa-coordinated** — **six lone pairs** are donated to them to form **six co-ordinate bonds** in an **octahedral** structure.
3) Four of the co-ordinate bonds come from a single **multidentate ligand**. Four **nitrogen atoms** from the **same** molecule co-ordinate around Fe^{2+} to form a **circle**. This part of the molecule is called **haem**.
4) The other two co-ordinate bonds come from a protein called **globin**, and either an **oxygen** or a **water** molecule — so the complex can **transport oxygen** to where it's needed, and then swap it for a water molecule — here's **how it works**:

- In the lungs, where the oxygen concentration is high, an **oxygen molecule** substitutes the water ligand and bonds co-ordinately to the Fe(II) ion to form **oxyhaemoglobin,** which is carried **around the body** in the blood.
- When the **oxyhaemoglobin** gets to a place where oxygen is needed, the **oxygen molecule is exchanged** for a **water molecule**. The haemoglobin then **returns to the lungs** and the whole process starts again.

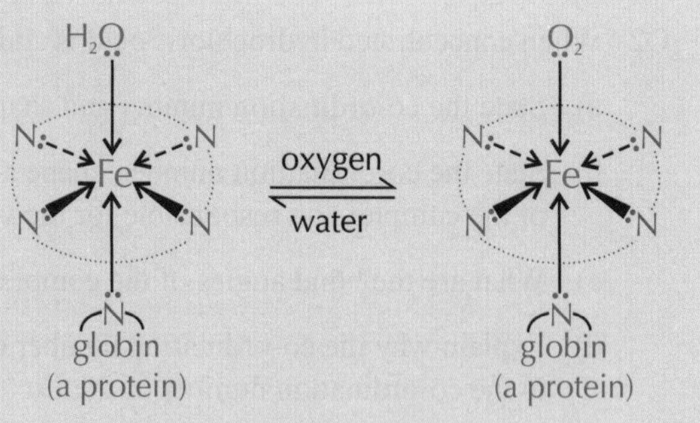

5) This process can be disrupted if **carbon monoxide** is inhaled. The **haemoglobin** swaps its **water** ligand for a **carbon monoxide** ligand, forming **carboxyhaemoglobin**. This is bad news because carbon monoxide is a **strong** ligand and **doesn't** readily exchange with oxygen or water ligands, meaning the haemoglobin **can't transport oxygen** any more. **Carbon monoxide poisoning** starves the organs of oxygen — it can cause **headaches, dizziness, unconsciousness** and even **death** if it's not treated.

More on Complex Ions

Complex Ions Can Show **Optical Isomerism**

1) **Optical isomerism** is a type of **stereoisomerism** (see page 131).
2) Complex ions can show **optical isomerism** — where an ion can exist in two forms that are **non-superimposable mirror images**. This happens with octahedral complexes when **three bidentate ligands**, such as ethane-1,2-diamine, $NH_2CH_2CH_2NH_2$, co-ordinately bond with a central metal ion, like **nickel**.

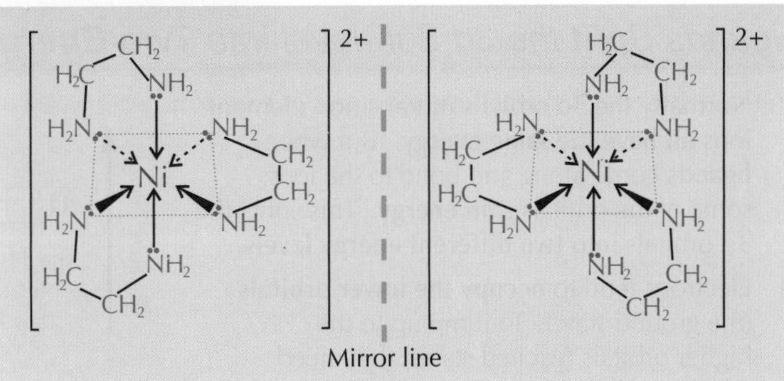

Cis-Trans Isomers Can Form in **Octahedral** and **Square Planar** Complexes

1) Cis-trans isomerism is another type of **stereoisomerism.** In fact, it's a special case of **E/Z isomerism** (see page 132).
2) **Octahedral** complexes with four monodentate ligands of one type and two monodentate ligands of another type can show **cis-trans isomerism.** If the **two odd** ligands are **opposite** each other, you've got the **trans isomer**. If they're **next** to each other, you've got the **cis isomer**. For example, the complex $[NiCl_2(H_2O)_4]$ has a trans and a cis isomer.

trans-$[NiCl_2(H_2O)_4]$ cis-$[NiCl_2(H_2O)_4]$

3) **Square planar** complex ions that have **two pairs** of ligands also show cis-trans isomerism. When two paired ligands are **opposite** each other it's the **trans isomer** and when they're **next** to each other it's the **cis isomer**.

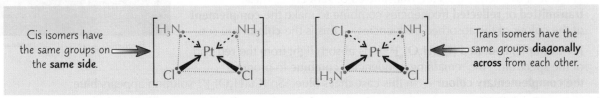

Cis isomers have the same groups on the **same side**.

Trans isomers have the same groups **diagonally across** from each other.

4) The molecule in the example above, is a complex of platinum(II) with two chloride ions and two ammonia molecules in a square planar shape. The **cis isomer** (cisplatin) is used as an anti-cancer drug (see page 189).

Practice Questions

Q1 Give an example of a monodentate ligand.

Q2 How many co-ordinate bonds can the ion $EDTA^{4-}$ form with a metal ion?

Q3 Name two types of stereoisomerism that an octahedral complex could show.

Exam Questions

Q1 Iron(III) can form the complex ion $[Fe(C_2O_4)_3]^{3-}$ with three ethanedioate ions. The ethanedioate ion is a bidentate ligand. Its structure is shown on the right.

a) What is a bidentate ligand? [1 mark]

b) i) Draw the two stereoisomers of $[Fe(C_2O_4)_3]^{3-}$. [2 marks]

ii) What type of stereoisomerism is this? [1 mark]

Q2 In the body, the chloride ion ligands of cisplatin are replaced by water ligands to form $[Pt(NH_3)_2(H_2O)_2]^{2+}$. Draw this compound, given that it is the cis isomer of a square planar complex. [1 mark]

Iron & $EDTA^{4-}$ used to be together, but now their status is 'it's complicated'...

Phew, whoever called them 'complex' ions really wasn't joking. Make sure you can look at a complex ion and figure out whether or not it will have isomers. Octahedral complexes form optical isomers, and even then only if they're bonded to three bidentate ligands. But both octahedral and square planar complexes can form cis-trans isomers.

Formation of Coloured Ions

Transition metal complex ions have distinctive colours, which is handy when it comes to identifying them. This page explains why they're so colourful.

*Ligands **Split** the 3d Sub-level into **Two Energy Levels***

1) Normally the 3d orbitals of transition element ions **all** have the **same energy**. But when **ligands** come along and bond to the ions, some of the orbitals **gain energy**. This splits the 3d orbitals into **two different energy levels**.
2) Electrons tend to **occupy the lower orbitals** (the ground state). To jump up to the higher orbitals (excited states) they need **energy** equal to the energy gap, ΔE. They get this energy from **visible light**.
3) The energy **absorbed** when electrons jump up from the ground state to an excited state can be worked out using this formula.

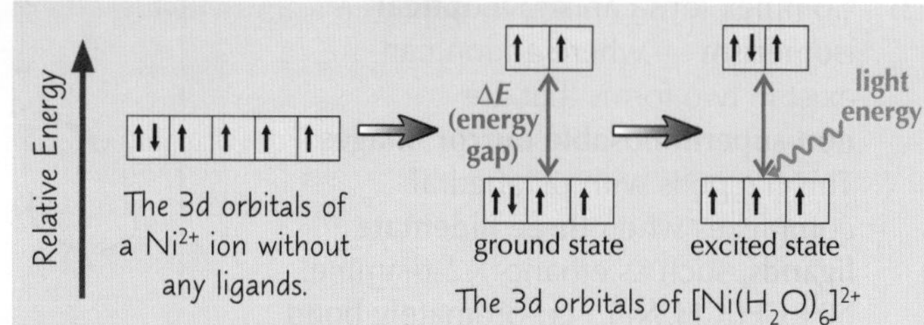

$$\Delta E = h\nu = \frac{hc}{\lambda}$$

ν = frequency of light absorbed (hertz/Hz)
h = Planck's constant (6.63×10^{-34} J s)
c = the speed of light (3.00×10^{8} m s^{-1})
λ = wavelength of light absorbed (m)

4) The amount of energy needed to make electrons jump depends upon the **central metal ion** and its **oxidation state**, the **ligands** and the **co-ordination number**, as these affect the **size of the energy gap** (ΔE).

*The **Colours** of Compounds are the **Complement** of Those That are **Absorbed***

1) When **visible light** hits a transition metal ion, some frequencies are **absorbed** when electrons jump up to the higher orbitals. The frequencies absorbed depend on the size of the **energy gap** (ΔE).
2) The rest of the frequencies are **transmitted** or **reflected**. These **transmitted** or **reflected** frequencies combine to make the **complement** of the colour of the absorbed frequencies — this is the **colour** you see.
3) For example, **hydrated $[Cu(H_2O)_6]^{2+}$** ions absorb light from the **red** end of the spectrum. The remaining frequencies **combine** to produce the **complementary colour** — in this case that's blue. So $[Cu(H_2O)_6]^{2+}$ solution appears **blue**.
4) If there are **no** 3d electrons or the 3d sub-level is **full**, then no electrons will jump, so **no energy** will be absorbed. If there's no energy absorbed, the compound will look **white** or **colourless**.

The larger the energy gap, the higher the frequency of light that is absorbed.

frequency increases →

Transition Metal Ions** can be **Identified** by their **Colour

It'd be nice if each transition metal formed ions or complexes with just one colour, but sadly it's not that simple. The **colour of a complex** can be altered by any of the factors that can affect the size of the **energy gap** (ΔE).

1) **Changes in oxidation state.**

Complex:	$[Fe(H_2O)_6]^{2+}_{(aq)}$	→	$[Fe(H_2O)_6]^{3+}_{(aq)}$		$[V(H_2O)_6]^{2+}_{(aq)}$	→	$[V(H_2O)_6]^{3+}_{(aq)}$
Oxidation state:	+2		+3	and	+2		+3
Colour:	pale green		yellow		violet		green

2) **Changes in co-ordination number** — this always involves a change of ligand too.

Complex:	$[Cu(H_2O)_6]^{2+} + 4Cl^-$	→	$[CuCl_4]^{2-} + 6H_2O$
Co-ordination number:	6		4
Colour:	blue		yellow

3) **Changes in ligand** — this can cause a colour change even if the oxidation state and co-ordination number remain the same.

Complex:	$[Co(H_2O)_6]^{2+} + 6NH_3$	→	$[Co(NH_3)_6]^{2+} + 6H_2O$
Oxidation state:	+2		+2
Colour:	pink		straw coloured

Formation of Coloured Ions

Spectroscopy can be used to Find **Concentrations** of Transition Metal Ions

Spectroscopy can be used to determine the **concentration of a solution** by measuring how much **light** it **absorbs**.

1) **White light** is shone through a **filter**, which is chosen to **only** let through the **colour of light** that is **absorbed** by the sample.
2) The light passes through the sample to a **colorimeter**, which calculates **how much light** was **absorbed** by the sample.
3) The more **concentrated** a coloured solution is, the more light it will absorb. So you can use this measurement to work out the **concentration** of a solution of transition metal ions.

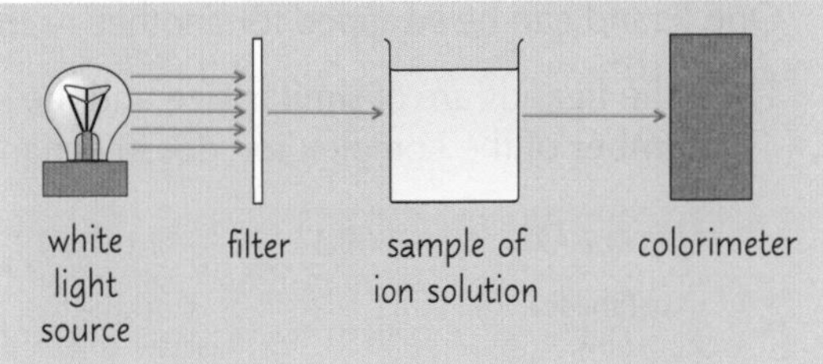

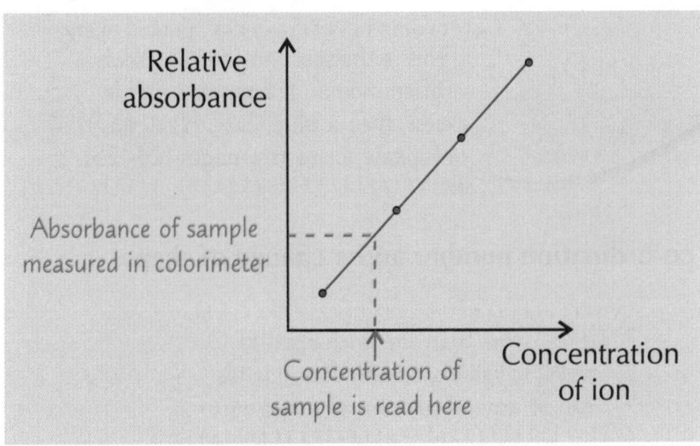

Before you can find the unknown concentration of a sample, you have to produce a **calibration curve** — like the lovely one on the left. This involves measuring the **absorbance** of **known concentrations** of solutions and plotting the results on a graph.

Once you've done this, you can measure the absorbance of your sample and read its **concentration** off the graph.

Stanley's concentration was strong — he'd show this cowboy who was boss.

Practice Questions

Q1 Which sub-level is split by the presence of ligands?

Q2 What two factors are changing in the following reaction that will cause a change in colour?

$$[Co(H_2O)_6]^{2+} + 4Cl^- \rightarrow [CoCl_4]^{2-} + 6H_2O$$

Q3 What is the purpose of the filter in spectroscopy?

Q4 What does a colorimeter measure?

Exam Questions

Q1 a) Explain why complex transition metal ions such as $[Fe(H_2O)_6]^{2+}$ are coloured. [1 mark]

b) State three changes to a complex ion that would result in a change in colour. [3 marks]

Q2 Colorimetry can be used to determine the concentration of a coloured solution. Briefly describe how you would construct a calibration graph, given a coloured solution of known concentration. [3 marks]

Q3 The frequency of light absorbed by a transition metal complex ion can be determined from the equation $\Delta E = h\nu$.

a) State what is meant by ΔE and what change this represents within the complex ion. [1 mark]

b) Using a noble gas core, [Ar], complete the electron arrangements for the following ions:

i) Cu^+ [1 mark]

ii) Cu^{2+} [1 mark]

c) Which one of the above ions has coloured compounds? State the feature of its electron arrangement that suggests this. [1 mark]

Blue's not my complementary colour — it clashes with my hair...

Transition metal ions are pretty colours, don't you think? The Romans did — they used iron, copper, cobalt and manganese compounds to add colour to glass objects. I'm not sure that they knew the colours were affected by variable oxidation states, ligands and co-ordination number, but it's pretty impressive even so.

Substitution Reactions

There are more equations on this page than the number of elephants you can fit in a Mini.

Ligands can Change Places with One Another

One ligand can be **swapped** for another — this is **ligand substitution** or **exchange**. It usually causes a **colour change**.

1) If the ligands are of **similar size** and the same **charge**, then the **co-ordination number** of the complex ion doesn't change, and neither does the **shape**.

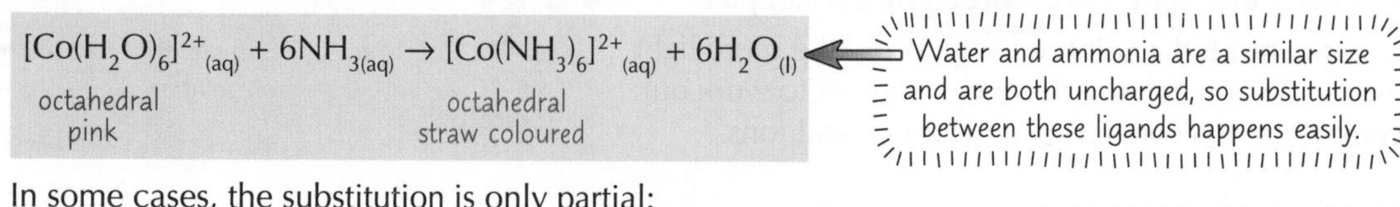

$$[Co(H_2O)_6]^{2+}_{(aq)} + 6NH_{3(aq)} \rightarrow [Co(NH_3)_6]^{2+}_{(aq)} + 6H_2O_{(l)}$$

octahedral pink → octahedral straw coloured

Water and ammonia are a similar size and are both uncharged, so substitution between these ligands happens easily.

In some cases, the substitution is only partial:

$$[Cu(H_2O)_6]^{2+}_{(aq)} + 4NH_{3(aq)} \rightarrow [Cu(NH_3)_4(H_2O)_2]^{2+}_{(aq)} + 4H_2O_{(l)}$$

octahedral blue → octahedral deep blue

This is the reaction with an excess of ammonia. If ammonia isn't in excess, then a blue $Cu(OH)_2(H_2O)_4$ precipitate forms (see pages 124-125).

2) If the ligands are **different sizes**, there's a **change of co-ordination number** and a **change of shape**.

$$[Co(H_2O)_6]^{2+}_{(aq)} + 4Cl^-_{(aq)} \rightleftharpoons [CoCl_4]^{2-}_{(aq)} + 6H_2O_{(l)}$$

octahedral pink → tetrahedral blue

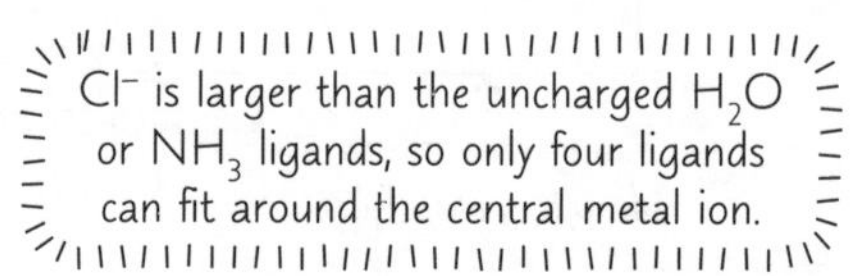

Cl^- is larger than the uncharged H_2O or NH_3 ligands, so only four ligands can fit around the central metal ion.

$$[Cu(H_2O)_6]^{2+}_{(aq)} + 4Cl^-_{(aq)} \rightleftharpoons [CuCl_4]^{2-}_{(aq)} + 6H_2O_{(l)}$$

octahedral pale blue → tetrahedral yellow

$$[Fe(H_2O)_6]^{3+}_{(aq)} + 4Cl^-_{(aq)} \rightleftharpoons [FeCl_4]^-_{(aq)} + 6H_2O_{(l)}$$

octahedral yellow → tetrahedral yellow

Climbing up was OK, but on the way down Tess had a sudden loss of co-ordination.

Have a quick look at the reactions on page 125 — these are **ligand exchange reactions** too. In these, **hydroxide precipitates** are formed when a little bit of **sodium hydroxide** or **ammonia solution** is added to metal-aqua ions. The hydroxide precipitates sometimes **dissolves** when excess sodium hydroxide or ammonia solution is added.

Different Ligands Form Different Strength Bonds

Ligand substitution reactions can be easily **reversed**, **UNLESS** the new complex ion is much **more stable** than the old one.

1) If the new ligands form **stronger** bonds with the central metal ion than the old ligands did, the change is **less easy** to reverse. E.g. **CN^- ions** form stronger co-ordinate bonds with Fe^{3+} ions than H_2O molecules, so it's hard to reverse this reaction:

$$[Fe(H_2O)_6]^{3+}_{(aq)} + 6CN^-_{(aq)} \rightarrow [Fe(CN)_6]^{3-}_{(aq)} + 6H_2O_{(l)}$$

Another example is carbon monoxide bonding to Fe^{2+} in haemoglobin much more strongly than oxygen does — see page 112 for more on this.

2) **Multidentate** ligands form more stable complexes than monodentate ligands, so a change like the one below is hard to reverse:

$$[Cu(H_2O)_6]^{2+}_{(aq)} + 3NH_2CH_2CH_2NH_{2(aq)} \rightarrow [Cu(NH_2CH_2CH_2NH_2)_3]^{2+}_{(aq)} + 6H_2O_{(l)}$$

This is a bidentate ligand.

This is explained on the next page...

Substitution Reactions

A Positive Entropy Change Makes a More Stable Complex

1) When a **ligand exchange reaction** occurs, co-ordinate bonds are **broken** and **formed**. The **strength** of the bonds being broken is often very **similar** to the strength of the new bonds being made. So the **enthalpy change** for a ligand exchange reaction is usually very **small**. For example, the reaction substituting ammonia with ethane-1,2-diamine in a nickel complex has a very **small** enthalpy change of reaction:

$$[Ni(NH_3)_6]^{2+} + 3NH_2CH_2CH_2NH_2 \rightarrow [Ni(NH_2CH_2CH_2NH_2)_3]^{2+} + 6NH_3 \quad \Delta H = -13 \text{ kJ mol}^{-1}$$

Break 6 co-ordinate bonds between Ni and N. (for $[Ni(NH_3)_6]^{2+}$)

Form 6 co-ordinate bonds between Ni and N. (for $[Ni(NH_2CH_2CH_2NH_2)_3]^{2+}$)

2) This is actually a **reversible** reaction, but the equilibrium lies so **far to the right** that it is thought of as being irreversible — $[Ni(NH_2CH_2CH_2NH_2)_3]^{2+}$ is **much more stable** than $[Ni(NH_3)_6]^{2+}$. This isn't accounted for by an enthalpy change.
3) Instead, the **increase in stability**, known as the **chelate effect**, explains why multidentate ligands always form much more stable complexes than monodentate ligands:

> When monodentate ligands are substituted with bidentate or multidentate ligands, the number of particles in solution increases — the **more particles**, the **greater the entropy**. Reactions that result in an increase in entropy are **more likely** to occur.

4) When the **hexadentate ligand EDTA^{4-}** replaces monodentate or bidentate ligands, the complex formed is **a lot more stable.**

$$[Cr(NH_3)_6]^{3+} + EDTA^{4-} \rightarrow [Cr(EDTA)]^{-} + 6NH_3 \quad \textbf{2 particles} \rightarrow \textbf{7 particles}$$

The enthalpy change for this reaction is almost zero and the entropy change is big and positive. This makes the free energy change ($\Delta G = \Delta H - T\Delta S$) negative, so the reaction is feasible (see p. 62).

5) It's difficult to reverse these reactions, because reversing them would cause a **decrease in entropy**.

Practice Questions

Q1 What colour is a solution of the complex ion $[Cu(NH_3)_4(H_2O)_2]^{2+}$?

Q2 When you add hydroxide ions to $[Cu(H_2O)_6]^{2+}$ you get a precipitate of $Cu(OH)_2(H_2O)_4$. Adding extra H_2O does not reverse the reaction. Which ligand bonds more strongly to Cu^{2+}, OH^- or H_2O?

Q3 What is the chelate effect?

Exam Questions

Q1 When a solution of EDTA^{4-} ions is added to an aqueous solution of $[Fe(H_2O)_6]^{3+}$ ions, a ligand substitution reaction occurs.

a) Write an equation for the reaction that takes place. [1 mark]

b) The new complex that is formed is more stable than $[Fe(H_2O)_6]^{3+}$. Explain why. [1 mark]

Q2 A scientist takes three samples of a solution that contains the complex ion $[Co(H_2O)_6]^{2+}$. She mixes the first sample with a solution of chloride ions, the second with an excess of ammonia solution, and the third with a solution containing ethane-1,2-diamine. She observes the reactions that occur.
Write an equation for a reaction the scientist observed where:

a) There was no change in the number of ligands surrounding the cobalt ion. [1 mark]

b) The number of ligands surrounding the cobalt ion changed, but the overall charge on the complex ion did not. [1 mark]

c) Both the number of ligands surrounding the cobalt ion and the overall charge on the complex ion changed. [1 mark]

Ligand exchange — the musical chairs of the molecular world...

Ligands generally don't mind swapping with other ligands, so long as they're not too tightly attached to the central metal ion. They also won't fancy changing if it means forming fewer molecules and having less entropy. It's kind of like you wouldn't want to swap a cow for a handful of beans. Unless they're magic beans. But that almost never happens...

Variable Oxidation States

One of the reasons why transition metal complexes have such a big range of colours is their variable oxidation states.

Transition Metals Can Exist in **Variable Oxidation States**

You learnt on page 109 that one of the properties of transition metals is that they can exist in variable oxidation states. When you switch between oxidation states, it's a **redox reaction** — the metal ions are either oxidised or reduced. For example, vanadium can exist in **four oxidation states** in solution — the +2, +3, +4 and +5 states. You can tell them apart by their colours:

Oxidation state of vanadium	Formula of ion	Colour of ion
+5	$VO_2^+{}_{(aq)}$	Yellow
+4	$VO^{2+}{}_{(aq)}$	Blue
+3	$V^{3+}{}_{(aq)}$	Green
+2	$V^{2+}{}_{(aq)}$	Violet

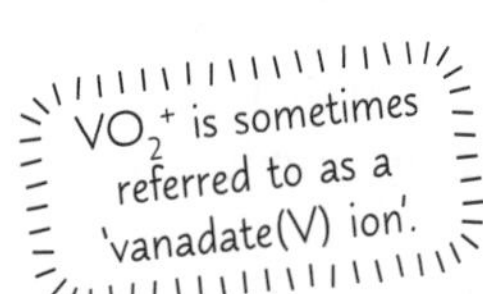

VO_2^+ is sometimes referred to as a 'vanadate(V) ion'.

Vanadium(V) ions can be reduced by adding them to **zinc metal** in an **acidic solution**. Here are the equations for each of the reduction reactions:

1) To begin with, the solution turns from **yellow** to **blue** as vanadium(V) is reduced to vanadium(IV):

$$2VO_2^+{}_{(aq)} + Zn_{(s)} + 4H^+{}_{(aq)} \rightarrow 2VO^{2+}{}_{(aq)} + Zn^{2+}{}_{(aq)} + 2H_2O_{(l)}$$

2) The solution then changes colour from **blue** to **green** as vanadium(IV) is reduced to vanadium(III):

$$2VO^{2+}{}_{(aq)} + Zn_{(s)} + 4H^+{}_{(aq)} \rightarrow 2V^{3+}{}_{(aq)} + Zn^{2+}{}_{(aq)} + 2H_2O_{(l)}$$

3) Finally, vanadium(III) is reduced to vanadium(II), and so the solution changes from **green** to **violet**.

$$2V^{3+}{}_{(aq)} + Zn_{(s)} \rightarrow 2V^{2+}{}_{(aq)} + Zn^{2+}{}_{(aq)}$$

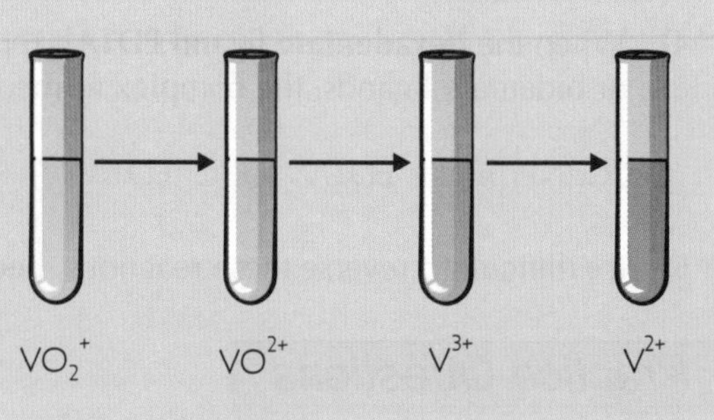

Redox Potentials Tell You How **Easy** it is to **Reduce** an Ion

1) The **redox potential** of an ion or atom tells you how easily it is **reduced** to a lower oxidation state. They're the same as electrode potentials (see page 76).
2) The **larger** the redox potential, the **less stable** the ion will be, and so the **more likely** it is to be **reduced**. For example, in the table on the right, copper(II) has a redox potential of +0.15 V, so is less stable and more likely to be reduced than chromium(III) which has a redox potential of –0.41 V.

Half equation	Standard electrode potential (V)
$Cr^{3+}{}_{(aq)} + e^- \rightleftharpoons Cr^{2+}{}_{(aq)}$	–0.41
$Cu^{2+}{}_{(aq)} + e^- \rightleftharpoons Cu^+{}_{(aq)}$	+0.15

3) The redox potentials in the table are **standard electrode potentials** — they've been measured with the reactants at a concentration of 1 mol dm^3 against a standard hydrogen electrode under standard conditions (see page 77).
4) The redox potential of an ion won't always be the same as its standard electrode potential. It can **vary** depending on the environment that the ion is in. For example:

- **Ligands:** Standard electrode potentials are measured in **aqueous solution**, so any aqueous ions will be surrounded by **water ligands**. Different **ligands** may make the redox potential larger or smaller depending on how well they bind to the metal ion in a particular oxidation state.
- **pH:** Some ions need H^+ ions to be present in order to be reduced, e.g. $2VO_2^+{}_{(aq)} + 4H^+{}_{(aq)} + 2e^- \rightleftharpoons 2VO^{2+}{}_{(aq)} + 2H_2O_{(l)}$
  Others **release** OH^- ions into solution when they are reduced, e.g. $CrO_4^{2-}{}_{(aq)} + 4H_2O_{(l)} + 3e^- \rightleftharpoons Cr(OH)_{3(s)} + 5OH^-{}_{(aq)}$

For reactions such as these, the **pH** of the solution affects the size of the redox potential. In general, redox potentials will be **larger** in more **acidic** solutions, making the ion more easily reduced.

Variable Oxidation States

Tollens' Reagent Contains a *Silver Complex*

Silver is a transition metal that is most commonly found in the +1 oxidation state (Ag^+). It is easily **reduced** to **silver metal**:

$$Ag^+_{(aq)} + e^- \rightarrow Ag_{(s)} \qquad \text{Standard electrode potential} = +0.80\ V$$

The standard electrode potential is large, so Ag^+ is easily reduced.

Steve had a serious silver complex.

Tollens' reagent uses this reduction reaction to distinguish between **aldehydes** and **ketones**. It's prepared by adding just enough ammonia solution to silver nitrate solution to form a colourless solution containing the complex ion $[Ag(NH_3)_2]^+$.

When added to **aldehydes**, Tollens' reagent reacts to give a **silver mirror** on the inside of the test tube. The aldehyde is **oxidised** to a **carboxylic acid**, and the Ag^+ ions are **reduced** to silver metal:

$$RCHO_{(aq)} + 2[Ag(NH_3)_2]^+_{(aq)} + 3OH^-_{(aq)} \rightarrow RCOO^-_{(aq)} + 2Ag_{(s)} + 4NH_{3(aq)} + 2H_2O_{(l)}$$

Tollens' reagent can't oxidise **ketones**, so it **won't react** with them, and no silver mirror will form.

Practice Questions

Q1 What colour changes are seen when zinc powder is added to an acidic solution of vanadium(V) ions?

Q2 What does the size of the redox potential tell you about a transition metal ion?

Q3 Name two things that can influence the size of the redox potential for a transition metal ion.

Q4 What would you observe if you were to react Tollens' reagent with an aldehyde?

Exam Questions

Q1 Tollens' reagent contains a transition metal complex ion.

a) i) What is the formula of the complex ion in Tollens' reagent? [1 mark]

ii) What is the oxidation state of the transition metal ion in Tollens' reagent? [1 mark]

b) Explain, using an appropriate equation, how Tollens' reagent can be used to distinguish between an aldehyde and a ketone. [3 marks]

Q2 The vanadate(V) ion, VO_2^+, can be reduced to vanadium(II), V^{2+}.
The reduction occurs in three steps, which can be observed as the yellow vanadium(V) solution first turns blue, then green and finally violet as a solution of V^{2+} ions is formed.

a) What are the reagent(s) and conditions needed to carry out this reaction? [2 marks]

b) Identify the ions responsible for the blue and green solutions. [2 marks]

c) Write reactions to show the three steps in the reduction of vanadium(V) to vanadium(II). [3 marks]

Q3 Which of the following statements about redox potentials is **not** correct?

A The redox potential of a transition metal ion tells you how easily it is transformed from a higher to a lower oxidation state.

B The larger the redox potential of a transition metal ion, the more stable it is.

C The ligands that co-ordinate to a transition metal ion can affect its redox potential.

D Changing the pH will change the redox potential of the following half equation:
$MnO_{4\ (aq)}^- + 2H_2O_{(l)} + 3e^- \rightleftharpoons MnO_{2(s)} + 4OH^-_{(aq)}$ [1 mark]

My brother often puts me in the +6 or +8 aggravation state...

This topic wasn't too bad — a few equations, some pretty colours and a nifty test to distinguish aldehydes from ketones. Have another look at the reduction reactions of vanadium. Now write them down. And write them down again. Now sing them to the tune of 'Dancing Queen'. And if you get bored of that, just sing 'Dancing Queen', or maybe 'Waterloo'.

Titrations with Transition Metals

These titrations are redox titrations. They're similar to acid-base titrations, but they're not exactly the same.

*Titrations Using **Transition Element Ions** are **Redox** Titrations*

You can use titrations to let you find out how much **oxidising agent** is needed to **exactly** react with a quantity of **reducing agent**. If you know the **concentration** of either the oxidising agent or the reducing agent, you can use the titration results to work out the concentration of the other.

Transition metals have **variable oxidation states** which means they are often present in either the oxidising or the reducing agent. Their **colour changes** also make them useful in titrations as it's easy to spot the **end point**.

1) First you measure out a quantity of **reducing agent** (e.g. aqueous Fe^{2+} ions or aqueous $C_2O_4^{2-}$ ions) using a pipette, and put it in a conical flask.
2) Using a **measuring cylinder**, you add about **20 cm³** of **dilute sulfuric acid** to the flask — this is an excess, so you don't have to be too exact.
3) Now you add the **oxidising agent** to the reducing agent using a **burette**, **swirling** the conical flask as you do so.
4) The **oxidising agent** that you add reacts with the reducing agent. This reaction will continue until **all** of the reducing agent is used up. The **very next drop** you add to the flask will give the mixture the **colour of the oxidising agent**. (You could use a coloured reducing agent and a colourless oxidising agent instead — then you'd be watching for the moment that the colour in the flask disappears.)
5) Stop when the mixture in the flask **just** becomes tainted with the colour of the oxidising agent (the **end point**) and record the volume of the oxidising agent added. This is the **rough titration**.
6) Now you do some **accurate titrations**. You need to do a few until you get **two or more** readings that are **within 0.10 cm³** of each other.

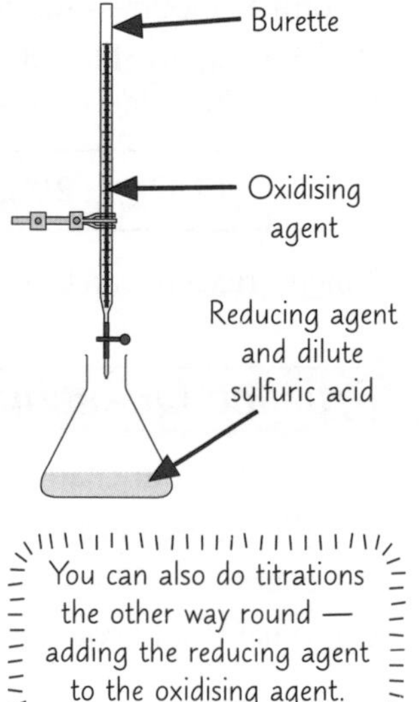

You can also do titrations the other way round — adding the reducing agent to the oxidising agent.

The main **oxidising agent** used is **aqueous potassium manganate(VII)**, which contains **purple** manganate(VII) ions. **Strong acidic** conditions are needed for the manganate(VII) ions to be reduced.

Use e^-, H^+ and H_2O to Balance Half-Equations

1) Before you do a titration calculation, you might need to **work out** the reaction equation first. You can do this by **balancing half-equations** — but that might take more than just adding a few electrons.
2) You might have to add some H^+ ions and H_2O to your half-equations to make them **balance**.

Example: Acidified manganate(VII) ions (MnO_4^-) can be reduced to Mn^{2+} by Fe^{2+} ions. Write the full equation for this reaction.

Iron is being oxidised. The half-equation for this is: $Fe^{2+}_{(aq)} \rightarrow Fe^{3+}_{(aq)} + e^-$

You can also use this method to balance other redox half-equations.

The second half-equation is a little bit trickier...

1) Manganate is being reduced. Start by writing this down: $MnO^-_{4\,(aq)} \rightarrow Mn^{2+}_{(aq)}$
2) To balance the **oxygens**, you need to add **water** to the right-hand side of the equation: $MnO^-_{4\,(aq)} + \rightarrow Mn^{2+}_{(aq)} + 4H_2O_{(l)}$
3) Now you need to add some H^+ **ions** to the left-hand side to balance the **hydrogens**: $MnO^-_{4\,(aq)} + 8H^+_{(aq)} \rightarrow Mn^{2+}_{(aq)} + 4H_2O_{(l)}$
4) Finally, balance the **charges** by adding some **electrons**: $MnO^-_{4\,(aq)} + 8H^+_{(aq)} + 5e^- \rightarrow Mn^{2+}_{(aq)} + 4H_2O_{(l)}$

There's an overall charge of +2 on each side of the equation.

Now you have to make sure the number of **electrons** produced in the **iron half-equation equal** the number of **electrons** used up in the **manganate half-equation**.

$$Fe^{2+}_{(aq)} \rightarrow Fe^{3+}_{(aq)} + e^- \xrightarrow{\times 5} 5Fe^{2+}_{(aq)} \rightarrow 5Fe^{3+}_{(aq)} + 5e^-$$

Now combine both half-equations to give a **full redox equation**.

$$MnO^-_{4\,(aq)} + 8H^+_{(aq)} + 5Fe^{2+}_{(aq)} \rightarrow Mn^{2+}_{(aq)} + 5Fe^{3+}_{(aq)} + 4H_2O_{(l)}$$

Titrations with Transition Metals

Use the Titration Results to Calculate the Concentration of a Reagent

Example: 27.5 cm^3 of 0.0200 mol dm^{-3} aqueous potassium manganate(VII) reacted with 25.0 cm^3 of acidified sodium ethanedioate solution. Calculate the concentration of $C_2O_4^{2-}$ ions in the solution.

$$2MnO_4^-{}_{(aq)} + 16H^+{}_{(aq)} + 5C_2O_4^{2-}{}_{(aq)} \rightarrow 2Mn^{2+}{}_{(aq)} + 8H_2O_{(l)} + 10CO_{2(g)}$$

$C_2O_4^{2-}$ ions are called ethanedioate or oxalate ions.

1) Work out the number of **moles of MnO_4^- ions** added to the flask.

$$\text{Number of moles } MnO_4^- \text{ added} = \frac{\text{concentration} \times \text{volume (cm}^3)}{1000} = \frac{0.0200 \times 27.5}{1000} = 5.50 \times 10^{-4} \text{ moles}$$

2) Look at the balanced equation to find how many moles of $C_2O_4^{2-}$ react with **every mole** of MnO_4^-. Then you can work out the **number of moles of $C_2O_4^{2-}$** in the flask.

5 moles of $C_2O_4^{2-}$ react with 2 moles of MnO_4^-.
So moles of $C_2O_4^{2-}$ = $(5.50 \times 10^{-4} \times 5) \div 2 = 1.38 \times 10^{-3}$ moles

Manganate 007, licensed to oxidise.

3) Work out the **number of moles of $C_2O_4^{2-}$** that would be in 1000 cm^3 (1 dm^3) of solution — this is the **concentration**.

25.0 cm^3 of solution contained 1.38×10^{-3} moles of $C_2O_4^{2-}$.

$$1000 \text{ cm}^3 \text{ of solution would contain } \frac{(1.38 \times 10^{-3}) \times 1000}{25.0} = 0.552 \text{ moles of } C_2O_4^{2-}$$

So the concentration of $C_2O_4^{2-}$ is **0.552 mol dm^{-3}**.

You can also work out the volume of one reagent if you know the volume of the other one and both concentrations.

Practice Questions

Q1 Outline how you would carry out a titration of Fe^{2+} with MnO_4^- in acid solution.

Q2 What colour marks the end point of a redox titration when potassium manganate(VII) is added from the burette?

Q3 Why is an excess of acid added to potassium manganate(VII) titrations?

Exam Questions

Q1 0.100 g of a sample of steel containing carbon and iron only was dissolved in sulfuric acid.
The resulting solution was titrated using 29.4 cm^3 of 0.0100 mol dm^{-3} potassium manganate(VII) solution.
The equation for the reaction is $MnO_4^- + 5Fe^{2+} + 8H^+ \rightarrow Mn^{2+} + 5Fe^{3+} + 4H_2O$

a) Calculate the number of moles of manganate ions present in the titre of potassium manganate(VII) solution. [1 mark]

b) Calculate the number of moles of Fe^{2+} that will have reacted with the manganate(VII) ions. [1 mark]

c) Calculate the mass of iron in the steel. [1 mark]

d) Calculate the percentage of iron in the steel. [1 mark]

Q2 A sample of sodium ethanedioate is dissolved in dilute sulfuric acid.
The solution is titrated using 18.3 cm^3 of 0.0200 mol dm^{-3} potassium manganate(VII).
The equation for the reaction is: $2MnO_4^- + 5C_2O_4^{2-} + 16H^+ \rightarrow 2Mn^{2+} + 10CO_2 + 8H_2O$
Calculate the mass of $Na_2C_2O_4$ in the initial sample. [4 marks]

Q3 A 0.100 mol dm^{-3} solution of iron(II) sulfate has been partly oxidised by the air to an iron(III) compound.
Explain how you could use a standard solution of potassium manganate(VII) to find the actual concentration of Fe^{2+} remaining. [4 marks]

And how many moles does it take to change a light bulb...

The example calculations on these two pages might look rather complicated, but if you break them down into a series of steps, they're really not that bad — especially as you've done titration calculations before. You know, the ones with acids and bases? Back on pages 90-91? What do you mean you don't remember? Go back and have another look.

Catalysts

Transition metals aren't just good — they're grrrreat.

Transition Metal Catalysts Work by Changing Oxidation States

Transition metals and their compounds make good catalysts because they can **change oxidation states** by gaining or losing electrons within their **d orbitals**. This means they can **transfer electrons** to **speed up** reactions.

For example, in the **Contact Process**, vanadium(V) oxide is able to **oxidise SO_2 to SO_3** because it can be **reduced** to vanadium(IV) oxide. It's then **oxidised** back to vanadium(V) oxide by oxygen ready to start all over again.

This example uses a heterogeneous catalyst (see below), but the principle also applies to homogeneous catalysts.

Vanadium oxidises SO_2 to SO_3 and is reduced itself.

$$V_2O_5 + SO_2 \rightarrow V_2O_4 + SO_3$$
vanadium(V) → vanadium(IV)

The reduced catalyst is then oxidised by oxygen gas back to its original state.

$$V_2O_4 + \tfrac{1}{2}O_2 \rightarrow V_2O_5$$
vanadium(IV) → vanadium(V)

Make sure you learn these equations showing how the V_2O_5 catalyst works.

Heterogeneous Catalysts are in a Different Phase From the Reactants

1) A heterogeneous catalyst is one that is in a **different phase** from the reactants — i.e. in a different **physical state**. For example, in the Haber Process (see below) **gases** are passed over a **solid iron catalyst**.
2) The **reaction** happens on active sites located on the **surface** of the **heterogeneous catalyst**. So, **increasing** the **surface area** of the catalyst increases the number of molecules that can **react** at the same time, **increasing the rate** of the reaction.
3) **Support mediums** are often used to make the **area** of a catalyst **as large as possible.** They help to **minimise** the **cost** of the reaction, because only a **small coating** of catalyst is needed to provide a large surface area.

You need to know these examples of **heterogeneous catalysts:**

1) **Iron** in the **Haber Process** for making **ammonia**: $N_{2(g)} + 3H_{2(g)} \xrightarrow{Fe_{(s)}} 2NH_{3(g)}$
2) **Vanadium(V) oxide** in the **Contact Process** for making **sulfuric acid**: $SO_{2(g)} + \tfrac{1}{2}O_{2(g)} \xrightarrow{V_2O_{5(s)}} SO_{3(g)}$

SO_3 is then reacted with water to produce sulfuric acid.

Impurities Can Poison Heterogeneous Catalysts

1) Heterogeneous catalysts often work by **adsorbing** reactants onto **active sites** located on their surfaces.
2) **Impurities** in the reaction mixture may also **bind** to the catalyst's surface and **block reactants** from being adsorbed. This process is called **catalyst poisoning**.
3) Catalyst poisoning **reduces the surface area** of the catalyst available to the reactants, **slowing down the reaction**.
4) Catalyst poisoning **increases the cost** of a chemical process because **less product** can be made in a certain **time** or with a certain amount of **energy**. The **catalyst** may even need **replacing or regenerating**, which costs money.

Sulfur poisons the iron catalyst in the Haber Process:
The **hydrogen** in the Haber process is produced from **methane**. The methane is obtained from natural gas, which contains impurities, including **sulfur** compounds. Any sulfur that is not removed is **adsorbed** onto the iron, forming iron sulfide, and stopping the iron from catalysing the reaction efficiently.

Homogeneous Catalysts are in the Same Phase as the Reactants

1) **Homogeneous catalysts** are in the **same physical state** as the reactants. Usually a **homogeneous** catalyst is an **aqueous catalyst** for a reaction between two **aqueous solutions**.
2) Homogeneous catalysts work by combining with the reactants to form an **intermediate species** which then reacts to form the **products** and **re-form the catalyst**.
3) The enthalpy profile for a **homogeneously catalysed** reaction has **two humps** in it, corresponding to the two steps in the reaction.
4) The activation energy needed to form the **intermediates** (and to form the products from the intermediates) is **lower** than that needed to make the products directly from the reactants.
5) The catalyst always **re-forms**, so it can carry on catalysing the reaction.

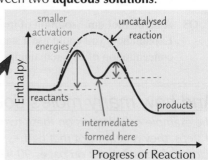

Catalysts

Fe^{2+} Catalyses the Reaction Between $S_2O_8^{2-}$ and I^-

The **redox** reaction between iodide ions and peroxodisulfate ($S_2O_8^{2-}$) ions takes place **annoyingly slowly** because both ions are **negatively charged**. The ions **repel** each other, so it's unlikely they'll **collide** and **react**.

$$S_2O_8^{2-}{}_{(aq)} + 2I^-{}_{(aq)} \rightarrow I_{2(aq)} + 2SO_4^{2-}{}_{(aq)}$$

The negative charges of the two ions is one reason why this reaction has high activation energy.

But if **Fe^{2+} ions** are added, things really **speed up** because each stage of the reaction involves a **positive and a negative ion**, so there's **no repulsion**.

1) First, the Fe^{2+} ions are **oxidised** to Fe^{3+} ions by the $S_2O_8^{2-}$ ions. $$S_2O_8^{2-}{}_{(aq)} + 2Fe^{2+}{}_{(aq)} \rightarrow 2Fe^{3+}{}_{(aq)} + 2SO_4^{2-}{}_{(aq)}$$
2) The newly formed intermediate Fe^{3+} ions now **easily oxidise** the I^- ions to iodine, and the **catalyst is regenerated**. $$2Fe^{3+}{}_{(aq)} + 2I^-{}_{(aq)} \rightarrow I_{2(aq)} + 2Fe^{2+}{}_{(aq)}$$

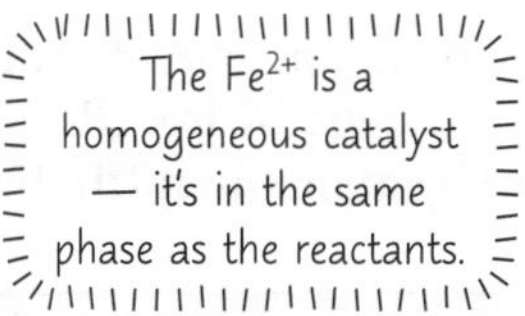

The Fe2+ is a homogeneous catalyst — it's in the same phase as the reactants.

You can test for **iodine** by adding **starch solution** — it'll turn **blue-black** if iodine is present.

Autocatalysis is when a **Product** Catalyses the Reaction

Another example of a **homogeneous catalyst** is Mn^{2+} in the reaction between $C_2O_4^{2-}$ and MnO_4^-.
It's an **autocatalysis reaction** because Mn^{2+} is a **product** of the reaction and **acts as a catalyst** for the reaction.
This means that as the reaction progresses and the **amount** of the **product increases**, the reaction **speeds up**.

The reactant ions are both negatively charged so repel each other and cause the rate of the uncatalysed reaction to be very slow.

$$2MnO_4^-{}_{(aq)} + 16H^+{}_{(aq)} + 5C_2O_4^{2-}{}_{(aq)} \rightarrow 2Mn^{2+}{}_{(aq)} + 8H_2O_{(l)} + 10CO_{2(g)}$$

1) Mn^{2+} catalyses the reaction by first reacting with MnO_4^- to form **Mn^{3+}** ions: $$MnO_4^-{}_{(aq)} + 4Mn^{2+}{}_{(aq)} + 8H^+{}_{(aq)} \rightarrow 5Mn^{3+}{}_{(aq)} + 4H_2O_{(l)}$$
2) The newly formed **Mn^{3+}** ions then react with $C_2O_4^{2-}$ ions to form carbon dioxide and **re-form** the Mn^{2+} catalyst ions: $$2Mn^{3+}{}_{(aq)} + C_2O_4^{2-}{}_{(aq)} \rightarrow 2Mn^{2+}{}_{(aq)} + 2CO_{2(g)}$$

Practice Questions

Q1 What property of transition elements makes them good catalysts?

Q2 What's the difference between a homogeneous and a heterogeneous catalyst?

Q3 Why is the rate of the uncatalysed reaction between iodide and peroxodisulfate ions so slow?

Q4 What term describes the process when a product catalyses a reaction?

Exam Questions

Q1 a) Using equations, explain how vanadium(V) oxide acts as a catalyst in the Contact Process. [2 marks]

b) i) Describe how heterogeneous catalysts can become poisoned and give an example. [2 marks]

ii) Give two consequences of catalytic poisoning. [2 marks]

Q2 A student is measuring the rate of the reaction between MnO_4^- ions and $C_2O_4^{2-}$ over time.
She predicts that the rate will decrease with time, but discovers instead that the rate of reaction over the first five minutes increases. Explain the student's results with use of appropriate equations. [5 marks]

What do you call an enthalpy profile diagram with no humps? Humphrey...

Hopefully you'll have realised from these two pages that transition metals are really useful. You should also know the difference between heterogeneous and homogeneous catalysts and some equations for how transition metals can catalyse certain reactions. If you do, then grab yourself a cup of tea— it speeds up the rate of revision, you know.

Metal-Aqua Ions

Metal-aqua ions are just complex ions where all of the ligands are water molecules. Pretty simple really. Unlike their reactions, which are just a tiny bit complicated. Sorry about that...

*Metal Ions Become **Hydrated** in Water*

When **transition metal compounds** dissolve in water, the water molecules form **co-ordinate bonds** with the **metal ions**. This forms **metal-aqua complex ions**. In general, **six water molecules** form co-ordinate bonds with each metal ion.

The water molecules do this by donating a **non-bonding pair of electrons** from their oxygen.

The diagrams show the metal-aqua ions formed by **iron(II)**, **$[Fe(H_2O)_6]^{2+}$**, and by **aluminium(III)**, **$[Al(H_2O)_6]^{3+}$**.

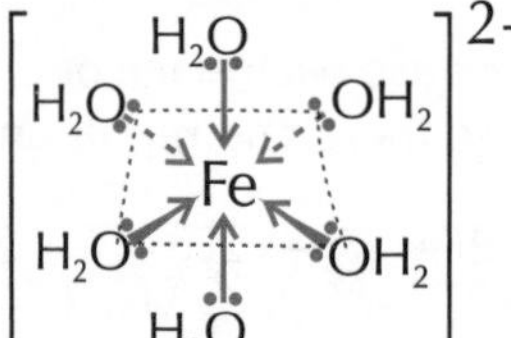

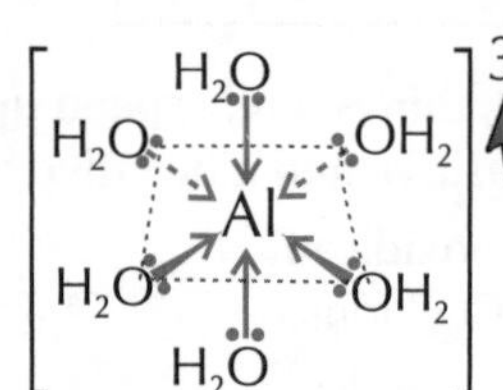

Water molecules are neutral so the overall charge of the complex is the same as the charge on the metal ion.

*Solutions Containing **Metal-Aqua Ions** are **Acidic***

In a solution containing metal-aqua **2+** ions, there's a reaction between the metal-aqua ion and the water — this is a **hydrolysis** or **acidity reaction**.

E.g.

$$Cu(H_2O)_6{}^{2+}{}_{(aq)} + H_2O_{(l)} \rightleftharpoons [Cu(OH)(H_2O)_5]^+{}_{(aq)} + H_3O^+{}_{(aq)}$$

Aqua-ironing — it keeps those flat fish smooth.

The metal-aqua **2+** ions release H^+ ions, so an **acidic** solution is formed. There's only **slight** dissociation though, so the solution is only **weakly acidic**.

Metal-aqua **3+** ions react in the same way. They dissociate **more** than 2+ ions, and so form **more acidic** solutions.

E.g.

$$Fe(H_2O)_6{}^{3+}{}_{(aq)} + H_2O_{(l)} \rightleftharpoons [Fe(OH)(H_2O)_5]^{2+}{}_{(aq)} + H_3O^+{}_{(aq)}$$

Here's why 3+ metal-aqua ions form more acidic solutions than 2+ metal-aqua ions:

Metal 3+ ions are pretty **small** but have a **big charge** — so they've got a **high charge density** (otherwise known as **charge/size ratio**). The metal 2+ ions have a **much lower** charge density.

This makes the 3+ ions much more **polarising** than the 2+ ions. More polarising power means that they attract **electrons** from the oxygen atoms of the co-ordinated water molecules more strongly, weakening the O–H bond.

So it's more likely that a **hydrogen ion** will be released. And more hydrogen ions means a **more acidic** solution.

*You Can **Hydrolyse** Metal-Aqua Ions **Further** to **Form Precipitates***

Adding **OH^- ions** to solutions of **metal-aqua ions** produces **insoluble metal hydroxides**. Here's why:

1) In water, **metal-aqua 3+ ions** such as Fe^{3+} or Al^{3+} form the equilibrium:

$$M(H_2O)_6{}^{3+}{}_{(aq)} + H_2O_{(l)} \rightleftharpoons [M(OH)(H_2O)_5]^{2+}{}_{(aq)} + H_3O^+{}_{(aq)}$$

If you add **OH^- ions** to the equilibrium, **H_3O^+ ions** are removed — this shifts the equilibrium to the **right**.

2) Now another equilibrium is set up in the solution:

$$[M(OH)(H_2O)_5]^{2+}{}_{(aq)} + H_2O_{(l)} \rightleftharpoons [M(OH)_2(H_2O)_4]^+{}_{(aq)} + H_3O^+{}_{(aq)}$$

Again the OH^- ions remove H_3O^+ ions from the solution, pulling the equilibrium to the right.

3) This happens one last time — now you're left with an **insoluble, uncharged metal hydroxide**:

$$[M(OH)_2(H_2O)_4]^+{}_{(aq)} + H_2O_{(l)} \rightleftharpoons M(OH)_3(H_2O)_{3(s)} + H_3O^+{}_{(aq)}$$

4) The same thing happens with **metal-aqua 2+ ions** (e.g. Fe^{2+} or Cu^{2+}), except this time there are only **two** steps:

$$M(H_2O)_6{}^{2+}{}_{(aq)} + H_2O_{(l)} \rightleftharpoons [M(OH)(H_2O)_5]^+{}_{(aq)} + H_3O^+{}_{(aq)}$$

$$[M(OH)(H_2O)_5]^+{}_{(aq)} + H_2O_{(l)} \rightleftharpoons M(OH)_2(H_2O)_{4(s)} + H_3O^+{}_{(aq)}$$

There are only two steps this time because only two of the water ligands need to be deprotonated to make the +2 complex uncharged (and so insoluble).

Metal-Aqua Ions

Metal Hydroxides That Can Act as an **Acid OR a Base** are **Amphoteric**

1) You saw on the last page that if you hydrolyse a metal-aqua ion in a base then you form an insoluble **metal hydroxide precipitate**.
2) **All** these metal hydroxide precipitates **will dissolve in acid**. They act as Brønsted-Lowry bases and **accept H^+ ions**. This **reverses** the hydrolysis reactions.
3) Some metal hydroxides are **amphoteric** — they can act as both acids and bases. This means they'll **dissolve in an excess of base** as well as in **acids**.

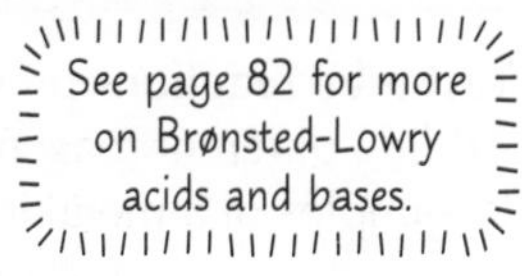

See page 82 for more on Brønsted-Lowry acids and bases.

4) **Aluminium hydroxide** is **amphoteric**. In the presence of a **base**, e.g. NaOH, it acts as a **Brønsted-Lowry acid** and **donates H^+ ions** to the OH^- ions, forming a **soluble compound**. It also acts as a **Brønsted-Lowry base** in the presence of an **acid** and **accepts H^+** ions from the H_3O^+ ions in solution.

$$[Al(H_2O)_6]^{3+}_{(aq)} + 3H_2O_{(aq)} \xleftarrow[\text{With acid}]{+3H_3O^+_{(aq)}} Al(OH)_3(H_2O)_{3(s)} \xrightarrow[\text{With base}]{+OH^-_{(aq)}} [Al(OH)_4(H_2O)_2]^-_{(aq)} + H_2O_{(l)}$$

Precipitates Form with **Ammonia Solution...**

1) The obvious way of adding hydroxide ions is to use a strong alkali, like **sodium hydroxide solution** — but you can use **ammonia solution** too. When ammonia dissolves in water this equilibrium occurs:

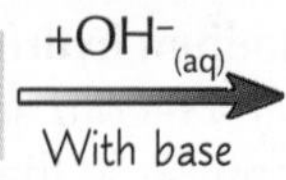

$$NH_{3(aq)} + H_2O_{(l)} \rightleftharpoons NH_4^+{}_{(aq)} + OH^-_{(aq)}$$

2) Because hydroxide ions are formed, adding a **small** amount of ammonia solution gives the same results as sodium hydroxide.
3) In some cases, such as with $Cu(OH)_2(H_2O)_4$, a further reaction happens if you add **excess** ammonia solution — the H_2O and OH^- ligands are **displaced** by NH_3 ligands. This forms a **charged complex** which is **soluble** in water, so the precipitate dissolves.

$$Cu(OH)_2(H_2O)_{4(s)} + 4NH_{3(aq)} \rightleftharpoons [Cu(NH_3)_4(H_2O)_2]^{2+}_{(aq)} + 2OH^-_{(aq)} + 2H_2O_{(l)}$$

...and **Sodium Carbonate** Too

Metal 2+ ions react with **sodium carbonate** to form **insoluble metal carbonates**, like this:

$$[M(H_2O)_6]^{2+}_{(aq)} + CO_3^{2-}{}_{(aq)} \rightleftharpoons MCO_{3(s)} + 6H_2O_{(l)}$$

But metal 3+ ions are **stronger acids** so there is a **higher concentration** of H_3O^+ ions in solution. Rather than displacing water from the metal ions, the carbonate ions react with **H_3O^+**, removing them from the solution and shifting the equilibria of the reactions on the previous page to the right. So the precipitate that forms is $M(OH)_3(H_2O)_3$ rather than $M_2(CO_3)_3$.

$$CO_3^{2-}{}_{(aq)} + 2H_3O^+_{(aq)} \rightleftharpoons CO_{2(g)} + 3H_2O_{(l)}$$

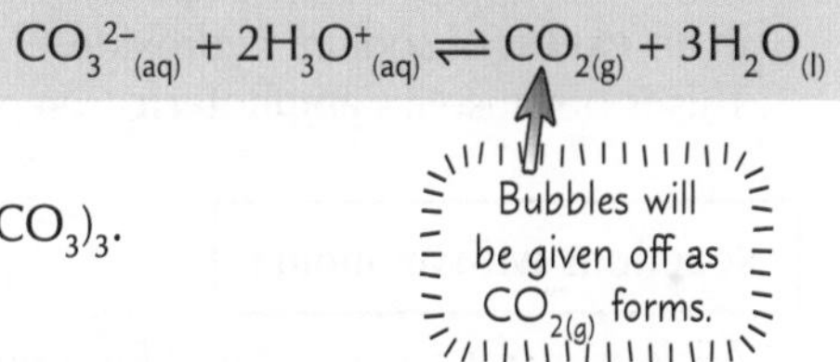

Bubbles will be given off as $CO_{2(g)}$ forms.

Practice Questions

Q1 Explain why 3+ metal-aqua ions form more acidic solutions than 2+ metal-aqua ions.

Q2 Write an equation to show what happens if you add an excess of ammonia to $Cu(OH)_2(H_2O)_{4(s)}$.

Q3 Show by equations how $Al(OH)_3$ can act as both a Brønsted-Lowry acid and a Brønsted-Lowry base.

Exam Questions

Q1 Why do separate solutions of iron(II) sulfate and iron(III) sulfate have different pH values? [2 marks]

Q2 a) Write an ionic equation to show the formation of the precipitate when sodium carbonate is added to aqueous iron(II) sulfate. [1 mark]

b) Explain how sodium carbonate reacts with aqueous iron(III) chloride to form a precipitate. Include equations in your answer. [2 marks]

I get grumpy when it rains — it's a precipitation reaction...

Quite a few more reactions to learn here, I'm afraid. Remember that the precipitates of these reactions will always be neutral compounds, and the soluble compounds will be charged. So if you know a precipitate forms but you can't remember the formula of the product, then it's a case of balancing out the charges until the compound is neutral.

More on Metal-Aqua Ions

It's true — learning about the reactions of metal-aqua ions is probably less exciting than reading the dictionary from cover to cover. Fortunately, you'll get to try them out for yourself in practical experiments. Oooo... pretty colours.

You Can Identify Metal Ions Using **Test Tube Reactions**

Test tube reactions provide a qualitative way of working out the identity of unknown metal ions in solution. Adding different reagents, such as sodium hydroxide solution, ammonia solution and sodium carbonate solution, to samples of a metal ion solution and recording what you see should help you identify what metal ion is present. Here's what you should do:

1) Measure out samples of the unknown metal ion solution into **three separate test tubes.**
2) To the first test tube, add **sodium hydroxide solution dropwise**, using a dropping pipette, and record any changes you see. Then add more NaOH dropwise so that it is in **excess**. Record any changes.
3) To the second test tube, add **ammonia solution** dropwise, using a dropping pipette, and record any changes you see. Keep adding ammonia so that it's in **excess**. Record any changes.
4) To the third test tube, add **sodium carbonate solution** dropwise. Record your observations.

Some of the solutions may irritate your skin and eyes, so make sure you wear gloves, a lab coat and goggles. Ammonia is very smelly and can make you cough if you breathe it in, so it's best to use it in a fume hood.

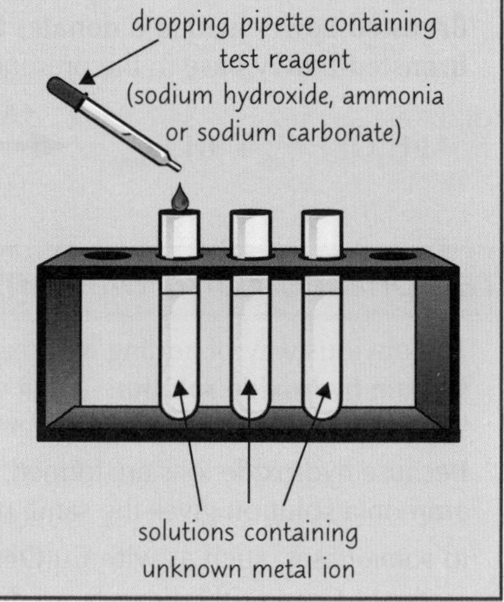

On pages 124-125, you learnt about some of the reactions of copper(II), iron(II), iron(III) and aluminium(III) aqua ions. If you had four unknown solutions, each of which contained one of these metal aqua ions, you could use the method above and the **differences** in their reactivities to distinguish between them. Here's how:

Reactions with sodium hydroxide

All four metal aqua ions will form precipitates with sodium hydroxide, but only the **aluminium hydroxide** precipitate will **dissolve** in an **excess** of sodium hydroxide. This is because it's **amphoteric** (see page 125).

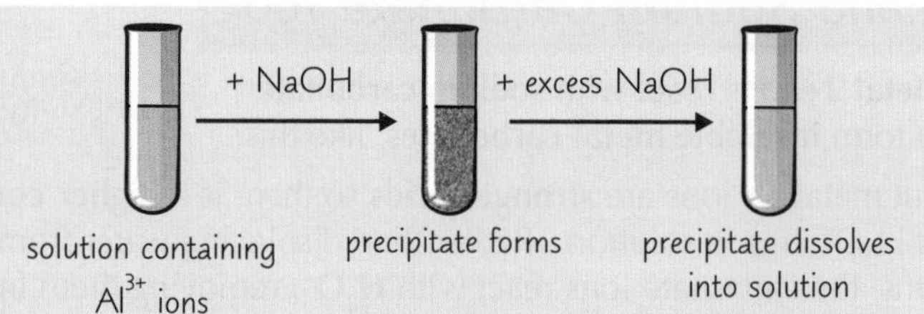

Reactions with ammonia

All four metal aqua ions will form precipitates with ammonia, but only the **copper hydroxide** precipitate will **dissolve** in an **excess** of ammonia. This is because it undergoes a **ligand exchange** reaction with excess ammonia (see page 125).

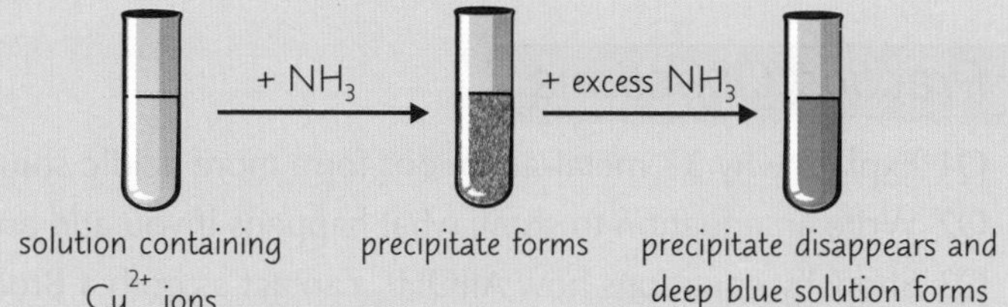

Reactions with sodium carbonate

All four metal aqua ions will form precipitates with sodium carbonate. The solutions containing Al^{3+} or Fe^{3+} will also form **bubbles** as CO_2 formed (see page 125). So if you're not sure whether a sample contains Fe^{2+} or Fe^{3+} ions, (which behave identically in the other two tests), you can use this test to decide — a sample containing Fe^{3+} ions will give off a **gas**, but a sample containing Fe^{2+} ions **won't**.

The solutions should be freshly made — if Fe^{2+} ions are left too long in contact with air they will oxidise to Fe^{3+} ions.

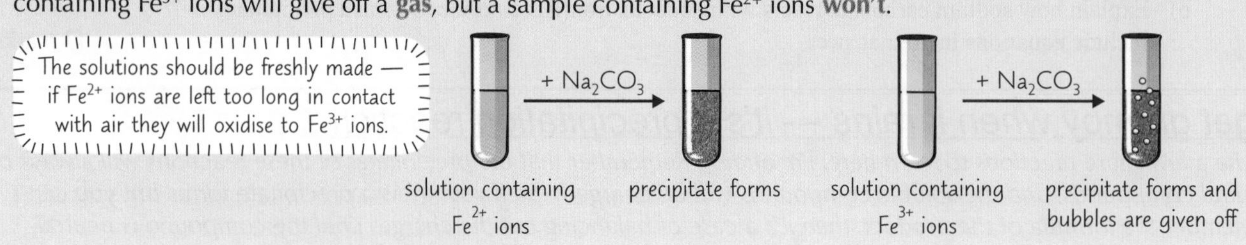

More on Metal-Aqua Ions

Learn the **Colours** of All the **Complex Ion Solutions** and **Precipitates**

This handy table summarises all the compounds that are formed in the reactions on these pages.
You need to know the **formulae** of all the complex ions, and their **colours**.

Metal-aqua ion	With $OH^-_{(aq)}$ or $NH_{3(aq)}$	With excess $OH^-_{(aq)}$	With excess $NH_{3(aq)}$	With $Na_2CO_{3(aq)}$
$[Cu(H_2O)_6]^{2+}$ blue solution	$Cu(OH)_2(H_2O)_4$ blue precipitate	no change	$[Cu(NH_3)_4(H_2O)_2]^{2+}$ deep blue solution	$CuCO_3$ green-blue precipitate
$[Fe(H_2O)_6]^{2+}$ green solution	$Fe(OH)_2(H_2O)_4$ green precipitate	no change	no change	$FeCO_3$ green precipitate
$[Al(H_2O)_6]^{3+}$ colourless solution	$Al(OH)_3(H_2O)_3$ white precipitate	$[Al(OH)_4(H_2O)_2]^-$ colourless solution	no change	$Al(OH)_3(H_2O)_3$ white precipitate
$[Fe(H_2O)_6]^{3+}$ yellow solution	$Fe(OH)_3(H_2O)_3$ brown precipitate	no change	no change	$Fe(OH)_3(H_2O)_3$ brown precipitate

Practice Questions

Q1 Outline an experiment you could carry out to distinguish between four unknown solutions which each contain either copper(II), iron(II), iron(III) or aluminium(III) ions.

Q2 Name any hazards associated with the experiment in Question 1. How would you minimise these hazards?

Q3 What differences would you observe if you were to add sodium carbonate solution to a solution containing iron(II) ions and to a solution containing iron(III) ions?

Q4 What colour solution is formed when you add excess ammonia to a solution containing $[Cu(H_2O)_6]^{2+}$ ions?

Exam Questions

Q1 Describe what you would see when ammonia solution is added slowly to a solution containing copper(II) sulfate until it is in excess. Write an equation for each reaction that occurs. [4 marks]

Q2 Aqueous sodium hydroxide was added to an aqueous solution of aluminium(III) chloride.

a) Identify the aluminium complex ion present in:

i) the aqueous aluminium(III) chloride. [1 mark]

ii) the white precipitate initially formed when aqueous sodium hydroxide is added. [1 mark]

iii) the colourless solution when an excess of aqueous sodium hydroxide is added. [1 mark]

b) Write an equation for the reaction in which the colourless solution is formed from the white precipitate. [1 mark]

Q3 a) Describe what you would observe when aqueous sodium carbonate is added to:

i) aqueous iron(III) chloride. [1 mark]

ii) freshly-prepared aqueous iron(II) sulfate. [1 mark]

b) Write an equation for the reaction of the iron(II)-aqua ion with the carbonate ion. [1 mark]

c) If iron(II) sulfate solution is left to stand overnight in an open beaker before the aqueous sodium carbonate is added, then a different reaction is observed.

i) Describe the new observation. [1 mark]

ii) Explain this change. [1 mark]

Test-tube reactions — proper Chemistry at last...

So many pretty colours. It's like walking into an exploded paint factory. Or my little sister's room when she's been doing arts and crafts. The only downside is that now you have to remember them all. You also need to know the reactions that make them. It's important that you learn them all, or come exam day you'll end up feeling blue. Or possibly blue-green...

Basic Stuff

Organic chemistry is all about carbon compounds. There are loads of them, and you're mainly made up of them, so they're fairly important. Chemists have organised them into families, which makes them a tad easier to cope with.

There are Loads of Ways of Representing Organic Compounds

TYPE OF FORMULA	WHAT IT SHOWS YOU	FORMULA FOR BUTAN-1-OL
General formula	An algebraic formula that can describe **any member** of a family of compounds.	$C_nH_{2n+1}OH$ (for all alcohols)
Empirical formula	The **simplest ratio** of atoms of each element in a compound (cancel the numbers down if possible). (So ethane, C_2H_6, has the empirical formula CH_3.)	$C_4H_{10}O$
Molecular formula	The **actual** number of atoms of each element in a molecule.	$C_4H_{10}O$
Structural formula	Shows the atoms **carbon by carbon**, with the attached hydrogens and functional groups.	$CH_3CH_2CH_2CH_2OH$
Skeletal formula	Shows the **bonds** of the carbon skeleton **only**, with any functional groups. The hydrogen and carbon atoms aren't shown. This is handy for drawing large complicated structures, like cyclic hydrocarbons.	OH
Displayed formula	Shows how all the atoms are **arranged**, and all the bonds between them.	H H H H / H–C–C–C–C–O–H / H H H H

Homologous Compounds have the Same General Formulas

1) A **homologous series** is a group of compounds that contain the same functional group. They can all be represented by the same **general formula**.
2) You can use a general formula to work out the **molecular formula** of any member of a homologous series.

A functional group is a reactive part of a molecule — it gives it most of its chemical properties.

Example: The alkanes have the general formula C_nH_{2n+2}. Give the formula of the alkane with six carbons.

n = 6, so the formula is $C_6H_{(2 \times 6) + 2}$ = $\mathbf{C_6H_{14}}$

3) Each **successive member** of a homologous series differs by a '$\mathbf{CH_2}$' group.

Example: Alcohols have the general formula $C_nH_{2n+1}OH$.
Give the formulas of the first six alcohols in the homologous series.

Each alcohol has one more **CH_2 group** than the one before:

Methanol	Ethanol	Propanol	Butanol	Pentanol	Hexanol
CH_3OH	C_2H_5OH	C_3H_7OH	C_4H_9OH	$C_5H_{11}OH$	$C_6H_{13}OH$

Don't worry if you don't recognise all of these yet — you'll meet them by the end of the course.

4) Here are the main **homologous series** that you need to know about:

Homologous series	Prefix or suffix	Example
alkanes	-ane	propane $CH_3CH_2CH_3$
branched alkanes	alkyl-	methylpropane $CH_3CH(CH_3)CH_3$
alkenes	-ene	propene $CH_3CH{=}CH_2$
halogenoalkanes	fluoro- / chloro- / bromo- / iodo-	chloroethane CH_3CH_2Cl
alcohols	-ol	ethanol CH_3CH_2OH
aldehydes	-al	ethanal CH_3CHO
ketones	-one	propanone CH_3COCH_3

Homologous series	Prefix or suffix	Example
carboxylic acids	-oic acid	ethanoic acid CH_3COOH
esters	alkyl- -oate	ethyl ethanoate $CH_3COOCH_2CH_3$
amines	-amine	methylamine CH_3NH_2
amides	-amide	ethanamide CH_3CONH_2
acyl chlorides	-oyl chloride	ethanoyl chloride CH_3COCl
cycloalkanes	cyclo- -ane	cyclohexane C_6H_{12}
arenes	-benzene (or phenyl-)	methylbenzene $C_6H_5CH_3$

Basic Stuff

Nomenclature is a Fancy Word for the Naming of Organic Compounds

You can name any organic compound using these **rules** of nomenclature:

1) Count the carbons in the **longest continuous chain** that contains the functional group— this gives you the stem:

Number of carbons	1	2	3	4	5	6
Stem	meth–	eth–	prop–	but–	pent–	hex–

Don't forget — the longest carbon chain may be bent.

2) The **main functional group** of the molecule usually gives you the end of the name (the **suffix**) — see the table on the previous page.

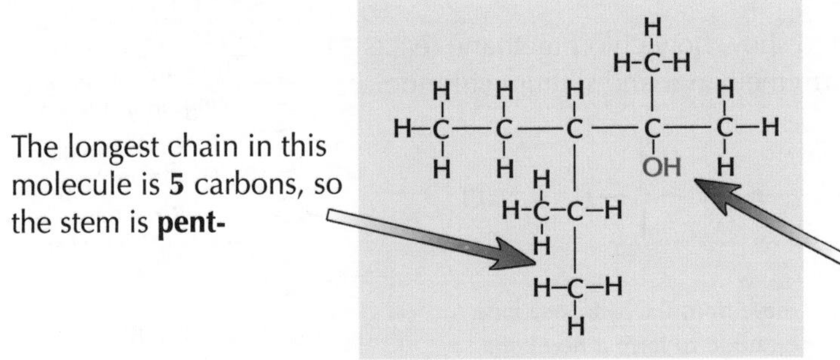

The longest chain in this molecule is **5** carbons, so the stem is **pent-**
OH

The main functional group is **-OH**, so the compound's name is going to be based on "**pentanol**".

3) Number the carbons in the **longest** carbon chain so that the carbon with the main functional group attached has the lowest possible number.
If there's more than one longest chain, pick the one with the **most side-chains**.

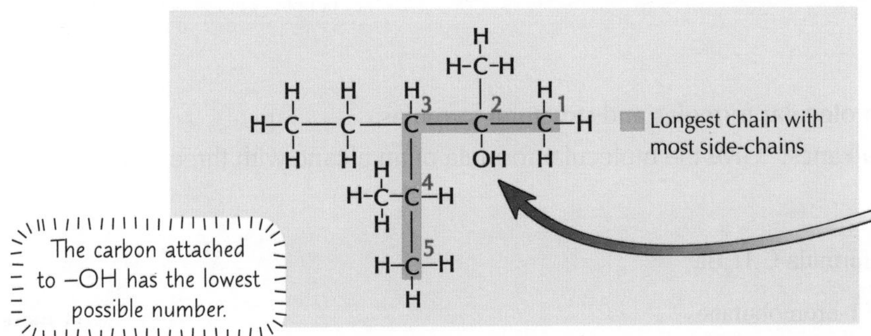

4) Write the carbon number that the functional group is on **before the suffix.**

–OH is on carbon-2, so it's some sort of "**pentan-2-ol**".

5) Any side-chains or less important functional groups are added as prefixes at the start of the name. Put them in **alphabetical** order, with the **number** of the carbon atom each is attached to.

6) If there's more than one **identical** side-chain or functional group, use **di-** (2), **tri-** (3) or **tetra-** (4) before that part of the name — but ignore this when working out the alphabetical order.

There's an ethyl group on carbon-3, and methyl groups on carbon-2 and carbon-4, so it's **3-ethyl-2,4-dimethylpentan-2-ol**

IUPAC Rules Help Avoid Confusion

Fancy a cup of tea?
Tea, you know... Brown stuff.
Oh come on, don't just sit
there looking confused..."

1) The **IUPAC system** for naming organic compounds is the agreed **international language** of chemistry. Years ago, organic compounds were given whatever names people fancied, such as acetic acid and ethylene. But these names caused **confusion** between different countries.

2) The IUPAC system means scientific ideas can be communicated **across the globe** more effectively. So it's easier for scientists to get on with testing each other's work, and either confirm or dispute new theories.

Basic Stuff

A Mechanism Breaks Down a Reaction into Individual Stages

1) It's all very well knowing the outcome of a reaction, but it can also be useful to know **how** a reaction happens.
2) **Mechanisms** are diagrams that break reactions down into individual stages.
   They show how molecules react together by using **curly arrows** to show which **bonds** are made or broken.

Curly Arrows Show How Electron Pairs Move Around

In order to make or break a bond in a reaction, **electrons** have to move around.
A **curly arrow** shows where a **pair** of electrons goes during a reaction. They look like this:

The arrow starts at the bond or lone pair where the electrons are at the beginning of the reaction.

The arrow points to where the new bond is formed at the end of the reaction.

Example: Draw a reaction mechanism to show how chloromethane reacts with sodium hydroxide to form methanol and sodium chloride.

Reaction:

$$CH_3Cl + NaOH \longrightarrow CH_3OH + NaCl$$

Mechanism:

$$CH_3Cl \;\; :OH^- \longrightarrow CH_3OH \;\; :Cl^-$$

Electrons move from the hydroxide lone pair to the carbon to form a new bond.

The carbon-chlorine bond breaks, and the electrons move onto the chlorine atom.

The overall charge of the reaction stays the same.

Na^+ doesn't get involved in the reaction, so you don't need to include it in the mechanism.

There are lots of mechanisms coming up in Unit 3, so if it all seems a bit strange now, don't worry. Before long you'll be a curly arrow wizard.

Practice Questions

Q1 Explain the difference between molecular formulas and structural formulas.

Q2 What is the general formula for alkanes? Give the molecular formula of an alkane with three carbons.

Exam Questions

Q1 1-bromobutane has the molecular formula C_4H_9Br.

a) Draw the displayed formula of 1-bromobutane. [1 mark]

b) Which homologous series does 1-bromobutane belong to? [1 mark]

c) 1-bromobutane can be made from the molecule shown on the right. Name the molecule. [1 mark]

$$CH_3CH_2CH=CH_2$$

Q2 a) Alkanes are an example of a homologous series. What is a homologous series? [1 mark]

b) i) Write down the molecular formula for the alkane molecule that has five carbon atoms. [1 mark]

ii) What is the IUPAC name of this molecule? [1 mark]

Q3 A molecule has the structural formula CH_2ClCH_2Cl. Give each of the following:

a) Its IUPAC name. [1 mark]

b) Its displayed formula. [1 mark]

c) Its skeletal formula. [1 mark]

I hope all that hasn't sent iu-packing...

I know it's a pain that there are about a gazillion different ways of describing the same molecule but examiners love asking you to draw molecules from their IUPAC name or formula and vice versa. You'll only be able to do this quickly if you practise. So choose a molecule and write down all its different representations. That's right. Every. Single. One.

Isomerism

Isomers are great fun — they're all about putting the same atoms together in different ways to make completely different molecules. It's like playing with plastic building bricks, but it hurts less when you tread on one by accident.

***Isomers** Have the Same **Molecular** Formula*

1) Two molecules are isomers of one another if they have the same **molecular formula** but the atoms are arranged **differently**.
2) There are two types of isomers you need to know about — **structural isomers** and **stereoisomers**.

***Structural Isomers** have different **Structural Arrangements** of Atoms*

In structural isomers the atoms are **connected** in different ways.
So they have the **same molecular formula** but different **structural formulas**.
There are **three types** of structural isomers:

Chain Isomers

Chain isomers have different arrangements of the **carbon skeleton**. Some are **straight chains** and others **branched** in different ways.

butane   methylpropane

Positional Isomers

Positional isomers have the **same skeleton** and the **same atoms or groups of atoms** attached.

The difference is that the atom or group of atoms is attached to a **different carbon atom**.

1-chlorobutane   2-chlorobutane

Functional Group Isomers

Functional group isomers have the same atoms arranged into **different functional groups**.

hex-1-ene   cyclohexane

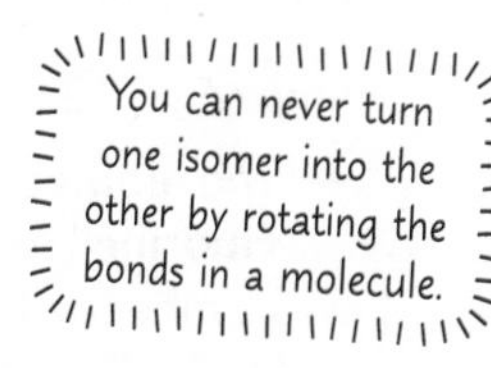

You can never turn one isomer into the other by rotating the bonds in a molecule.

Stereoisomers** Have Different Arrangements in **Space

Stereoisomers have the same structural formula but a **different arrangement** in space.
(Just bear with me for a moment... that will become clearer, I promise.)

Double Bonds** Can't **Rotate

1) Carbon atoms in a C=C double bond and the atoms bonded to these carbons all lie in the **same plane** (they're **planar**).
2) Another important thing about C=C double bonds is that atoms **can't rotate** around them like they can around single bonds. In fact, double bonds are fairly **rigid** — they don't bend much either.
3) The **restricted rotation** around the C=C double bond causes a type of stereoisomerism called **E/Z isomerism** (more about this on the next page).

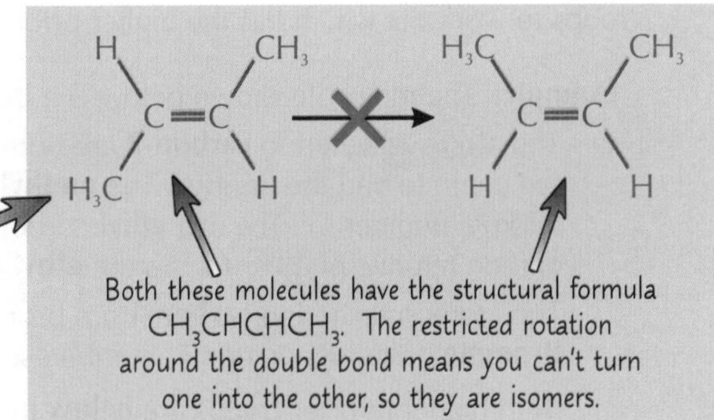

Both these molecules have the structural formula $CH_3CHCHCH_3$. The restricted rotation around the double bond means you can't turn one into the other, so they are isomers.

Isomerism

Alkenes Show E/Z Isomerism

(You might see E/Z isomerism being called geometric isomerism.)

1) **Alkenes** have **restricted rotation** around their carbon-carbon double bonds.
2) This means that if both of the double-bond carbons have **different atoms** or **groups** attached to them, the arrangement of those groups around the double bond becomes important — you get two **stereoisomers.**
3) One of these isomers is called the **'E-isomer'** and the other is called the **'Z-isomer'**.
4) The **Z-isomer** has the same groups either **both above** or **both below** the double bond.
5) The **E-isomer** has the same groups positioned **across** the double bond.

E stands for 'entgegen', the German for 'opposite'. Z stands for 'zusammen', the German for 'together'.

For example, here are the two stereoisomers of **but-2-ene**:

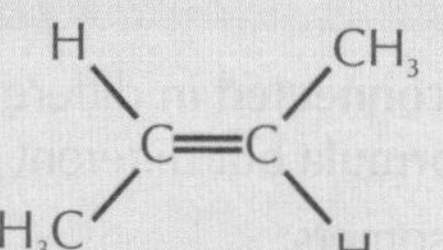

Here, the same groups are both **above** the double bond so it's the **Z-isomer**. This molecule is **Z-but-2-ene**.

Here, the same groups are **across** the double bond so it's the **E-isomer**. This molecule is **E-but-2-ene**.

One way to remember which isomer is which is to say that in the Z-isomer the matching groups are on 'ze zame zide', but in the E-isomer, they are 'enemies'.

The E/Z System Works Even When the Groups Are Different

A molecule that has a C=C bond surrounded by **four different groups** still has an E- and a Z-isomer — it's just harder to work out which is which. Fortunately, you can solve this problem using the **Cahn-Ingold-Prelog (CIP) priority rules**:

Example: Using the Cahn-Ingold-Prelog Priority Rules to Identify an E/Z Isomer

1) The first step is to **assign a priority** to the two atoms attached to each carbon in the double bond.
2) To do this, you look at the atoms that are **directly bonded** to each of the C=C carbon atoms. The atom with the higher **atomic number** on each carbon is given the higher **priority**.

Here's one of the stereoisomers of 1-bromo-1-chloro-2-fluoro-ethene:

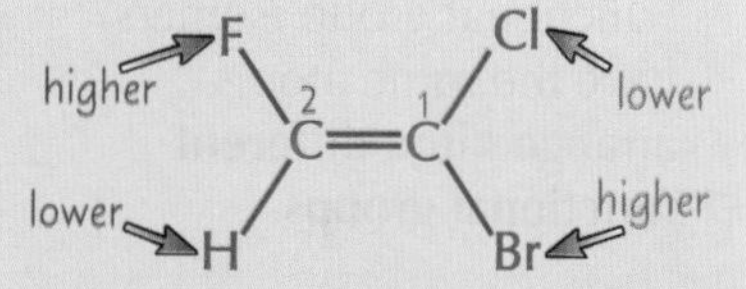

- The atoms directly attached to **carbon-2** are fluorine and hydrogen. **Fluorine** has an atomic number of **9** and **hydrogen** has an atomic number of **1**. So **fluorine** is the higher priority group.
- The atoms directly attached to **carbon-1** are bromine and chlorine. **Bromine** has an atomic number of **35** and **chlorine** has an atomic number of **17**. So **bromine** is the higher priority group.

3) Now to work out which isomer you have, just look at how the **two higher priority groups** are arranged.

In the stereoisomer shown above, the two higher priority groups (**Br** and **F**) are positioned **across** the double bond from one another — so this is the **E-isomer** of 1-bromo-1-chloro-2-fluoro-ethene.

You May Have to Look Further Along a Chain to work out Priorities

If the atoms **directly bonded** to the carbon are the **same**, then you have to look at the **next** atom in the groups to work out which has the higher priority.

Example: The molecule shown below is a branched alkene. State whether it is an E-isomer or a Z-isomer.

1) The atoms attached to **carbon-2** are both carbons, so go a step **along** the chain to find the priority. The **methyl** carbon is joined to hydrogen (atomic number 1). The first **ethyl** carbon is joined to another carbon (atomic number 6). So the **ethyl group** has higher priority.
2) The atoms attached to **carbon-1** are bromine and chlorine. **Bromine** has a higher atomic number, so it is the higher priority group.
3) Both higher priority groups are **below** the double bond — so this molecule is a **Z-isomer.**

Isomerism

Don't be Fooled — What Looks Like an Isomer Might **Not** Be

Beware — sometimes what looks like an isomer, isn't. If you can **switch** between two drawings of a molecule, either by rotating the **C-C single bonds** or rotating the **entire molecule**, then you've drawn the same isomer twice.

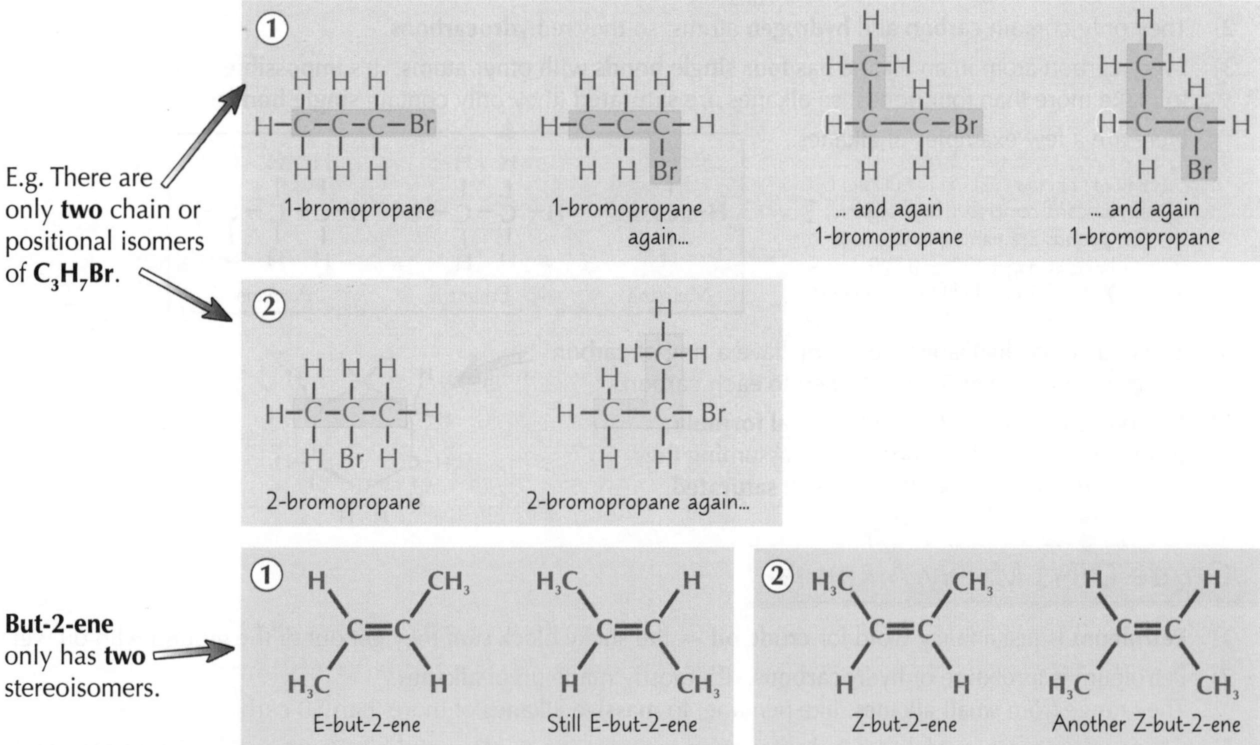

Practice Questions

Q1 What are structural isomers?

Q2 Define the term 'stereoisomers'.

Q3 What property of alkenes gives rise to E/Z isomerism?

Q4 Which group would have a higher priority under the CIP priority rules: a bromine atom or an –OH group?

Exam Questions

Q1 There are four halogenoalkanes with the molecular formula C_4H_9Cl.

a) Give the names of all four of these halogenoalkanes. [4 marks]

b) Identify a pair of positional isomers from your answer to part a). [1 mark]

c) Identify a pair of chain isomers from your answer to part a). [1 mark]

Q2 1-bromopropene has the structural formula $CH_3CH{=}CHBr$.

a) Draw the structure of E-1-bromopropene. [1 mark]

b) Draw the structures of two isomers of 1-bromopropene that do not exhibit E/Z isomerism. [2 marks]

Q3 a) Draw and name the E/Z isomers of pent-2-ene. [2 marks]

b) Explain why alkenes can have E/Z isomers but alkanes cannot. [2 marks]

I just love my new surround-sound stereoisomer...

IMPORTANT FACT: If the two groups connected to one of the double-bonded carbons in an alkene are the same, then it won't have E/Z isomers. For example, neither propene nor but-1-ene have E/Z isomers. Try drawing them out if you're not sure. And then draw out all the structural isomers of butene. Just to prove you've got this completely sussed.

Alkanes and Petroleum

Alkanes are the first set of organic chemicals you need to know about. They're what petroleum's mainly made of.

Alkanes are **Saturated Hydrocarbons**

1) Alkanes have the **general formula C_nH_{2n+2}**.
2) They only contain **carbon** and **hydrogen** atoms, so they're **hydrocarbons**.
3) Every carbon atom in an alkane has **four single bonds** with other atoms. It's **impossible** for carbon to make more than four bonds, so alkanes are **saturated** (they only contain **single bonds**).

   Here are a few examples of alkanes:

   If you can't remember how organic compounds are named, have a look back at pages 128 and 129.

```
   H            H  H             H  H  H
   |            |  |             |  |  |
H— C —H     H— C — C —H      H— C — C — C —H
   |            |  |             |  |  |
   H            H  H             H  H  H
Methane        Ethane           Propane
```

4) You can get **cycloalkanes** too. They have a ring of carbon atoms with two hydrogens attached to each carbon.
5) Cycloalkanes have a **different general formula** from that of normal alkanes (C_nH_{2n}, assuming they have only one ring), but they are still **saturated**.

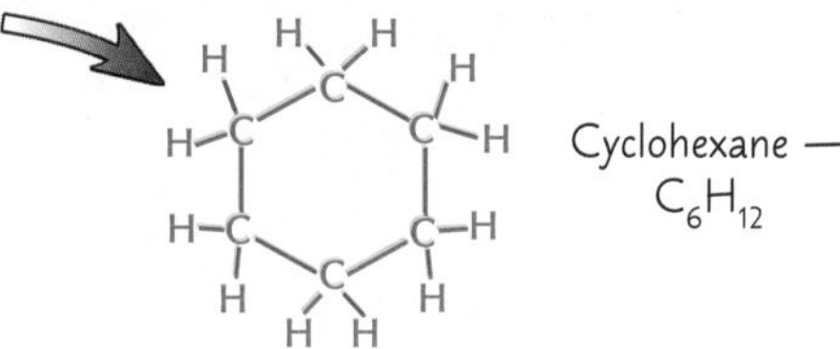

Crude Oil is Mainly **Alkanes**

1) **Petroleum** is just a fancy word for **crude oil** — the sticky black stuff they get out of the ground with oil wells.
2) Petroleum is a mixture of **hydrocarbons**. It's mostly made up of **alkanes**.
   They range from **small alkanes**, like pentane, to **massive alkanes** of more than 50 carbons.
3) Crude oil isn't very useful as it is, but you can **separate** it out into useful bits (**fractions**) by **fractional distillation**.

Here's how fractional distillation works — don't try this at home.

You might do fractional distillation in the lab, but if you do you'll use a safer crude oil substitute instead.

1) First, the crude oil is **vaporised** at about 350 °C.
2) The vaporised crude oil goes into a **fractionating column** and rises up through the trays. The largest hydrocarbons don't **vaporise** at all, because their boiling points are too high — they just run to the bottom and form a gooey **residue**.
3) As the crude oil vapour goes up the fractionating column, it gets **cooler**.
   Because the alkane molecules have different chain lengths, they have different **boiling points**, so each fraction **condenses** at a different temperature. The fractions are **drawn off** at different levels in the column.
4) The hydrocarbons with the **lowest boiling points** don't condense.
   They're drawn off as **gases** at the top of the column.

fractionating column — tray — Heater 350 °C — crude oil

	Fraction	Number of Carbons	Uses
	Gases	1 - 4	liquefied petroleum gas (LPG), camping gas
40 °C	Petrol (gasoline)	5 -12	petrol
110 °C	Naphtha	7 - 14	processed to make petrochemicals
180 °C	Kerosene (paraffin)	11 - 15	jet fuel, petrochemicals, central heating fuel
250 °C	Gas Oil (diesel)	15 - 19	diesel fuel, central heating fuel
340 °C	Mineral Oil (lubricating)	20 - 30	lubricating oil
	Residue: Fuel Oil	30 - 40	ships, power stations
	Residue: Wax, grease	40 - 50	candles, lubrication
	Residue: Bitumen	50+	roofing, road surfacing

Alkanes and Petroleum

Heavy Fractions can be *'Cracked'* to Make *Smaller Molecules*

1) People want loads of the **light** fractions of crude oil, like petrol and naphtha. They don't want so much of the **heavier** stuff like bitumen though. Stuff that's in high demand is much more **valuable** than the stuff that isn't.
2) To meet this demand, the less popular heavier fractions are **cracked**. Cracking is **breaking** long-chain alkanes into **smaller** hydrocarbons (which can include alkenes). It involves breaking the **C–C bonds.**

For example, **decane** could crack like this:

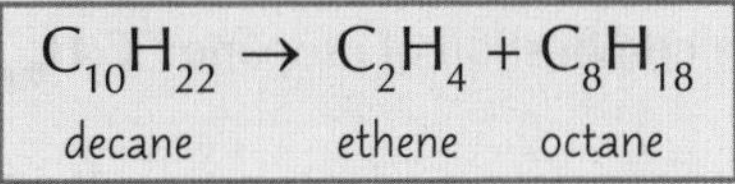

$$C_{10}H_{22} \rightarrow C_2H_4 + C_8H_{18}$$

decane ethene octane

Where the chain breaks is random, so you'll get a different mixture of products every time you crack a hydrocarbon.

There are **two types** of **cracking** you need to know about:

Thermal cracking:
- This takes place at **high temperature** (up to 1000 °C) and **high pressure** (up to 70 atm).
- It produces a lot of **alkenes.**
- These **alkenes** are used to make heaps of valuable products, like **polymers** (plastics). A good example is **poly(ethene)**, which is made from ethene.

Catalytic cracking:
- Catalytic cracking uses something called a **zeolite catalyst** (**hydrated aluminosilicate**), at a **slight pressure** and **high temperature** (about 450 °C).
- It mostly produces **aromatic** hydrocarbons and **motor fuels.**

  Aromatic compounds contain benzene rings. Benzene rings have six carbon atoms with three double bonds. They're pretty stable because the electrons are delocalised around the carbon ring.
- Using a catalyst **cuts costs**, because the reaction can be done at a **low** pressure and a **lower** temperature. The catalyst also **speeds** up the reaction, saving time (and time is money).

Practice Questions

Q1 What is the general formula for alkanes?

Q2 Alkanes are saturated hydrocarbons — explain what the term 'saturated' means.

Q3 What is cracking?

Q4 What type of organic chemical does thermal cracking produce?

Exam Questions

Q1 Crude oil contains many different alkane molecules.
These are separated using a process called fractional distillation.

a) Why do the components of crude oil need to be separated? *[1 mark]*

b) What physical property of the molecules is used to separate them? *[1 mark]*

c) A typical alkane found in the petrol fraction has 8 carbon atoms.

i) Give the molecular formula for this alkane. *[1 mark]*

ii) Would you find the petrol fraction near the top or the bottom of the fractionating column? Explain your answer. *[2 marks]*

Q2 Crude oil is a source of fuels and petrochemicals. It can be vaporised and separated into fractions by fractional distillation. Some heavier fractions are processed using cracking.

a) Give one reason why cracking is carried out. *[1 mark]*

b) An alkane containing 12 carbon atoms is cracked to produce hexane, butene and ethene. Write a balanced equation for this reaction. *[1 mark]*

Crude oil — not the kind of oil you could take home to meet your mother...

This isn't the most exciting topic in the history of the known universe. Although in a galaxy far, far away there may be lots of pages on more boring topics. But, that's neither here nor there, because you've got to learn the stuff anyway. Get fractional distillation and cracking straight in your brain and make sure you know why people bother to do it.

Alkanes as Fuels

Alkanes are absolutely fantastic as fuels. Except for the fact that they produce loads of nasty pollutant gases.

Alkanes** are Useful **Fuels

1) If you burn (**oxidise**) alkanes (and other hydrocarbons) with **plenty of oxygen**, you get **carbon dioxide** and water — it's a **combustion reaction**.

For example, here's the equation for the combustion of propane: $C_3H_{8(g)} + 5O_{2(g)} \rightarrow 3CO_{2(g)} + 4H_2O_{(g)}$

2) This is **complete combustion** — the only products are **water** and **carbon dioxide**.
(There's also incomplete combustion, which is really bad — see below.)
3) Alkanes make great **fuels** — burning just a small amount releases a humongous amount **of energy**.
4) They're burnt in power stations, central heating systems and, of course, to power car engines.
5) They do have a downside though — burning alkanes also produces lots of **pollutants**.
Handily, that's what the rest of this topic is all about...

Incomplete Combustion** Happens When There's Not Enough **Oxygen

If there's not enough oxygen around, hydrocarbons combust **incompletely**.
This changes the **products** of the reaction and can lead to some nasty side effects.

Carbon Monoxide

- When a hydrocarbon undergoes incomplete combustion, you can get **carbon monoxide** gas instead of, or as well as, carbon dioxide. For example:

$CH_{4(g)} + 1\frac{1}{2}O_{2(g)} \rightarrow CO_{(g)} + 2H_2O_{(g)}$ $\quad$ $C_8H_{18(g)} + 10\frac{1}{2}O_{2(g)} \rightarrow 4CO_{2(g)} + 4CO_{(g)} + 9H_2O_{(g)}$

- This is bad news because carbon monoxide gas is **poisonous**.
Carbon monoxide molecules bind to the same sites on **haemoglobin molecules** in red blood cells as oxygen molecules. So **oxygen** can't be carried around the body.
- Luckily, carbon monoxide can be removed from exhaust gases by **catalytic converters** on cars.

Carbon

- **Carbon particles (soot)** can also be formed by incomplete combustion. For example:

$CH_4 + O_2 \rightarrow C + 2H_2O$

- Soot is thought to cause **breathing problems**.
It can also build up in **engines**, meaning they don't work properly.

*Burning Fossil Fuels Contributes to **Global Warming***

1) Burning fossil fuels produces **carbon dioxide**. Carbon dioxide is a **greenhouse gas**.
2) The greenhouse gases in our atmosphere are really good at absorbing **infrared energy** (**heat**). They emit some of the energy they absorb back towards the Earth, keeping it warm. This is called the **greenhouse effect**.
3) Most scientists agree that by **increasing** the amount of carbon dioxide in our atmosphere, we are making the Earth **warmer**.
4) This process is known as **global warming**.

Marjorie loved her new green-house effect.

Alkanes as Fuels

Unburnt Hydrocarbons and *Oxides of Nitrogen* Can Cause *Smog*

1) Engines **don't burn** all of the fuel molecules. Some of them will come out as **unburnt hydrocarbons**.
2) **Oxides of nitrogen** (**NO_x**) are produced when the high **pressure** and **temperature** in a car engine cause the nitrogen and oxygen atoms from the air to react together.
3) Hydrocarbons and nitrogen oxides react in the presence of sunlight to form **ground-level ozone** (O_3), which is a major component of **smog**. **Ground-level ozone** irritates people's eyes, aggravates respiratory problems and even causes lung damage (ozone isn't nice stuff, unless it is high up in the atmosphere as part of the ozone layer).
4) **Catalytic converters** on cars remove unburnt hydrocarbons and oxides of nitrogen from the exhaust.

As if That's Not Bad Enough...Acid Rain is Caused by Sulfur Dioxide

1) Some fossil fuels contain **sulfur**. When they are burnt, the sulfur reacts to form **sulfur dioxide gas** (SO_2).
2) Sulfur dioxide is a bit of a nasty beast. If it gets into the atmosphere, it dissolves in the moisture and is converted into **sulfuric acid**. This is what causes **acid rain**.

The same process occurs when nitrogen oxides escape into the atmosphere — nitric acid is produced.

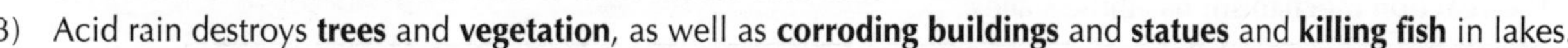

3) Acid rain destroys **trees** and **vegetation**, as well as **corroding buildings** and **statues** and **killing fish** in lakes.

Fortunately, sulfur dioxide can be **removed** from power station flue gases before it gets into the atmosphere. Powdered **calcium carbonate** (limestone) or **calcium oxide** is mixed with water to make an **alkaline slurry**. When the flue gases mix with the alkaline slurry, the **acidic sulfur dioxide** gas reacts with the calcium compounds to form a harmless salt (**calcium sulfate**) — see page 99 for more on this.

Practice Questions

Q1 Which two compounds are produced when an alkane burns completely?

Q2 Under what conditions does incomplete combustion of a fuel take place?

Q3 Name four pollutants that can be produced when a fuel is burnt.
For each pollutant you have named, give one environmental problem that it causes.

Q4 Describe how burning fossil fuels causes acid rain.

Exam Questions

Q1 Heptane, C_7H_{16}, is an alkane present in some fuels.

a) Write a balanced equation for the complete combustion of heptane. [1 mark]

b) Fuels often contain compounds called oxygenates which are added to ensure that the fuel burns completely.

i) What poisonous compound is produced by incomplete combustion of alkanes like heptane? [1 mark]

ii) Apart from adding oxygenates, how else can this compound be removed from exhaust gases? [1 mark]

Q2 Burning fossil fuels produces a variety of gaseous pollutants, such as oxides of nitrogen and sulfur dioxide.

a) Explain how oxides of nitrogen are produced in car engines. [2 marks]

b) Explain how sulfur dioxide can be removed from power station flue gases using calcium carbonate. [2 marks]

Burn, baby, burn — so long as the combustion is complete...

Don't you just hate it when you come up with a great idea, then everyone picks holes in it? Just imagine you were the one who thought of burning alkanes for fuel... it seemed like such a good idea at the time. Despite all the problems, we still use them — and until we find some suitable alternatives, we all have to deal with the negative consequences.

Chloroalkanes and CFCs

For a while, everyone thought CFCs were the bee's knees. That was until they started destroying the ozone layer...

Free Radicals *Have* ***Unpaired*** *Electrons*

1) A **free radical** is a particle with an **unpaired electron**.
2) Free radicals form when a covalent bond splits **equally**, giving **one electron** to each atom.
3) The unpaired electron makes them **very reactive**.
4) You can show something's a free radical in a mechanism by putting a dot next to it, like this 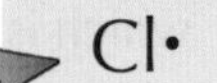 The dot represents the unpaired electron.

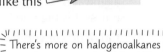

chlorine free radical

Cl•

There's more on halogenoalkanes coming up on pages 140-142.

Halogens React with ***Alkanes*** *to form* ***Halogenoalkanes***

1) Halogens react with alkanes in **photochemical** reactions — reactions that are started by **ultraviolet** light.
2) A hydrogen atom is **substituted** (replaced) by chlorine or bromine. This is a **free-radical substitution reaction**.

Example: *Reacting* ***Chlorine*** *With* ***Methane***

Chlorine and **methane** react with a bit of a bang to form **chloromethane**: $CH_4 + Cl_2 \xrightarrow{UV} CH_3Cl + HCl$

The **reaction mechanism** has three stages.

Stage 1: Initiation reactions — free radicals are produced.

1) Sunlight provides enough energy to break the Cl-Cl bond — this is **photodissociation.**

$$Cl_2 \xrightarrow{UV} 2Cl\bullet$$

2) The bond splits **equally** and each atom gets to keep one electron. The atom becomes a highly reactive **free radical**, Cl•, because of its **unpaired electron.**

Stage 2: Propagation reactions — free radicals are used up and created in a chain reaction.

1) Cl• attacks a **methane** molecule: $Cl\bullet + CH_4 \rightarrow CH_3\bullet + HCl$
2) The new **methyl free radical**, $CH_3\bullet$, can attack another Cl_2 molecule: $CH_3\bullet + Cl_2 \rightarrow CH_3Cl + Cl\bullet$
3) The new Cl• can attack **another** CH_4 molecule, and so on, until all the Cl_2 or CH_4 molecules are used up.

Stage 3: Termination reactions — free radicals are mopped up.

1) If two free radicals join together, they make a **stable molecule.** The two unpaired electrons form a covalent bond.
2) There are **heaps** of possible termination reactions. Here's a couple of them to give you the idea:

$$Cl\bullet + CH_3\bullet \rightarrow CH_3Cl \qquad CH_3\bullet + CH_3\bullet \rightarrow C_2H_6$$

What happens after this depends on whether there's more **chlorine** or **methane** around:

1) If the **chlorine's** in excess, Cl• free radicals will start attacking chloromethane, producing **dichloromethane** CH_2Cl_2, **trichloromethane** $CHCl_3$, and even **tetrachloromethane** CCl_4.
2) **But** if the **methane's** in excess, then the product will mostly be **chloromethane.**

Chlorofluorocarbons *Contain No* ***Hydrogen***

Chlorofluorocarbons (**CFCs**) are halogenoalkane molecules where all of the hydrogen atoms have been replaced by **chlorine** and **fluorine** atoms. 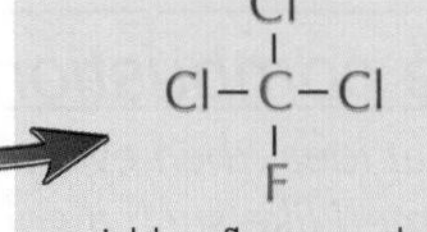

```
    Cl              F
    |               |
Cl—C—Cl          F—C—F
    |               |
    F               Cl
```

trichlorofluoromethane   chlorotrifluoromethane

Chloroalkanes and CFCs

Chlorine Atoms are Destroying The Ozone Layer

1) Ozone (O_3) in the upper atmosphere acts as a **chemical sunscreen**. It absorbs a lot of **ultraviolet radiation** from the Sun, stopping it from reaching us. Ultraviolet radiation can cause **sunburn** or even **skin cancer**.
2) Ozone's **formed naturally** when an oxygen molecule (O_2) is **broken down** into two **free radicals** by **ultraviolet radiation**. The free radicals **attack** other oxygen molecules forming **ozone**. Just like this:

$$O_2 \xrightarrow{UV} O\cdot + O\cdot \quad \text{then} \quad O_2 + O\cdot \rightarrow O_3$$

You've probably heard that **CFCs** are creating **holes** in the **ozone layer**. Well, here's what's happening.

1) **Chlorine free radicals** ($Cl\cdot$) are formed in the upper atmosphere when C–Cl bonds in **CFCs** are broken down by **ultraviolet radiation.** E.g. $CCl_3F \xrightarrow{UV} CCl_2F\cdot + Cl\cdot$
2) These free radicals are **catalysts.** They react with **ozone** to form an **intermediate** ($ClO\cdot$) and an oxygen molecule.

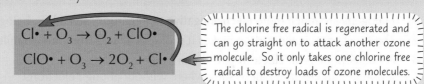

$$Cl\cdot + O_3 \rightarrow O_2 + ClO\cdot$$
$$ClO\cdot + O_3 \rightarrow 2O_2 + Cl\cdot$$

The chlorine free radical is regenerated and can go straight on to attack another ozone molecule. So it only takes one chlorine free radical to destroy loads of ozone molecules.

3) So the **overall reaction** is... $2O_3 \rightarrow 3O_2$ ... and $Cl\cdot$ is the **catalyst.**

CFCs Are Now Banned

1) CFCs are pretty **unreactive**, **non-flammable** and **non-toxic**. They used to be used as coolant gas in fridges, as solvents, and as propellants in aerosols.
2) In the 1970s research by several different scientific groups demonstrated that CFCs were causing **damage** to the **ozone layer**. The **advantages** of CFCs couldn't outweigh the **environmental problems** they were causing, so they were **banned**.
3) Chemists have developed safer **alternatives** to CFCs which contain no **chlorine** such as **HFCs (hydrofluorocarbons)** and **hydrocarbons.**

Practice Questions

Q1 What is a free radical?

Q2 Write an overall equation for the reaction of methane with chlorine in the presence of UV light.

Q3 Explain why having ozone in the upper atmosphere is beneficial to humans.

Q4 Name two past uses for CFCs. Name one type of compound that is used as an alternative to CFCs today.

Exam Questions

Q1 In the upper atmosphere, chlorine free radicals are produced by the action of ultraviolet light on CFC molecules. Chlorine free radicals can catalyse the breakdown of ozone (O_3), damaging the ozone layer.
Write two equations to show how a chlorine free radical catalyses the breakdown of an ozone molecule. [2 marks]

Q2 The alkane ethane (C_2H_6) will react with chlorine in a photochemical reaction.

a) Give the name for this type of reaction. [1 mark]

b) Write equations to show the mechanism for the reaction of ethane (C_2H_6) with chlorine.
Your mechanism should include an initiation reaction, propagation reactions and an example of one possible termination reaction that produces an organic compound. [4 marks]

This stuff is like...totally radical, man...

Mechanisms are an absolute pain in the neck to learn, but unfortunately reactions are what Chemistry's all about. If you don't like it, you should have taken art — no mechanisms in that, just pretty pictures. Ah well, there's no going back now. You've just got to sit down and learn the stuff. Keep hacking away at it, till you know it all off by heart.

Halogenoalkanes

Don't worry if you see halogenoalkanes called haloalkanes. It's a government conspiracy to confuse you.

Halogenoalkanes are Alkanes with Halogen Atoms

A **halogenoalkane** is an alkane with at least one **halogen atom** in place of a hydrogen atom.

E.g.

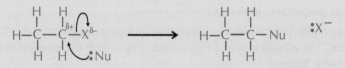

trichloromethane — 2-iodopropane — 2-bromo-2-chloro-1, 1, 1-trifluoroethane

There's more about how to name organic compounds, including halogenoalkanes, on pages 128-129.

The Carbon–Halogen Bond in Halogenoalkanes is Polar

1) Halogens are much more **electronegative** than carbon, so carbon-halogen bonds are **polar**.
2) The δ+ charge on the carbon makes it prone to attacks from **nucleophiles**.
3) A nucleophile is an **electron-pair donor**.
   It donates an electron pair to somewhere without enough electrons.
4) **OH⁻**, **CN⁻** and **NH_3** are all **nucleophiles** that can react with halogenoalkanes.

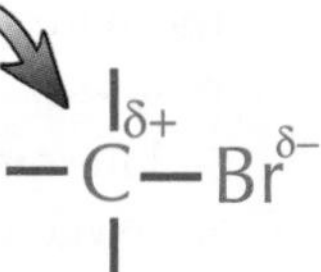

Halogenoalkanes Can Undergo Nucleophilic Substitution Reactions

1) A nucleophile can react with a polar molecule by kicking out the functional group and taking its place.
2) This is called a **nucleophilic substitution reaction**. It works like this:

Curly arrows show the movement of electron pairs. If you can't remember how they work, look back at page 130.

The lone pair of electrons on the nucleophile attacks the δ+ carbon. The C–X bond breaks.

The halogen leaves, taking both electrons with it. A new bond forms between the carbon and the nucleophile.

X stands for one of the halogens (F, Cl, Br or I). **Nu** stands for a nucleophile.

3) The **product** of these reactions depends on what the **nucleophile** is...

Halogenoalkanes React with Hydroxides to Form Alcohols

For example, bromoethane can react to form ethanol. You have to use **warm aqueous sodium** or **potassium hydroxide** — it's a **nucleophilic substitution reaction**.

Here's how it happens:

The δ+ carbon attracts a lone pair of electrons from the OH⁻ ion. The C–Br bond breaks.

The bromine leaves, taking both electrons to become Br⁻. A new bond forms between the carbon and the OH⁻ ion, making an alcohol.

This is sometimes called hydrolysis, because exactly the same reaction will happen with water. (Hydrolysis means splitting a molecule apart by reacting it with water.)

Nitriles Are Formed by Reacting Halogenoalkanes with Cyanide

Nitriles have –C≡N groups.

If you **warm** a halogenoalkane with **ethanolic potassium cyanide** (potassium cyanide dissolved in ethanol) you get a **nitrile**. It's yet another **nucleophilic substitution reaction** — the **cyanide ion**, CN⁻, is the **nucleophile**.

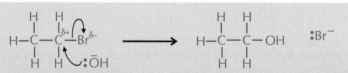

reflux, ethanol

This follows the same pattern — the lone pair of electrons on the CN⁻ ion attacks the δ+ carbon, the C–Br bond breaks and the bromine leaves.

Halogenoalkanes

Reacting Halogenoalkanes with **Ammonia** Forms **Amines**

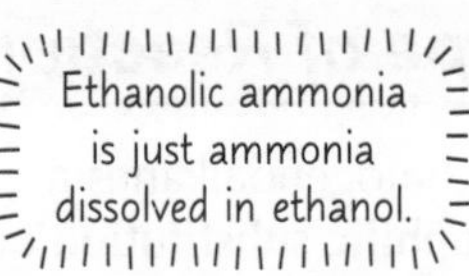

Ethanolic ammonia
is just ammonia
dissolved in ethanol.

1) If you **warm** a halogenoalkane with excess **ethanolic** ammonia, the **ammonia** swaps places with the **halogen** — yes, it's another one of these **nucleophilic substitution reactions**.

The first step is the same as the mechanism on the previous page, except this time the nucleophile is NH_3.

In the second step, an ammonia molecule removes a hydrogen from the NH_3 group, forming an amine and an ammonium ion (NH_4^+).

2) The ammonium ion can react with the bromine ion to form ammonium bromide. So the overall reaction looks like this:

$$CH_3CH_2Br + 2NH_3 \xrightarrow{\text{ethanol}} CH_3CH_2NH_2 + NH_4Br$$

3) The **amine group** in the product still has a lone pair of electrons. This means that it can also act as a **nucleophile** — so it may react with halogenoalkane molecules itself, giving a mixture of products.

Iodoalkanes React **Fastest**, Fluoroalkanes React **Slowest**

1) The **carbon-halogen bond strength** (or enthalpy) decides **reactivity**. For any reaction to occur the carbon-halogen bond needs to **break**.
2) The **C–F bond** is the **strongest** — it has the highest **bond enthalpy**. So **fluoroalkanes** undergo nucleophilic substitution reactions **more slowly** than other halogenoalkanes.
3) The **C–I bond** has the **lowest bond enthalpy**, so it's easier to break. This means that **iodoalkanes** are substituted more **quickly**.

bond	bond enthalpy kJ mol^{-1}
C–F	467
C–Cl	346
C–Br	290
C–I	228

Faster substitution as bond enthalpy decreases (the bonds are getting weaker).

Halogenoalkanes also Undergo **Elimination Reactions**

The conditions are anhydrous (there's no water).

1) If you warm a halogenoalkane with hydroxide ions dissolved in **ethanol** instead of water, an **elimination reaction** happens, and you end up with an **alkene**.
2) You have to heat the mixture **under reflux** or you'll lose volatile stuff.
3) Here's how the reaction works:

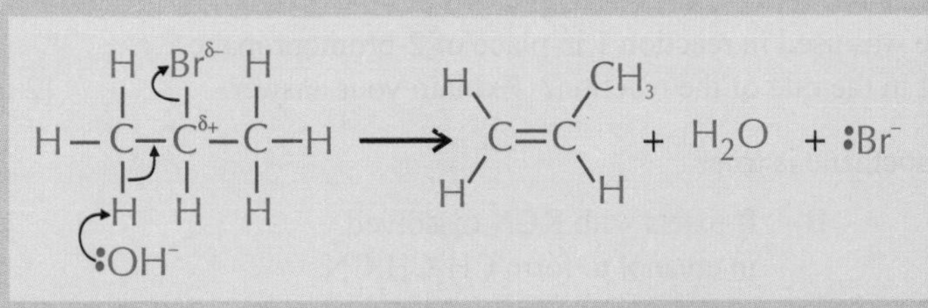

1) OH^- acts as a base and takes a proton, H^+, from the carbon on the left. This makes water.
2) The left carbon now has a spare electron, so it forms a double bond with the middle carbon.
3) To form the double bond, the middle carbon has to let go of the Br, which drops off as a Br^- ion.

4) If you use 2-bromopropane and potassium hydroxide, the equation for the reaction will look like this:

$$CH_3CHBrCH_3 + KOH \xrightarrow[\text{reflux}]{\text{ethanol}} CH_2CHCH_3 + H_2O + KBr$$

5) This is an example of an **elimination reaction**. In an elimination reaction, a **small group** of atoms breaks away from a molecule. This group is **not replaced** by anything else (whereas it would be in a substitution reaction). For example, in the reaction above, H and Br have been eliminated from $CH_3CHBrCH_3$ to leave CH_2CHCH_3.

Halogenoalkanes

The Type of Reaction Depends on the Conditions

1) When halogenoalkanes are reacted with **hydroxides**, they may undergo either **nucleophilic substitution** or **elimination**. The two reactions are said to be **competing**.
2) You can influence which reaction will happen the most by **changing the conditions**.

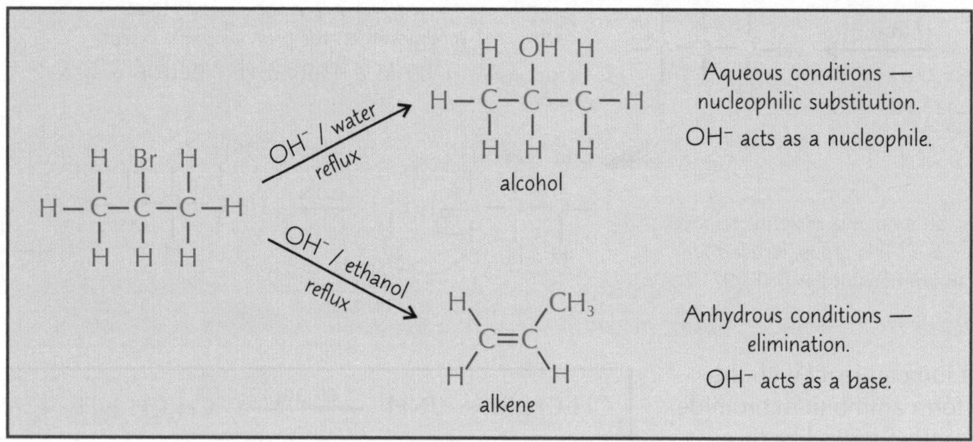

Bob's reaction to the sight of a golf course was not affected by the conditions.

3) If you use a **mixture** of water and ethanol as the solvent, **both** reactions will happen and you'll get a mixture of the two products.

Practice Questions

Q1 Why are carbon-halogen bonds polar?

Q2 What is a nucleophile?

Q3 Name the type of organic compound formed by the reaction of a halogenoalkane with:
a) aqueous potassium hydroxide b) potassium cyanide c) ethanolic ammonia

Q4 Why does iodoethane react faster than chloroethane with nucleophiles?

Q5 Draw the mechanism for the reaction of bromoethane with potassium hydroxide dissolved in ethanol.

Exam Questions

Q1 Three reactions of 2-bromopropane, $CH_3CHBrCH_3$, are shown below.

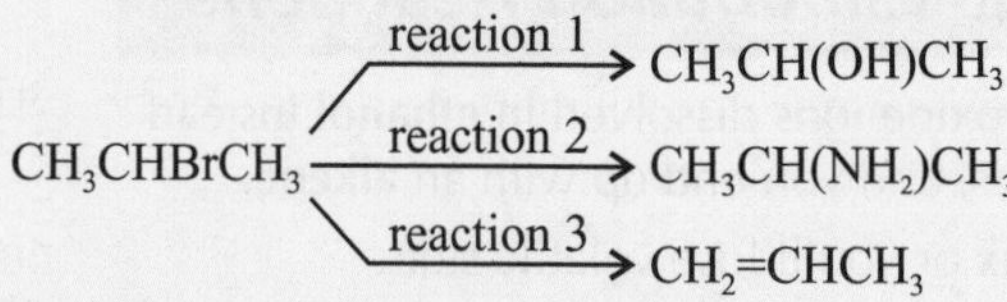

a) For each reaction, name the reagent and the solvent used. [6 marks]

b) Under the same conditions, 2-iodopropane was used in reaction 1 in place of 2-bromopropane. What difference (if any) would you expect in the rate of the reaction? Explain your answer. [2 marks]

Q2 Which of the following statements about bromoethane is true?

A The carbon attached to the bromine atom has a partial negative charge.

B It reacts with KCN dissolved in ethanol to form CH_3CH_2CN.

C It reacts with KOH in aqueous conditions to form $CH_2{=}CH_2$.

D It reacts with nucleophiles at a slower rate than chloroethane. [1 mark]

If you don't learn this — you will be eliminated. Resistance is nitrile...

Polar bonds get into so many bits of chemistry. If you still think they're something to do with bears or mints, flick back to page 30 and have a read. Make sure you learn all these reactions and mechanisms, as well as which halogenoalkanes react fastest. This stuff's always coming up in exams, so make sure it's filed away in your brain for when you need it.

Alkenes

I'll warn you now — some of this stuff gets a bit heavy — but stick with it, as it's pretty important.

Alkenes are Unsaturated Hydrocarbons

Alkenes with more than one double bond have fewer Hs than the general formula suggests.

1) Alkenes have the **general formula C_nH_{2n}**. They're just made of carbon and hydrogen atoms, so they're **hydrocarbons**.
2) Alkene molecules **all** have at least one **C=C double covalent bond**. Molecules with C=C double bonds are **unsaturated** because they can make more bonds with extra atoms in **addition** reactions.
3) Because there are two pairs of electrons in the C=C double bond, it has a really **high electron density**. This makes alkenes pretty reactive. Here are a few pretty diagrams of **alkenes**:

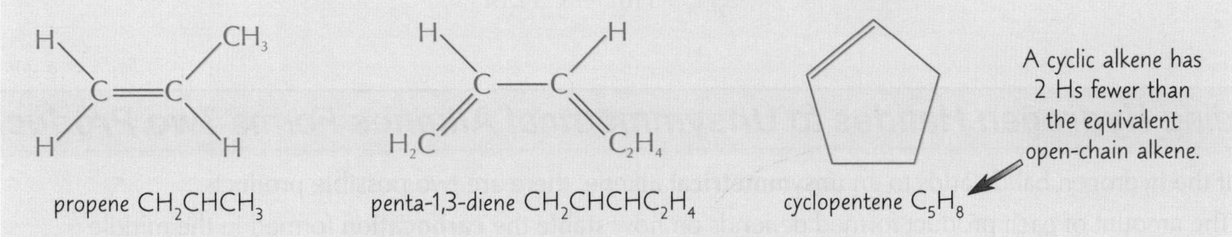

propene CH_2CHCH_3
penta-1,3-diene $CH_2CHCHC_2H_4$
cyclopentene C_5H_8
A cyclic alkene has 2 Hs fewer than the equivalent open-chain alkene.

Electrophilic Addition Reactions Happen to Alkenes

Electrophilic addition reactions aren't too complicated...

1) The **double bonds** open up and atoms are **added** to the carbon atoms.
2) Electrophilic addition reactions happen because the double bond has got plenty of **electrons** and is easily attacked by **electrophiles**.

Electrophiles are **electron-pair acceptors** — they're usually a bit short of electrons, so they're **attracted** to areas where there are lots of them about.
Here are a few examples:

- **Positively charged ions**, like H^+ and NO_2^+.
- **Polar molecules** — the δ+ atom is attracted to places with lots of electrons.

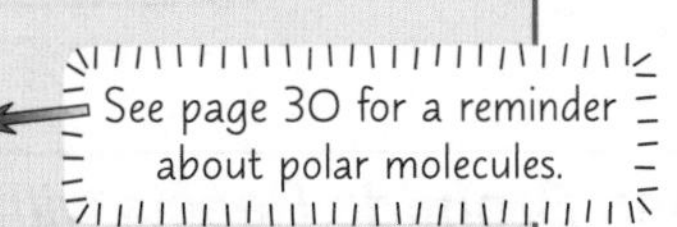

See page 30 for a reminder about polar molecules.

Eric Trofill — double bond hunter. Between you and me, he's a few electrons short of an atom.

Use Bromine Water to Test for Unsaturation

When you shake an alkene with **orange bromine water**, the solution quickly **decolourises** (goes from **orange** to **colourless**). Bromine is added across the C=C double bond to form a colourless **dibromoalkane** — this happens by **electrophilic addition**. Here's the mechanism...

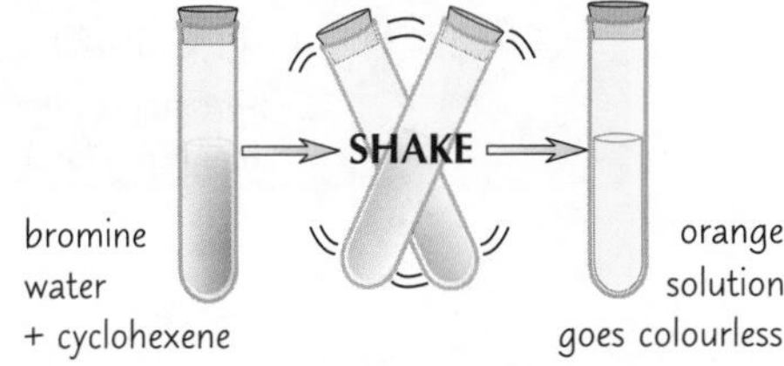

bromine water + cyclohexene
orange solution goes colourless

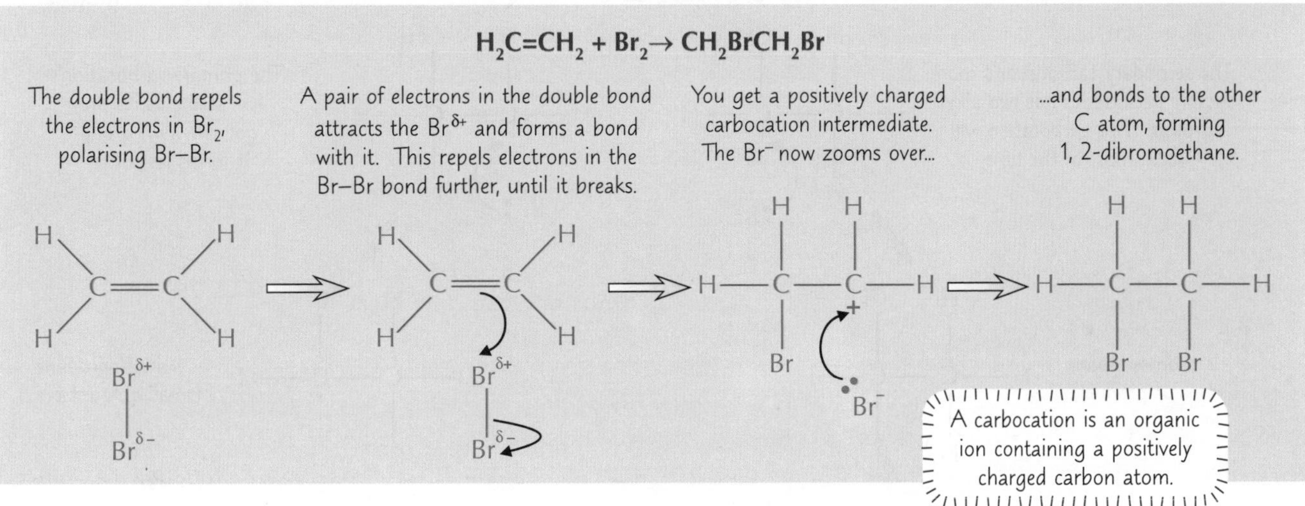

$H_2C=CH_2 + Br_2 \rightarrow CH_2BrCH_2Br$
The double bond repels the electrons in Br_2, polarising Br–Br.
A pair of electrons in the double bond attracts the $Br^{\delta+}$ and forms a bond with it. This repels electrons in the Br–Br bond further, until it breaks.
You get a positively charged carbocation intermediate. The Br^- now zooms over...
...and bonds to the other C atom, forming 1, 2-dibromoethane.
A carbocation is an organic ion containing a positively charged carbon atom.

Alkenes

Alkenes also Undergo **Addition** with **Hydrogen Halides**

Alkenes also undergo **electrophilic addition** reactions with hydrogen halides — to form **halogenoalkanes**. This is the reaction between **ethene** and **hydrogen bromide**, to form **bromoethane**.

$$H_2C{=}CH_2 + HBr \longrightarrow H-CH_2-CH_2Br$$

$$C_2H_4 + HBr \rightarrow C_2H_5Br$$

Other alkenes react in a similar way.

Adding **Hydrogen Halides** to **Unsymmetrical Alkenes** Forms **Two Products**

1) If the hydrogen halide adds to an **unsymmetrical** alkene, there are two possible products.
2) The amount of each product formed depends on how **stable** the **carbocation** formed in the middle of the reaction is — this is known as the **carbocation intermediate**.
3) Carbocations with more **alkyl groups** are more stable because the alkyl groups feed **electrons** towards the positive charge. The **more stable carbocation** is much more likely to form.

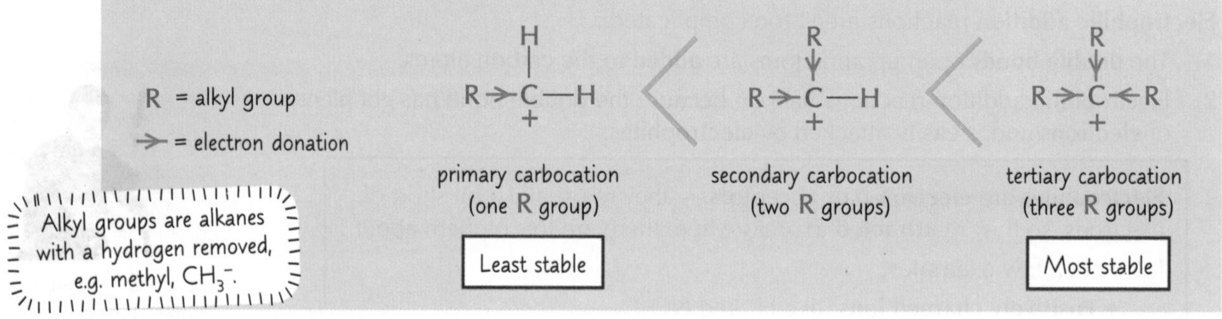

Alkyl groups are alkanes with a hydrogen removed, e.g. methyl, CH_3–.

Here's how **Hydrogen Bromide** Reacts with **Propene**

$H_2C{=}CHCH_3 \rightarrow CH_3CHBrCH_3$
2-bromopropane (major product)

$H_2C{=}CHCH_3 \rightarrow CH_2BrCH_2CH_3$
1-bromopropane (minor product)

$H^{\delta+}$ — $Br^{\delta-}$

The secondary carbocation's more stable because it's got two alkyl groups. This carbocation will form most of the time.

The primary carbocation's less stable as it's only got one alkyl group. It forms less often.

$:Br^-$

2-bromopropane (major product)

1-bromopropane (small amount only)

Alkenes

Alkenes also Undergo **Electrophilic Addition Reactions** with H_2SO_4

Alkenes will react with **cold concentrated sulfuric acid** to form **alkyl hydrogen sulfates.** You can then convert the alkyl hydrogen sulfates formed into **alcohols** by adding water and warming the reaction mixture. For example:

1) Cold concentrated **sulfuric acid** reacts with an alkene in an **electrophilic addition** reaction. → $H_2C{=}CH_2 + H_2SO_4 \rightarrow CH_3CH_2OSO_2OH$ (ethene + sulfuric acid → ethyl hydrogen sulfate)
2) If you then add cold **water** and warm the product, it **hydrolyses** to form an alcohol. → $CH_3CH_2OSO_2OH + H_2O \rightarrow CH_3CH_2OH + H_2SO_4$ (ethyl hydrogen sulfate → ethanol)
3) The **sulfuric acid** isn't used up — it acts as a **catalyst**.

$+ H_2O$, warm

This step is a hydrolysis reaction (see p.140).

If you're using this reaction to produce ethanol, then the equation for the overall reaction looks like this:

$$H_2C = CH_2 + H_2O \xrightarrow{H_2SO_4} C_2H_5OH$$

Just as with hydrogen halides on the previous page, if you do this reaction with an **unsymmetrical alkene**, you get a **mixture of products**. The one that's formed via the **most stable** carbocation intermediate will be the **major product**.

Practice Questions

Q1 What is the general formula for an alkene?

Q2 Why do alkenes react with electrophiles?

Q3 What is an electrophile?

Q4 What is a carbocation?

Exam Questions

Q1 But-1-ene is an alkene. Alkenes contain at least one C=C double bond.

a) Describe how bromine water can be used to test for C=C double bonds. [2 marks]

b) Name the reaction mechanism involved in the above test. [1 mark]

c) Hydrogen bromide will react with but-1-ene by this mechanism, producing two isomeric products.

i) Draw a mechanism for the reaction of HBr with $CH_2{=}CHCH_2CH_3$, showing the formation of the major product only. Name the product. [4 marks]

ii) Explain why it is the major product for this reaction. [2 marks]

Q2 Cold concentrated sulfuric acid is mixed with propene. Cold water is then added to the mixture, and the mixture is warmed. What are the end products of this procedure?

A $C_3H_7OH + H_2SO_4$ B $C_3H_7SO_4H + H_2O$

C $C_3H_8 + H_2SO_4 + H_2O$ D $C_3H_8OH + H_2SO_4$ [1 mark]

Got an unsymmetrical alkene? I'd get that looked at...

Wow... these pages really are packed. There's not one, not two, but three mechanisms to learn. They mightn't be as handy in real life as a tin opener, but you won't need a tin opener in the exam. Get the book shut and scribble them out. Make sure your arrows start at the electron pair and finish exactly where the electrons are going.

Addition Polymers

Polymers are long, stringy molecules made by joining lots of little molecules together. They're made up of one unit repeated over and over and over and over and over and over and over and over again. Get the idea? Let's get started.

Polymers are Formed from Monomers

1) **Polymers** are long chain molecules formed when lots of small molecules, called **monomers**, join together.
2) Polymers can be **natural**, like DNA and proteins, or **synthetic** (man-made), like poly(ethene).
3) People have been using **natural polymers**, like silk, cotton and rubber, for hundreds of years.
4) In the **19th century**, research concentrated on modifying the properties of natural polymers — for example **hardening rubber**, to make it more suitable for machine parts and car tyres.
5) The **20th century** saw the invention and production of **synthetic polymers**, like **nylon** and **Kevlar®**.
6) Scientists are still developing **new polymers**, with new properties, all the time.

Addition Polymers are Formed from Alkenes

Alkenes can act as monomers and form polymers because their double bonds can open up and join together to make long chains. These type of polymers are called **addition polymers**.

Addition polymers made from alkenes are called polyalkenes.

Poly(phenylethene) is formed from **phenylethene**.

the double bond opens up

'n' means there are lots of these units

phenylethene monomer — poly(phenylethene) polymer

This is what a section of the chain would look like:

section of poly(phenylethene) polymer

Polyalkene chains are **saturated** molecules (they only contain single bonds in the carbon chain). The main carbon chain of a polyalkene is also **non-polar**. These factors result in addition polymers being very **unreactive**.

The Properties of Polyalkenes Depend on Their Intermolecular Forces

1) Polyalkene chains are usually **non-polar** — so the chains are only held together by **Van der Waals forces** (see page 31).
2) The **longer** the polymer chains are and the **closer** together they can get, the **stronger** the Van der Waals forces between the chains will be.
3) This means that polyalkenes made up of **long**, **straight chains** tend to be **strong** and **rigid**, while polyalkenes that are made up of **short**, **branched chains** tend to be **weaker** and more **flexible**.

Polymers that contain electronegative atoms (like Cl) can be polar. These polymers will have permanent dipole-dipole forces.

You Can Modify The Properties of Polymers Using Plasticisers

flexibility

Adding a **plasticiser** to a polymer makes it more flexible. The plasticiser molecules get **between** the polymer chains and push them apart. This **reduces** the strength of the **intermolecular forces** between the chains — so they can slide around more, making the polymer easy to **bend**.

Poly(chloroethene), PVC is formed from **chloroethene**.

chloroethene monomer — poly(chloroethene) polymer

1) PVC has long, closely packed polymer chains, making it hard but brittle at room temperature. **Rigid PVC** is used to make **drainpipes** and **window frames**.
2) **Plasticised PVC** is much more flexible than rigid PVC. It's used to make **electrical cable insulation, flooring tiles** and **clothing**.

Pretty Polymer.

Addition Polymers

You can Draw the **Repeating Unit** from a **Monomer** or **Polymer**

Polymers are made up of **repeating units** (a bit of molecule that repeats over and over again). The repeating units of addition polymers look very similar to the monomer, but with the double bond opened out.

1) To draw the **repeating unit** of an addition polymer from its **monomer**, first draw the two alkene carbons. Replace the double bond with a single bond. Add a single bond coming out from each of the carbons.
2) Then just fill in the rest of the groups in the same way they surrounded the double bond.

Example: Draw the repeating unit of the polymer formed from ethene, C_2H_4.

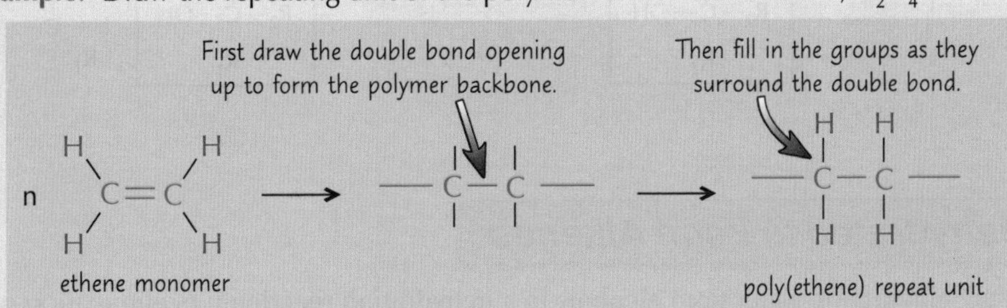

First draw the double bond opening up to form the polymer backbone.
Then fill in the groups as they surround the double bond.
ethene monomer
poly(ethene) repeat unit

To write the name of an addition polymer, you write the name of the alkene monomer, put brackets around it and stick 'poly' in front. For example, the polymer made from ethene is poly(ethene).

3) To draw the **repeating unit** from its **polymer**, you just need to look at the chain and work out which part is repeating. For an addition polymer, the repeating unit should be **two** carbons long (so it looks like the alkene monomer). Once you have the repeating unit, you can easily draw the **monomer** by adding a double bond.

Example: Draw the repeating unit and monomer of the polymer chain Teflon® (poly(tetrafluoroethene)).

section of Teflon® polymer → Teflon® repeat unit → Teflon® monomer

The polymer is only made up of carbon and fluorine atoms, so the repeating unit is just two carbons, surrounded by fluorines.

Practice Questions

Q1 What is a polymer?

Q2 What type of polymers do alkenes form?

Q3 Explain why polyalkenes with long, straight chains are stronger than those with short, branched chains.

Q4 Give two typical uses of rigid PVC.

Exam Questions

Q1 The diagram on the right shows a section of the polymer poly(propene).

a) Draw the structure of the monomer that forms this polymer. [1 mark]

b) Draw the structure of the repeating unit of poly(propene). [1 mark]

Q2 The polymer poly(chloroethene) (PVC) is made from chloroethene, shown right.

a) Draw the structure of the repeating unit of poly(chloroethene). [1 mark]

b) Explain how the properties of PVC change when you add a plasticiser. [3 marks]

c) Give two typical uses of plasticised PVC. [2 marks]

Never miss your friends again — form a polymer...

These polymers are really useful — drainpipes, clothing, electrical cable insulation... you couldn't do without them. And someone had to invent them. Just think, you could be the next mad inventor, working for the biggest secret agency in the world. And you'd have a really fast car, which obviously would turn into a yacht with the press of a button...

Alcohols

Alcohol — evil stuff, it is. I could start preaching, but I won't, because this page is enough to put you off alcohol for life...

Alcohols are **Primary**, **Secondary** or **Tertiary**

1) The **functional group** in the **alcohol** homologous series is the **hydroxyl** group, **-OH**.
2) An alcohol is **primary**, **secondary** or **tertiary**, depending on which carbon atom the **-OH** group is bonded to...

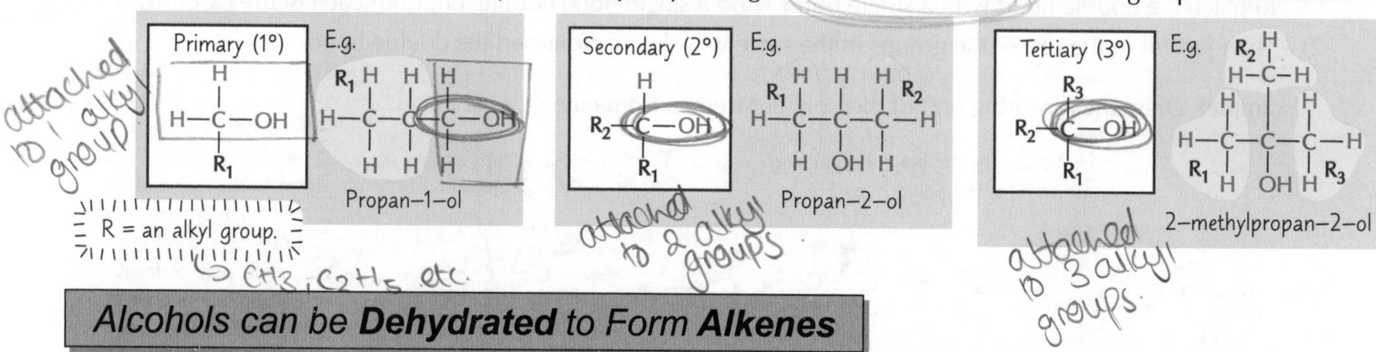

Alcohols can be **Dehydrated** to Form **Alkenes**

1) You can make alkenes by **eliminating** water from **alcohols** in a **dehydration reaction** (i.e. elimination of **water**).

$$C_nH_{2n+1}OH \rightarrow C_nH_{2n} + H_2O$$

In an elimination reaction, a small group of atoms breaks away from a larger molecule, and isn't replaced by anything else.

2) This reaction allows you to produce alkenes from **renewable** resources.
3) This is important, because it means that you can produce **polymers** (poly(ethene), for example) **without** needing **oil**.
4) One of the main industrial uses for alkenes is as a **starting material** for **polymers**.
   Here's how you dehydrate **ethanol** to form **ethene**:

Because you can produce ethanol by fermentation of glucose, which you can get from plants.

Ethanol is heated with a **concentrated sulfuric acid** catalyst: $C_2H_5OH \xrightarrow{H_2SO_4} C_2H_4 + H_2O$

Phosphoric acid (H_3PO_4) can also be used as a catalyst for this reaction. Often, the two are used together.

The product is usually in a **mixture** with water, acid and reactant in it so the alkene has to be **separated** out.

And here's the mechanism for the elimination of water from ethanol...

$$CH_3CH_2OH + H^+ \rightarrow CH_3CH_2OH_2^+ \rightarrow CH_3CH_2^+ + H_2O \rightarrow CH_2{=}CH_2 + H_2O + H^+$$

A **lone pair of electrons** from the oxygen bonds to an **H⁺** from the acid. The alcohol is **protonated**, giving the oxygen a **positive charge**.

The positively charged oxygen **pulls** electrons away from the carbon. An **H_2O molecule** leaves, creating an **unstable carbocation intermediate**.

The carbocation **loses an H⁺**...

...and the **alkene** is formed.

Dehydration of more complicated alcohols can have **more than one possible product** — the double bond can form either side of the carbon with the –OH group attached. For example:

$$CH_3CH(OH)CH_2CH_3 \xrightarrow{+H_2SO_4} CH_2{=}CHCH_2CH_3 \text{ and } CH_3CH{=}CHCH_3$$

Alcohols

Distillation is Used to Separate Chemicals

1) The products of organic reactions are often **impure** — so you've got to know how to **purify** them.
2) In the dehydration of alcohols to form alkenes, the **mixture** at the end contains the **product**, the **reactant**, **acid** and **water**. To get a **pure alkene**, you need a way to **separate** it from the other substances.
3) Distillation is a technique which uses the fact that different chemicals have **different boiling points** to separate them. Often further **separation and purification** is needed though.
Here's an example of how to produce an **alkene** from an **alcohol** and separate out the product.

Producing cyclohexene from cyclohexanol

Stage 1 — distillation

1) Add concentrated H_2SO_4 and H_3PO_4 to a **round-bottom flask** containing cyclohexanol. Mix the solution by swirling the flask and add 2-3 **carborundum boiling chips** (these make the mixture boil more calmly).
2) The mixture should be gently heated to around 83 °C (the boiling point of cyclohexene) using a **water bath** or **electric heater.**
3) Chemicals with boiling points up to 83 °C will evaporate. The warm gas will rise out of the flask and into the **condenser**, which has cold water running through the outside, turning it into a liquid.
4) The product can then be collected in a **cooled flask**.

water out
condenser
reaction mixture
impure product
water in
heat
cooled flask

Stage 2 — separation

1) The product collected after distillation will still contain impurities.
2) Transfer the product mixture to a **separating funnel** and add water to **dissolve** water soluble impurities and create an aqueous solution.
3) Allow the mixture to settle into layers. Drain the aqueous lower layer, leaving the impure cyclohexene.

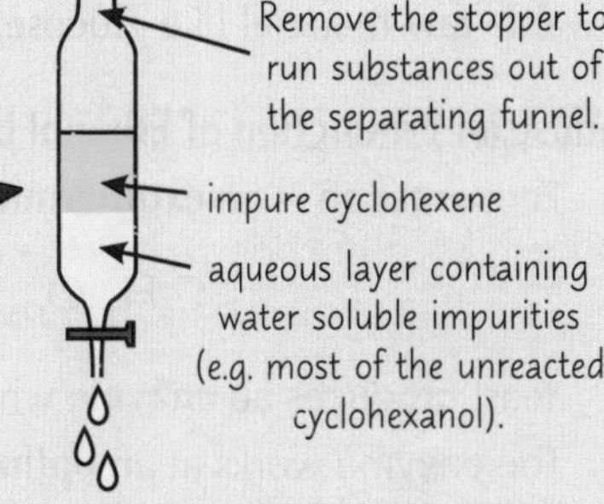
Remove the stopper to run substances out of the separating funnel.
impure cyclohexene
aqueous layer containing water soluble impurities (e.g. most of the unreacted cyclohexanol).

Stage 3 — purification

1) Drain the cyclohexene into a round-bottomed flask.
2) Add anhydrous $CaCl_2$ (a drying agent) and stopper the flask. Let the mixture dry for at least 20 minutes with occasional swirling.
3) The cyclohexene will still have small amounts of impurities so **distil** the mixture one last time.

Practice Questions

Q1 What is the general formula for an alcohol?

Q2 Draw out the mechanism for the elimination of water from ethanol.

Exam Questions

Q1 Butanol C_4H_9OH has four chain and positional isomers.

a) Class each one as a primary, secondary or tertiary alcohol.

```
i)    H  H  H  H          ii)      H                 iii)   H  H  H  H           iv)      OH
      |  |  |  |                   |                        |  |  |  |                     |
   H—C—C—C—C—OH                 H—C—H                   H—C—C—C—C—H                 H—C—H
      |  |  |  |                 H   |   H                  |  |  |  |                 H   |   H
      H  H  H  H                 |   |   |                  H  H  OH H                 |   |   |
                              H—C——C——C—H                                          H—C——C——C—H
                                 |   |   |                                             |   |   |
                                 H   OH  H                                             H   H   H
```

[4 marks]

b) Draw the organic product formed when each of the molecules above reacts with concentrated H_2SO_4. [4 marks]

Q2 Explain how distillation can help separate a liquid product from liquid by-products. [2 marks]

Euuurghh, what a page... I think I need a drink...

Not much to learn here — a few basic definitions, an industrial process and a three-step experimental process... Like I said, not much here at all. Think I'm going to faint. *[THWACK]*

Ethanol Production

Humans have been making ethanol for thousands of years. We've gotten pretty good at it.

*Alcohols are Produced by **Hydration** of **Alkenes***

1) The standard industrial method for producing alcohols is to **hydrate** an **alkene** using **steam** in the presence of an **acid catalyst**.

$$C_nH_{2n} + H_2O \overset{H^+}{\rightleftharpoons} C_nH_{2n+1}OH$$

2) Here's the general mechanism for this type of reaction. (It's the reverse of the dehydration mechanism from page 148.)

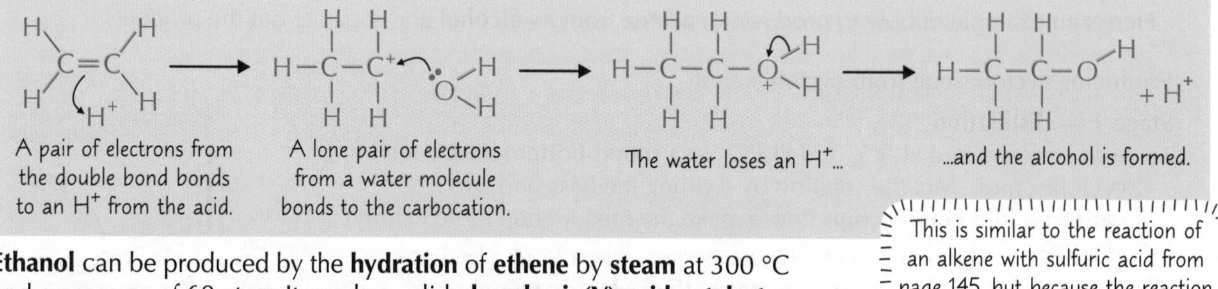

A pair of electrons from the double bond bonds to an H⁺ from the acid.
A lone pair of electrons from a water molecule bonds to the carbocation.
The water loses an H⁺...
...and the alcohol is formed.

3) **Ethanol** can be produced by the **hydration** of **ethene** by **steam** at 300 °C and a pressure of 60 atm. It needs a solid **phosphoric(V) acid catalyst.**

This is similar to the reaction of an alkene with sulfuric acid from page 145, but because the reaction conditions are different the mechanism's slightly different too.

*Ethanol can be Produced Industrially by **Fermentation***

At the moment, **steam hydration of ethene** is the most widely used technique in the industrial production of ethanol. The ethene comes from cracking heavy fractions of crude oil. But in the future, when crude oil supplies start **running out**, petrochemicals like ethene will be expensive — so producing ethanol by **fermentation**, using a renewable raw material like glucose, will become much more important...

Industrial Production of Ethanol by Fermentation of Glucose

1) Fermentation is an **exothermic** process, carried out by **yeast** in **anaerobic conditions** (without oxygen).

$$C_6H_{12}O_{6(aq)} \xrightarrow[\text{yeast}]{30\text{-}40\ ^\circ C} 2C_2H_5OH_{(aq)} + 2CO_{2(g)}$$

2) Yeast produces an **enzyme** which converts sugars, such as glucose, into **ethanol** and **carbon dioxide**.
3) The enzyme works at an **optimum** (ideal) temperature of **30-40 °C**. If it's too cold, the reaction is **slow** — if it's too hot, the enzyme is **denatured** (damaged).

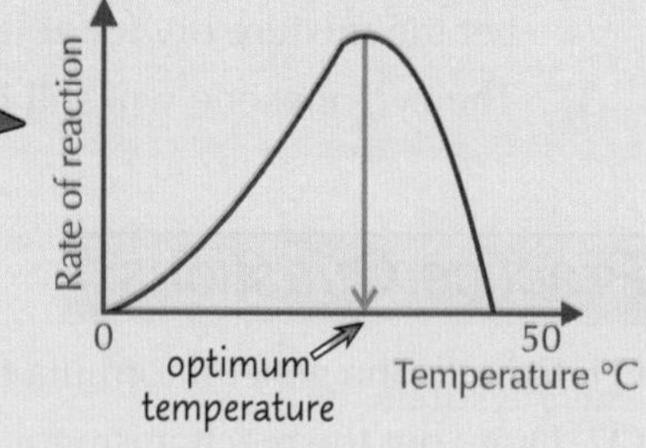

4) Once formed, ethanol is **separated** from the rest of the mixture by **fractional distillation**.
5) Fermentation is **low-tech**. It uses cheap equipment and **renewable resources**. But the fractional distillation step that is needed to **purify** the ethanol produced using this method takes extra time and money.

*Ethanol is a **Biofuel***

Ethanol is increasingly being used as a **fuel**, particularly in countries with few oil reserves. For example, in Brazil, **sugars** from sugar cane are **fermented** to produce alcohol, which is added to petrol. Ethanol made in this way is a **biofuel** (and is sometimes called **bioethanol**).

A biofuel is a fuel that's made from biological material that's recently died.

Biofuels have some advantages over fossil fuels (coal, oil and gas) and some potential drawbacks.

1) Biofuels are **renewable** energy sources. Unlike fossil fuels, biofuels won't run out, so they're more **sustainable**.
2) Biofuels do produce CO_2 when they're burnt, but it's CO_2 that the plants **absorbed** while growing, so **biofuels** are usually still classed as **carbon neutral** (see next page).
3) But one problem with switching from fossil fuels to biofuels in transport is that **petrol car engines** would have to be **modified** to use fuels with high ethanol concentrations.
4) Also, when you use land used to grow crops for fuel, that land can't be used to grow **food**. If countries start using land to grow biofuel crops instead of food, they may be unable to feed everyone in the country.

Ethanol Production

Bioethanol Production is **Almost Carbon Neutral**... But **Not Quite**

1) Just like burning the hydrocarbons from fossil fuels, burning ethanol produces **carbon dioxide** (CO_2). Carbon dioxide is a **greenhouse gas** — it contributes to **global warming** (see pages 136 and 159).
2) But the plants that are grown to produce bioethanol **take in carbon dioxide** from the **atmosphere** when they're growing. As they grow, they take in the same amount of carbon dioxide as burning the bioethanol you produce from them gives out. So it could be argued that burning ethanol as a fuel is **carbon neutral**.

   Here are the **chemical equations** to support that argument...

Plants take in **carbon dioxide** from the atmosphere to produce **glucose** by **photosynthesis**...

$$6CO_2 + 6H_2O \rightarrow C_6H_{12}O_6 + 6O_2$$

6 moles of carbon dioxide are **taken** from the atmosphere to produce **1 mole of glucose.**

In the **fermentation** process, **glucose** is converted into **ethanol**...

$$C_6H_{12}O_6 \rightarrow 2C_2H_5OH + 2CO_2$$

2 moles of carbon dioxide are released into the atmosphere when **1 mole of glucose** is converted to **2 moles of ethanol.**

When **ethanol** is **burned**, **carbon dioxide** and water are produced...

$$2C_2H_5OH + 6O_2 \rightarrow 4CO_2 + 6H_2O$$

4 moles of carbon dioxide are released into the atmosphere when **2 moles** of **ethanol** are burned completely.

If you **combine** all three of these equations, you'll find that exactly **6 moles of CO_2** are taken in...
...and exactly **6 moles of CO_2** are given out.

3) However, **fossil fuels** will need to be burned to power the machinery used to make **fertilisers** for the crops, **harvest the crops** and **refine and transport** the bioethanol. Burning the fuel to power this machinery produces carbon dioxide. So using bioethanol made by fermentation **isn't completely carbon neutral**.

Practice Questions

Q1 What is the standard method of producing alcohols from alkenes?

Q2 Why should the fermentation of glucose not be carried out at more than 40 °C?

Q3 What is a biofuel?

Q4 Write down one advantage and one disadvantages of replacing fossil fuels with biofuels.

Exam Questions

Q1 Industrially, ethanol can be produced by fermentation of glucose, $C_6H_{12}O_6$.

a) Write a balanced equation for this reaction. [1 mark]

b) State the conditions used for the industrial fermentation of glucose. [3 marks]

b) How is the ethanol separated from the reaction mixture? [1 mark]

Q2 In a classroom debate on the future of fuels, one student states that bioethanol is "carbon neutral". Another student argues that bioethanol is not really a carbon neutral fuel.

a) Explain why bioethanol is sometimes described as a carbon neutral fuel. [1 mark]

b) Explain why bioethanol cannot really be considered a carbon neutral fuel [2 marks]

Steam hydration or fermentation? Come on, everyone has a favourite...

The hydration reaction here is just the opposite of the elimination reaction a couple of pages back. Make sure that you know the fermentation reaction (including the conditions) inside out. Biofuels are a relatively simple idea to get your head round — then it's just a matter of learning about the pros and cons of the hydration and fermentation processes.

Oxidation of Alcohols

Another couple of pages of alcohol reactions. Probably not what you wanted for Christmas... But at least you're almost at the end of the section... and your wits, probably.

How Much an Alcohol can be **Oxidised** Depends on its **Structure**

1) The simple way to oxidise alcohols is to burn them. But you don't get the most exciting products by doing this. If you want to end up with something more interesting, you need a more sophisticated way of oxidising...
2) You can use the **oxidising agent acidified potassium dichromate(VI)**, $K_2Cr_2O_7$, to **mildly** oxidise alcohols. The **orange** dichromate(VI) ion, $Cr_2O_7^{2-}$, is reduced to the **green** chromium(III) ion, Cr^{3+}.

- **Primary** alcohols are oxidised to **aldehydes** and then to **carboxylic acids.**
- **Secondary** alcohols are oxidised to **ketones** only.
- **Tertiary** alcohols aren't oxidised.

Learn What **Aldehydes, Ketones** and **Carboxylic Acids** are

Aldehydes and **ketones** are **carbonyl** compounds — they have the functional group C=O. Their general formula is $\mathbf{C_nH_{2n}O}$. **Carboxylic acids** have the functional group COOH and have the general formula $\mathbf{C_nH_{2n+1}COOH}$.

ALDEHYDES

1) Have a **hydrogen** and **one alkyl group** attached to the carbonyl carbon atom.
2) Their suffix is **-al**. You don't have to say which carbon the functional group is on — it's always on carbon-1.

propanal
CH_3CH_2CHO

KETONES

1) Have **two alkyl groups** attached to the carbonyl carbon atom.
2) Their suffix is **-one**. For ketones with five or more carbons, you always have to say which carbon the functional group is on.

propanone
CH_3COCH_3

pentan-2-one
$CH_3COC_3H_7$

CARBOXYLIC ACIDS

1) Have a **COOH group** at the end of their carbon chain.
2) Their suffix is **-oic acid**.

propanoic acid
CH_3CH_2COOH

Primary Alcohols will Oxidise to **Aldehydes** and **Carboxylic Acids**

[O] = oxidising agent

$$R{-}CH_2{-}OH + [O] \longrightarrow R{-}CHO + H_2O$$

primary alcohol → aldehyde

then

$$R{-}CHO + [O] \xrightarrow{\text{reflux}} R{-}COOH$$

aldehyde → carboxylic acid

You can control how **far** the alcohol is oxidised by controlling the **reaction conditions**:

1) Gently heating ethanol with potassium dichromate(VI) and sulfuric acid in a test tube produces **ethanal** (an aldehyde). However, it's **tricky** to control the heat and the aldehyde is usually oxidised to form **ethanoic acid**.
2) To get just the **aldehyde**, you need to get it out of the oxidising solution **as soon** as it forms. You do this using **distillation apparatus**, so the aldehyde (which boils at a lower temperature than the alcohol) is distilled off **immediately**.
3) To produce the **carboxylic acid**, the alcohol has to be **vigorously oxidised**. The alcohol is mixed with excess oxidising agent and heated under **reflux**.
4) Heating under reflux means you can increase the **temperature** of an organic reaction without losing **volatile** solvents, reactants or products. Any vaporised compounds cool, condense and drip back into the reaction mixture. So the aldehyde stays in the reaction mixture and is oxidised to carboxylic acid.

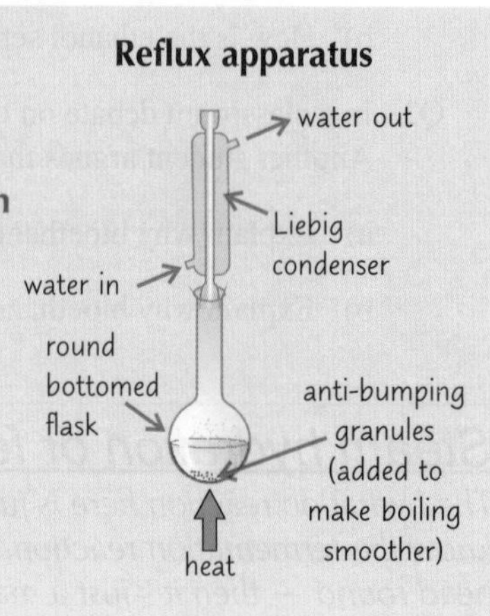

Oxidation of Alcohols

Secondary *Alcohols will Oxidise to* ***Ketones***

$$R_1-\underset{R_2}{\overset{H}{C}}-OH + [O] \xrightarrow{\text{reflux}} \begin{matrix} R_1 \\ R_2 \end{matrix}\!\!>C=O + H_2O$$

1) Refluxing a secondary alcohol, e.g. propan-2-ol, with acidified dichromate(VI) will produce a **ketone**.
2) Ketones can't be oxidised easily, so even prolonged refluxing won't produce anything more.

Tertiary *Alcohols* ***can't*** *be Oxidised Easily*

Tertiary alcohols don't react with potassium dichromate(VI) at all — the solution stays orange. The only way to oxidise tertiary alcohols is by **burning** them.

Use Oxidising Agents to ***Distinguish*** *Between* ***Aldehydes*** *and* ***Ketones***

Aldehydes and ketones can be distinguished using **oxidising agents**. Aldehydes are easily oxidised but ketones aren't.

1) **Fehling's solution** and **Benedict's solution** are both deep blue Cu^{2+} complexes, which reduce to brick-red Cu_2O when warmed with an aldehyde, but stay blue with a ketone.
2) **Tollens' reagent** is $[Ag(NH_3)_2]^+$ — it's reduced to **silver** when warmed with an aldehyde, but not with a ketone. The silver will coat the inside of the apparatus to form a **silver mirror**.

There's more details on test you can use to distinguish between aldehydes and ketones on pages 154-155.

Practice Questions

Q1 What is the colour change when potassium dichromate(VI) is reduced?

Q2 What's the difference between the structure of an aldehyde and a ketone?

Q3 What will acidified potassium dichromate(VI) oxidise secondary alcohols to?

Q4 Describe two tests you can use to distinguish between a sample of an aldehyde and a sample of a ketone.

Exam Question

Q1 A student wanted to produce the aldehyde propanal from propanol, and set up a reflux apparatus using a suitable oxidising agent.

a) i) Suggest an oxidising agent that the student could use. [1 mark]

ii) Draw the displayed formula of propanal. [1 mark]

b) The student tested his product and found that he had not produced propanal.

i) Describe a test for an aldehyde. [1 mark]

ii) What is the student's product? [1 mark]

iii) Write equations to show the two-stage reaction. Use [O] to represent the oxidising agent. [2 marks]

iv) What technique should the student have used to form propanal? [1 mark]

c) The student also tried to oxidise 2-methylpropan-2-ol, unsuccessfully.

i) Draw the skeletal formula for 2-methylpropan-2-ol. [1 mark]

ii) Why is it not possible to oxidise 2-methylpropan-2-ol with an oxidising agent? [1 mark]

So many organic compounds... eyes glazing over... brain freezing...

Don't give up now. Only as a fully-trained Chemistry warrior can you take on the Examiner and emerge victorious. If you quit now, if you choose the easy path, then all the marks you've fought for will be lost. Be strong, brave student.

Tests for Functional Groups

There's no escaping all these alcohols, aldehydes, alkenes, carboxylic acids. As well as knowing all about them and their reactions, you need to be able to carry out tests to distinguish between them. These pages should help...

You Can **Test** Whether You've Got a **Primary**, **Secondary** or **Tertiary** Alcohol

You've already seen how to **oxidise alcohols** using **acidified potassium dichromate (VI)** on pages 152 to 153. You can use that reaction to test which sort of alcohol you've got — **primary**, **secondary**, or **tertiary**. Here's what you do:

1) Add 10 drops of the alcohol to 2 cm^3 of acidified potassium dichromate solution in a test tube.
2) Warm the mixture gently in a hot water bath.
3) Then watch for a colour change:

PRIMARY – the **orange** solution slowly turns **green** as an aldehyde forms. (If you carry on heating, the aldehyde will be oxidised further to give a carboxylic acid.)
SECONDARY – the **orange** solution slowly turns **green** as a ketone forms.
TERTIARY – nothing happens — boring, but easy to remember.

You can also test for alcohols using sodium metal. If you add a small piece of sodium to a pure alcohol, it will fizz as it gives off H_2 gas.

The colour change is the orange dichromate(VI) ion ($Cr_2O_7^{2-}$) being reduced to the green chromium(III) ion (Cr^{3+}).

The problem with this test is that it shows the **same result** for **primary** and **secondary alcohols**.
To find out which one you started with, you'll have to repeat the experiment and collect some of the **product**:

1) Add excess alcohol to 2 cm^3 of acidified potassium dichromate solution in a round bottomed flask.
2) Set up the flask as part of distillation apparatus (see the picture on the right).
3) Gently heat the flask. The alcohol will be oxidised and the product will be distilled off immediately so you can collect it.

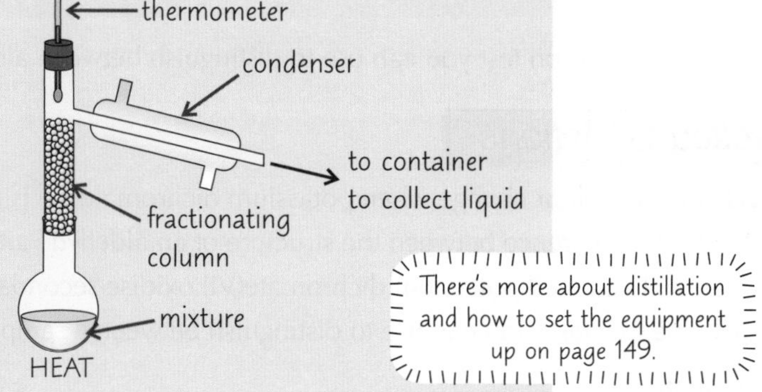

There's more about distillation and how to set the equipment up on page 149.

Once you've collected the product, you'll need to test it to find out if it's an **aldehyde** or a **ketone**. Handily, what's coming up next is how to do just that...

You Can **Test** Whether You've Got an **Aldehyde** or a **Ketone**

There are three main reagents you can use to distinguish between **aldehydes** and **ketones** — Fehling's solution, Benedict's solution and Tollens' reagent.

Fehling's Solution and **Benedict's Solution**

Fehling's solution and Benedict's solution work in exactly the same way.

1) Add 2 cm^3 of Fehling's or Benedict's solution to a test tube. (Whichever one you use, it should be a clear blue solution.)
2) Add 5 drops of the aldehyde or ketone to the test tube.
3) Put the test tube in a hot water bath to warm it for a few minutes.

ALDEHYDE – the blue solution will give a **brick red precipitate**.
KETONE – nothing happens.

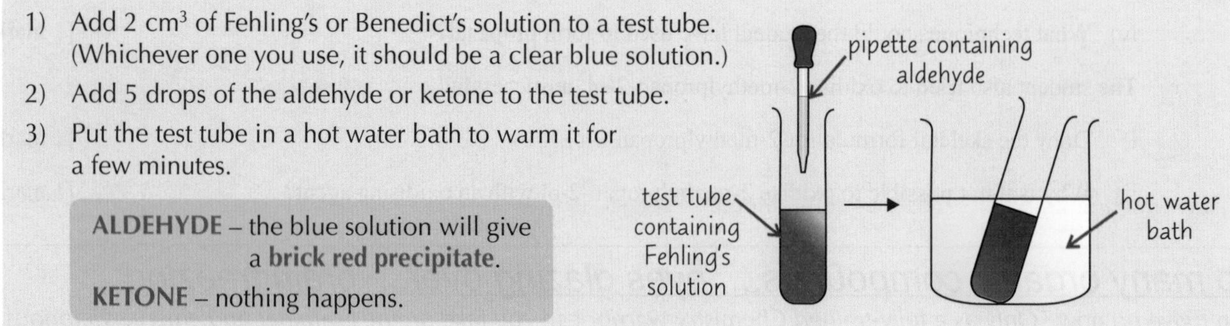

Tests for Functional Groups

Tollens' Reagent

You'd think that two reagents you could use to test for aldehydes and ketones would be enough, but there's one more to go. This one's a bit more tricky, because you have to start by making up the Tollens' reagent yourself.

1) Put 2 cm^3 of 0.10 mol dm^{-3} silver nitrate solution in a test tube.

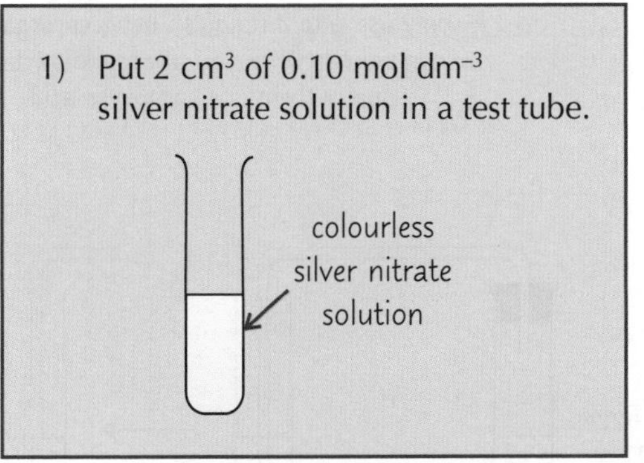

2) Add a few drops of dilute sodium hydroxide solution. A light brown precipitate should form.

light brown precipitate

3) Add drops of dilute ammonia solution until the brown precipitate dissolves completely.

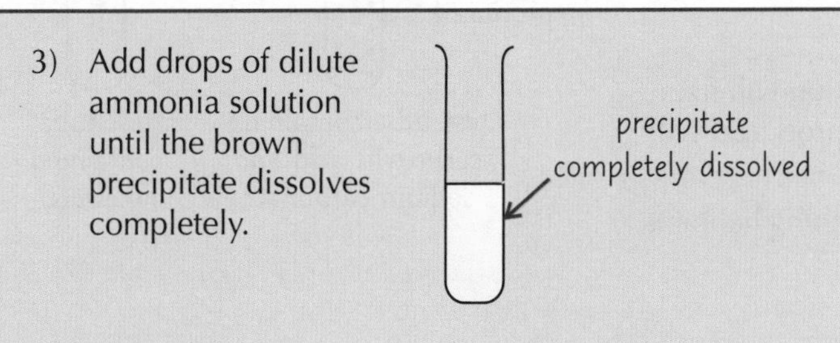

The solution you've made at the end of step 3) is Tollens' reagent.

4) Place the test tube in a hot water bath and add 10 drops of aldehyde or ketone. Wait for a few minutes.

ALDEHYDE – a **silver mirror** (a thin coating of silver) forms on the walls of the test tube.
KETONE – nothing happens.

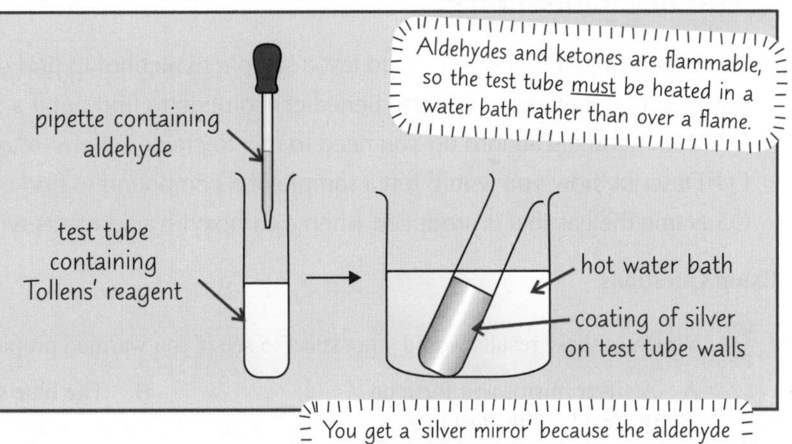

Aldehydes and ketones are flammable, so the test tube _must_ be heated in a water bath rather than over a flame.

You get a 'silver mirror' because the aldehyde reduces the Ag^+ ions to silver atoms.

Use *Bromine Water* to Test for *Alkenes*

Here's another test that you've come across before (on page 143). This one allows you to test a solution to find out if it's an **alkene** — what you're actually testing for is the presence of **double bonds**. Here's what you do:

1) Add 2 cm^3 of the solution that you want to test to a test tube.
2) Add 2 cm^3 of bromine water to the test tube.
3) Shake the test tube.

ALKENE – the solution will decolourise (go from **orange** to **colourless**).
NOT ALKENE – nothing happens.

SHAKE

test tube containing alkene and bromine water

solution is decolourised

Tests for Functional Groups

Use Sodium Carbonate to Test for Carboxylic Acids

1) Another thing you might be asked to test for is **carboxylic acids**. You met carboxylic acids on page 152, so have a look back if you can't remember what they are.
2) Carboxylic acids react with **carbonates** to form a salt, carbon dioxide and water. You can use this reaction to test whether a solution is a carboxylic acid, like this:

Be careful though — this test will give a positive result with any acid, so you can only use it to distinguish between organic compounds when you already know that one of them is a carboxylic acid.

1) Add 2 cm^3 of the solution that you want to test to a test tube.
2) Add 1 small spatula of solid sodium carbonate (or 2 cm^3 of sodium carbonate solution).
3) If the solution begins to fizz, bubble the gas that it produces through some limewater in a second test tube.

CARBOXYLIC ACID – the solution will fizz. The carbon dioxide gas that is produced will turn limewater cloudy.
NOT CARBOXYLIC ACID – nothing happens.

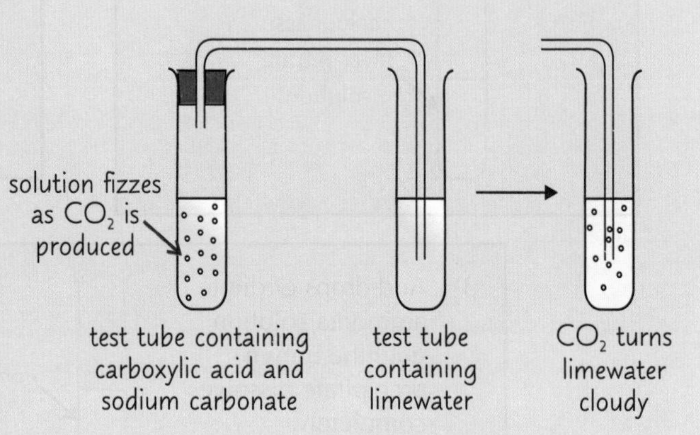

Practice Questions

Q1 What reagent could you use to test a sample of alcohol to find out whether it was a tertiary alcohol?

Q2 Describe how you could use Benedict's solution to find out if a solution was an aldehyde or a ketone.

Q3 Which three solutions do you need to mix to create Tollens' reagent?

Q4 Describe how you would test a sample of a compound to find out if it was an alkane or an alkene.

Q5 Name the gas that is produced when a carboxylic acid reacts with sodium carbonate.

Exam Questions

Q1 Which of these results would you expect to see if you warmed propanone with Fehling's solution?

A A silver mirror will form on the inside of the test tube.
B The blue solution will give a brick red precipitate.
C The orange solution will decolourise.
D Nothing will happen. [1 mark]

Q2 Which of these statements about cyclohexene is correct?

A It produces a brick red precipitate with Fehling's solution.
B It decolourises bromine water.
C It turns limewater cloudy.
D It forms a silver mirror with Tollens' reagent. [1 mark]

Q3 Describe a chemical test you could use to show that a solution is a carboxylic acid.
Include any reagents, conditions and expected observations in your answer. [3 marks]

Q4 A student has a sample of alcohol. He is told that it is a primary or secondary alcohol.
Describe a procedure he could carry out to test which of these it is.
Your answer should include the result you would expect to see in each case. [5 marks]

"Testing, testing, 1,2,3..."

Fehling, Tollens, Benedict... lots of people all busy coming up with different ways to test for ketones and aldehydes. You'd think they could've found something better to do with their time. Unfortunately for you, they didn't, and you could actually be asked to do these tests yourself. So make sure you learn how to do each one, you won't regret it. Honest.

Analytical Techniques

Get ready for the thrilling climax of the section — and watch out for the big twist at the end...

Mass Spectrometry Can Help to Identify Compounds

1) You saw on page 7 that **mass spectrometry** can be used to find the **relative molecular mass** (M_r) of a compound.
2) In the mass spectrometer, a **molecular ion** is formed when a molecule loses an **electron**.
3) The molecular ion produces a **molecular ion peak** on the mass spectrum of the compound.
4) For any compound, the **mass/charge** (m/z) value of the molecular ion peak will be the same as the **molecular mass** of the compound.

Assuming the ion has a 1+ charge, which it normally will have.

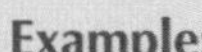

Example: The mass spectrum of a straight chain alkane contained a molecular ion peak with m/z = 72.0. Identify the compound.

1) The m/z value of the molecular ion peak is **72.0** — so the M_r of the compound must be 72.0.
2) If you calculate the molecular masses of the first few straight-chain alkanes, you'll find that the one with a molecular mass of 72.0 is pentane (C_5H_{12}):
   M_r of pentane = (5 × 12.0) + (12 × 1.0) = 72.0
3) So the compound must be **pentane**.

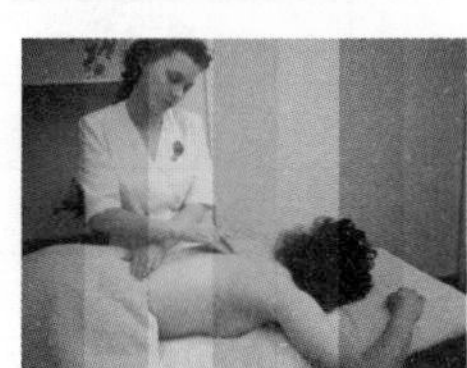

A massage spectrum

High Resolution Mass Spectrometry

1) Some mass spectrometers can measure atomic and molecular masses **extremely accurately** (to several decimal places). These are known as **high resolution mass spectrometers**.
2) This can be useful for identifying compounds that appear to have the **same M_r** when they're **rounded** to the nearest whole number.
3) For example, propane (C_3H_8) and ethanal (CH_3CHO) both have an M_r of 44 to the nearest whole number. But on a high resolution mass spectrum, propane has a molecular ion peak with m/z = 44.0624 and ethanal has a molecular ion peak with m/z = 44.0302.

Example: On a high resolution mass spectrum, a compound had a molecular ion peak of 98.0448.
What was its molecular formula?

A $C_5H_{10}N_2$
B $C_6H_{10}O$
C C_7H_{14}
D $C_5H_6O_2$

Use these precise atomic masses to work out your answer:
^{1}H — 1.0078 ^{12}C — 12.0000 ^{14}N — 14.0064 ^{16}O — 15.9990

1) Work out the precise molecular mass of each compound:
   $C_5H_{10}N_2$: M_r = (5 × 12.0000) + (10 × 1.0078) + (2 × 14.0064) = 98.0908
   $C_6H_{10}O$: M_r = (6 × 12.0000) + (10 × 1.0078) + 15.9990 = 98.0770
   C_7H_{14}: M_r = (7 × 12.0000) + (14 × 1.0078) = 98.1092
   $C_5H_6O_2$: M_r = (5 × 12.0000) + (6 × 1.0078) + (2 × 15.9990) = 98.0448
2) So the answer is **D**, $C_5H_6O_2$.

On a normal (low resolution) mass spectrum, all of these molecules would show up as having an M_r of 98.

Analytical Techniques

Infrared Spectroscopy Helps You *Identify Organic Molecules*

1) In infrared (IR) spectroscopy, a beam of **IR radiation** is passed through a sample of a chemical.
2) The IR radiation is absorbed by the **covalent bonds** in the molecules, increasing their **vibrational** energy.
3) **Bonds between different atoms** absorb **different frequencies** of IR radiation. Bonds in different **places** in a molecule absorb different frequencies too — so the O–H group in an **alcohol** and the O–H in a **carboxylic acid** absorb different frequencies. This table shows what **frequencies** different bonds absorb:

Bond	Where it's found	Wavenumber (cm^{-1})
N–H (amines)	amines (e.g. methylamine, CH_3NH_2)	3300 - 3500
O-H (alcohols)	alcohols	3230 - 3550
C–H	most organic molecules	2850 - 3300
O–H (acids)	carboxylic acids	2500 - 3000
C≡N	nitriles (e.g. ethanenitrile, CH_3CN)	2220 - 2260
C=O	aldehydes, ketones, carboxylic acids, esters	1680 - 1750
C=C	alkenes	1620 - 1680
C–O	alcohols, carboxylic acids	1000 - 1300
C–C	most organic molecules	750 - 1100

Wavenumber is the measure used for the frequency — it's just $\frac{1}{\text{wavelength (cm)}}$

You don't need to learn this data (but you do need to understand how to use it).

4) An infrared spectrometer produces a **graph** that shows you what frequencies of radiation the molecules are absorbing. So you can use it to identify the **functional groups** in a molecule:

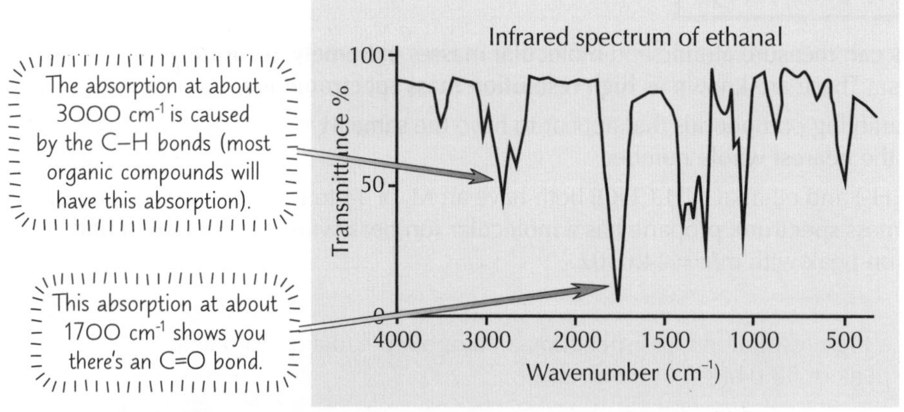

Infrared spectrum of ethanal
100
50
0
Transmittance %
4000 3000 2000 1500 1000 500
Wavenumber (cm⁻¹)
The absorption at about 3000 cm⁻¹ is caused by the C–H bonds (most organic compounds will have this absorption).
This absorption at about 1700 cm⁻¹ shows you there's an C=O bond.

The peaks show you where radiation is being absorbed.

The 'peaks' on IR spectra point downwards.

The *Fingerprint* Region *Identifies* a Molecule

1) The region between **500 cm^{-1} and 1500 cm^{-1}** is called the **fingerprint** region. It's **unique** to a **particular compound**. You can use a computer database to check this region of an unknown compound's IR spectrum against those of known compounds. If it **matches** one of them, you know what the molecule is.

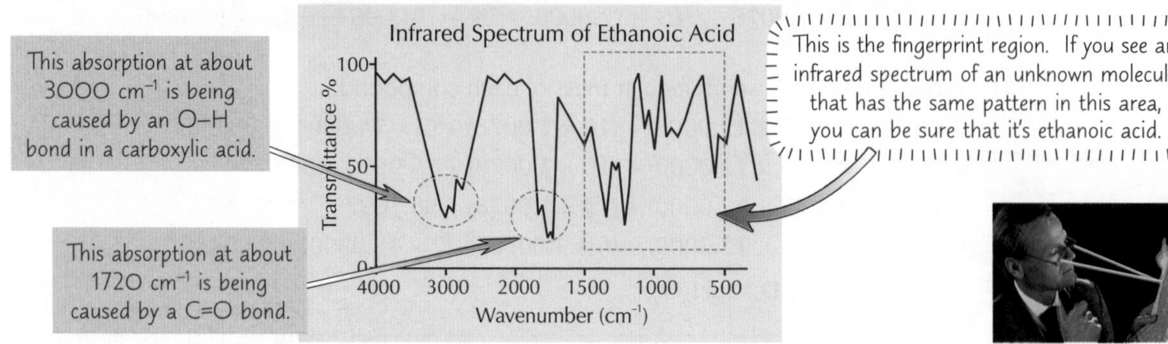

Infrared Spectrum of Ethanoic Acid
100
50
0
Transmittance %
4000 3000 2000 1500 1000 500
Wavenumber (cm⁻¹)
This absorption at about 3000 cm⁻¹ is being caused by an O–H bond in a carboxylic acid.
This absorption at about 1720 cm⁻¹ is being caused by a C=O bond.
This is the fingerprint region. If you see an infrared spectrum of an unknown molecule that has the same pattern in this area, you can be sure that it's ethanoic acid.

Clark began to regret having an infrared mechanism installed in his glasses.

2) Infrared spectroscopy can also be used to find out how **pure** a compound is, and identify any impurities. Impurities produce **extra peaks** in the fingerprint region.

Analytical Techniques

Infrared Radiation **Absorption** is Linked to **Global Warming**

1) Some of the electromagnetic radiation emitted by the **Sun** reaches the Earth and is absorbed. The Earth then re-emits some of it as **infrared radiation** (heat).
2) Molecules of **greenhouse gases**, like **carbon dioxide**, **methane** and **water vapour**, in the atmosphere absorb this infrared radiation. Then they re-emit some of it back towards the Earth, keeping us warm. This is called the '**greenhouse effect**'.
3) Human activities, such as burning fossil fuels and leaving rubbish to rot in landfill sites, have caused a **rise** in **greenhouse gas concentrations**.
4) This means more heat is being trapped and the Earth is getting warmer — this is **global warming**.

It's the bonds of these molecules that absorb the IR radiation.

Practice Questions

Q1 In mass spectrometry, what is meant by a molecular ion?

Q2 Explain how you could find the molecular mass of a compound by looking at its mass spectrum.

Q3 Which parts of a molecule absorb infrared radiation?

Q4 On an infrared spectrum, what is meant by the 'fingerprint region'?

Q5 Explain how increasing the amount of carbon dioxide in the atmosphere leads to global warming.

Exam Questions

Q1 Use the following precise atomic masses to answer this question:
^{1}H — 1.0078 ^{12}C — 12.0000 ^{14}N — 14.0064 ^{16}O — 15.9990

a) The high resolution mass spectrum of a compound has a molecular ion peak with m/z = 74.0908. Which of the following could be the molecular formula of the compound?
A $C_3H_6O_2$ B $C_4H_{10}O$ C $C_3H_{10}N_2$ D $C_2H_6N_2O$ [1 mark]

b) Explain why low resolution mass spectrometry would not allow you to distinguish between the options given in part a). [1 mark]

c) The high resolution mass spectrum of 2-fluoroethanol (CH_2FCH_2OH) produced a molecular ion peak with m/z = 64.0364. Use this information to find the precise atomic mass of fluorine. [1 mark]

Q2 An organic molecule with a molecular mass of 74 produces the IR spectrum shown below.

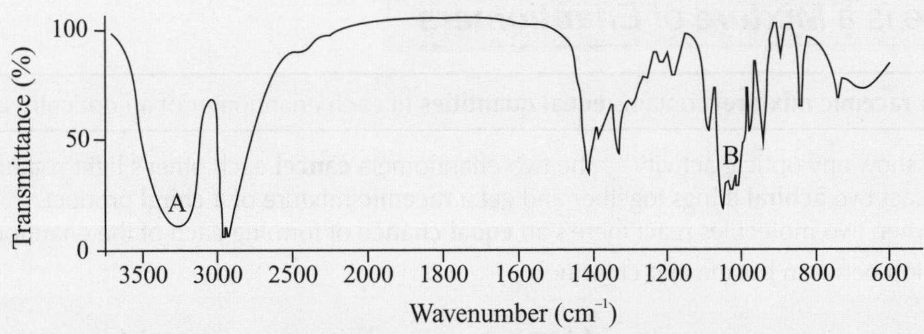

a) Identify the bonds that are responsible for causing the peaks labelled A and B. Use the data table on page 158 to help. [2 marks]

b) Suggest the molecular formula of this molecule. Explain your answer. [3 marks]

I wonder what the infrared spectrum of a fairy cake would look like...

Very squiggly I imagine. Luckily you don't have to be able to remember what any infrared spectrum graphs look like. But you definitely need to know how to interpret them. And don't worry, I haven't forgotten I said there was twist at the end... erm... hydrogen was my sister all along... and all the elements went to live in Jamaica. The End.

Optical Isomerism

Optical isomerism isn't something your optician and a pair of glasses can sort — though you may well be cross-eyed after looking at these two pages of mirrors and molecules.

Optical Isomers are Mirror Images of Each Other

1) **Optical isomerism** is a type of stereoisomerism. Stereoisomers have the **same structural formula**, but have their atoms arranged differently in **space**.
2) A **chiral** (or **asymmetric**) carbon atom is one that has **four different groups** attached to it. It's possible to arrange the groups in two different ways around the carbon atom so that two different molecules are made — these molecules are called **enantiomers** or **optical isomers**.
3) The enantiomers are **mirror images** and no matter which way you turn them, they can't be **superimposed**.
4) You have to be able to **draw** optical isomers. But first you have to identify the chiral centre...

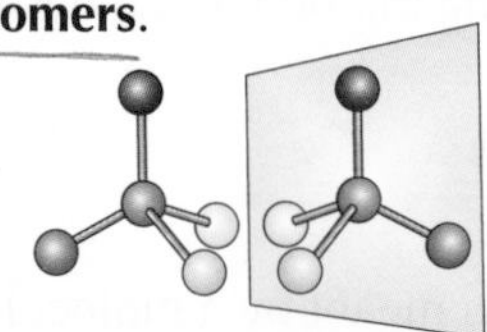

'Superimposed' means that you can put one thing on top of another and they will match up completely.

If molecules can be superimposed, they're achiral — and there's no optical isomerism.

1) **Locating the chiral centre:**
Look for the carbon atom with **four different groups** attached. Here it's the carbon with the four groups H, OH, COOH and CH_3 attached.

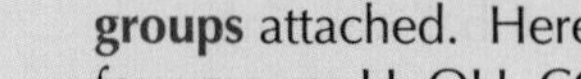

chiral centre

2-hydroxypropanoic acid

2) **Drawing isomers:**
Once you know the **chiral carbon**, draw one enantiomer in a **tetrahedral shape**. Don't try to draw the full structure of each group — it gets confusing. Then draw a **mirror image** beside it.

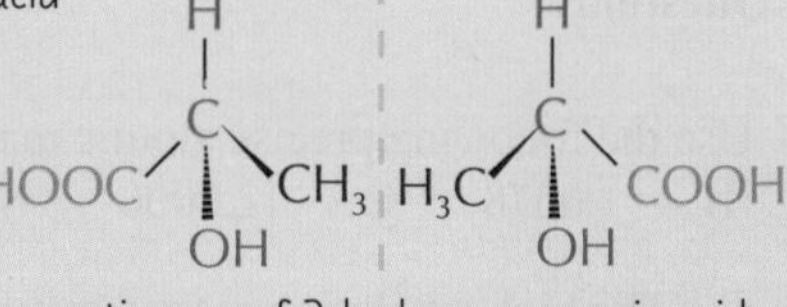

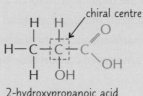

enantiomers of 2-hydroxypropanoic acid

Optical Isomers Rotate Plane-Polarised Light

1) Normal light vibrates in all directions. **Plane-polarised light** only vibrates in one direction.
2) Optical isomers are **optically active** — they **rotate plane-polarised light**.
3) One enantiomer rotates it in a **clockwise** direction, and the other rotates it in an **anticlockwise** direction.

Christmas is a time to embrace your choral centre.

A Racemate is a Mixture of Enantiomers

A **racemate** (or **racemic mixture**) contains **equal quantities** of each enantiomer of an optically active compound.

Racemates **don't** show any optical activity — the two enantiomers **cancel** each other's light-rotating effect. Chemists often react two **achiral** things together and get a **racemic** mixture of a **chiral** product. This is because when two molecules react there's an **equal chance** of forming each of the enantiomers.
Look at the reaction between butane and chlorine:

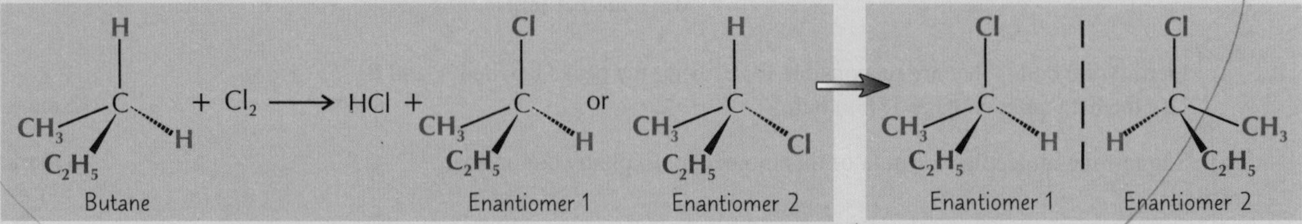

A **chlorine atom** replaces one of the **H atoms**, to give **2-chlorobutane**.
Either of the H atoms can be replaced, so the reaction produces a **mixture** of the **two possible enantiomers**.
Each hydrogen has a **fifty-fifty chance** of being replaced, so the two optical isomers are formed in **equal amounts**.

You can modify a reaction to produce a **single enantiomer** using chemical methods, but it's **difficult** and **expensive**.

Optical Isomerism

Reactions Involving Planar Bonds often Produce Racemates

Double bonds, such as C=O and C=C bonds, are **planar** (flat). The products of reactions that happen at the carbonyl group of **aldehydes** and **unsymmetrical ketones** are often **enantiomers** present as a **racemic mixture**:

Example: The reaction of propanal (C_3H_6O) with acidified potassium cyanide (KCN).

This reaction is covered in more detail on page 163.

1) The reaction of propanal with potassium cyanide involves a CN^- ion attacking the δ^+ carbon of propanal's planar C=O group.
2) The CN^- ion can attack from **two directions** — from **above** the plane of the molecule, or from **below** it.
3) Depending on which direction the nucleophilic attack happens from, one of two **enantiomers** is formed.
4) Because the C=O bond is **planar**, there is an **equal chance** that the nucleophile will attack from either of these directions. This means that an **equal amount** of each enantiomer will be formed.
5) So when propanal reacts with acidified potassium cyanide, you get a **racemic mixture** of products.

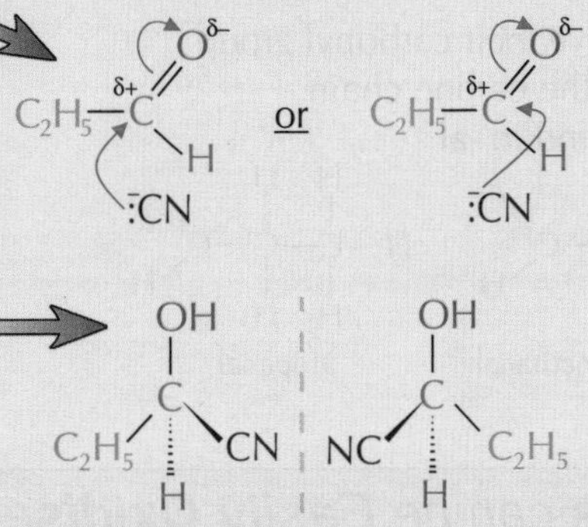

C_2H_5—C(δ^+)=O(δ^-), H, :CN or C_2H_5—C(δ^+)=O(δ^-), H, :CN

OH, C, C_2H_5, CN, H | OH, C, NC, C_2H_5, H

If you start with a **symmetrical ketone** instead, you'll make a product that **doesn't** have a chiral centre, so it **won't** display optical isomerism.

Practice Questions

Q1 What's a chiral molecule?

Q2 The displayed formula of 2-methylbutan-1-al is shown on the right. Explain why the carbon atom marked with a * is a chiral centre.

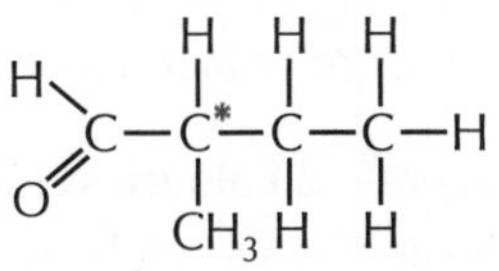

Q3 What's a racemic mixture?

Exam Questions

Q1 There are sixteen possible structural isomers of the compound $C_3H_6O_2$, four of which show stereoisomerism.

a) Explain the meaning of the term stereoisomerism. [2 marks]

b) i) There are two chiral isomers of $C_3H_6O_2$. Draw the enantiomers of one of the chiral isomers. [2 marks]

ii) State how you could distinguish between the enantiomers. [2 marks]

Q2 Parkinson's disease involves a deficiency of dopamine. It is treated by giving patients a single pure enantiomer of DOPA (dihydroxyphenylalanine), as shown on the right, a naturally occurring amino acid, which is converted to dopamine in the brain.

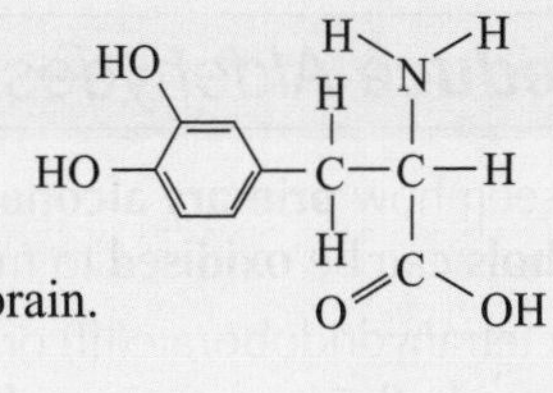

a) DOPA is a chiral molecule. Mark the structure's chiral centre. [1 mark]

b) DOPA was first synthesised as a racemate in 1911. Explain the meaning of the term racemate. [1 mark]

Q3 Pentanal reacts with potassium cyanide to produce an optically inactive product.

a) Explain why the product of the reaction is optically inactive. [4 marks]

b) Predict whether the reaction of pentan-3-one would produce an optically active product. Explain your answer. [2 marks]

Some quiet time for reflection...

This isomer stuff's not bad — well, not all bad... If you're having difficulty picturing them as 3D shapes, you could always make some models of them. It's easier to see the mirror image structure with a solid version in front of you.

Aldehydes and Ketones

You've already met aldehydes and ketones but it may have been a while ago, so here's a quick recap and then some more, meaty organic chemistry to learn. It's just the gift that never stops giving, isn't it?

Aldehydes and Ketones contain a Carbonyl Group

Aldehydes and ketones are both **carbonyl compounds** so they both contain the **carbonyl** functional group, **C=O**. The difference is, they've got their carbonyl groups in **different positions**.

Aldehydes have their carbonyl group at the **end** of the carbon chain. Their names end in **-al**.

$R-CHO$ (R = carbon chain), $H-CHO$ (methanal), CH_3CH_2CHO (propanal)

Ketones have their carbonyl group in the middle of the carbon chain. Their names end in **-one**, and often have a number to show which carbon the carbonyl group is on.

$R-CO-R'$, CH_3COCH_3 (propanone), $CH_3COCH_2CH_2CH_3$ (pentan-2-one)

Aldehydes can be Easily Oxidised but Ketones Can't

Oxidising agents will react with **aldehydes** to produce carboxylic acids, but not with **ketones**.

aldehyde $R-CHO + [O] \rightarrow R-COOH$ carboxylic acid

ketone $R-CO-R' + [O] \rightarrow$ ✗ nothing happens

As an aldehyde is oxidised, another compound is **reduced**. You can use compounds that change colour when they're reduced as a test for whether something could be an **aldehyde** or a **ketone**.

Tollens' reagent is a **colourless** solution of **silver nitrate** dissolved in **aqueous ammonia**.

- When heated in a test tube with an **aldehyde**, the Ag^+ ions in Tollens' reagent are reduced to Ag atoms and a **silver mirror** forms after a few minutes.
- Ketones can't be oxidised by Tollens' reagent, so with ketones there's no reaction and no colour change.

Fehling's solution is a **blue** solution of **copper(II) ions** dissolved in **sodium hydroxide**.

- If it's heated with an **aldehyde** the copper(II) ions are reduced to a **brick-red precipitate** of **copper(I) oxide**.
- As with Tollens' reagent, ketones don't react with Fehling's solution, so no precipitate is formed.

You can Reduce Aldehydes and Ketones Back to Alcohols

You've already seen how **primary alcohols** can be **oxidised** to produce **aldehydes** and **carboxylic acids**, and how **secondary alcohols** can be **oxidised** to make **ketones**. Using a **reducing agent**, you can **reverse** these reactions. $NaBH_4$ (sodium tetrahydridoborate(III) or sodium borohydride) dissolved in **water with methanol** is usually the reducing agent used. But in equations, **[H]** is often used to indicate a hydrogen from a reducing agent.

1) Reducing an **aldehyde** to a **primary alcohol**.

$$R-CHO + 2[H] \rightarrow R-CH_2-OH$$

2) Reducing a **ketone** to a **secondary alcohol**.

$$R-CO-R' + 2[H] \rightarrow R-CH(OH)-R'$$

The reaction mechanism for ketones is the same as for aldehydes.

Here's the reaction mechanism: (It's shown with an aldehyde here, but the mechanism works just the same for a ketone.)

The H^- ions come from the reducing agent.

$R-\overset{\delta+}{C}H=O^{\delta-}$ + $:H^-$ $\rightarrow$ $R-CH_2-O^-$ (+ H^+) $\rightarrow$ H_3C-CH_2-OH

The H^+ ions usually come from water. Sometimes a weak acid is added as a source of H^+.

These are **nucleophilic addition** reactions. The H^- ion acts as a **nucleophile** and **adds** on to the δ^+ carbon atom.

Aldehydes and Ketones

Potassium Cyanide will React with Carbonyls by Nucleophilic Addition

Potassium cyanide reacts with carbonyl compounds to produce **hydroxynitriles** (molecules with a CN and an OH group). It's a **nucleophilic addition reaction** — a **nucleophile** attacks the molecule, and **adds** itself as an extra group.

Potassium cyanide dissociates in water to form K^+ ions and CN^- ions: $KCN \rightarrow K^+ + CN^-$

1) The CN^- group **attacks** the partially positive carbon atom and **donates** a pair of electrons. Both electrons from the double bond transfer to the oxygen.

2) H^+ ions add to the oxygen to form the **hydroxyl group** (OH). Acidified KCN is usually used so there's a source of H^+ ions.

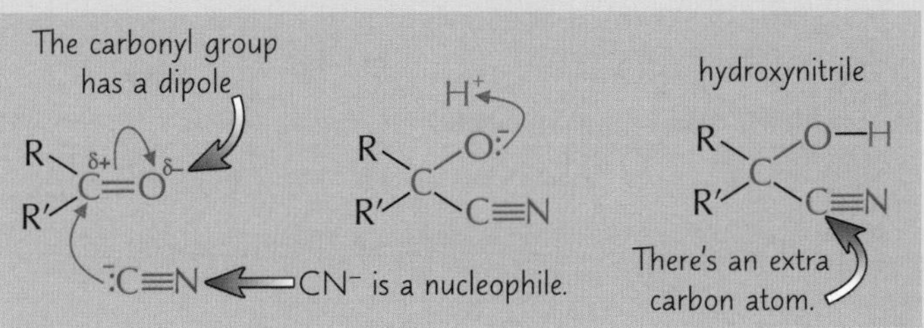

3) The overall reaction for an **aldehyde** is: $RCHO_{(aq)} + KCN_{(aq)} \xrightarrow{H^+_{(aq)}} RCH(OH)CN_{(aq)} + K^+_{(aq)}$

And for a **ketone**: $RCOR'_{(aq)} + KCN_{(aq)} \xrightarrow{H^+_{(aq)}} RCR'(OH)CN_{(aq)} + K^+_{(aq)}$

4) If you start with an **unsymmetrical ketone** or any **aldehyde** (except methanal), you will produce a mixture of **enantiomers** (see page 161).

Hydrogen cyanide (HCN) reacts with carbonyls in the same way, except you don't need the acid. Just swap KCN for HCN and drop the K^+ ion. E.g.: $RCHO + HCN \rightarrow RCH(OH)CN$

Potassium Cyanide is a Dangerous Chemical to Work With

Here's a risk assessment for reacting potassium cyanide with a carbonyl (as above): **Potassium cyanide** is an **irritant** and is also extremely dangerous if it's **ingested** (eaten) or **inhaled**. It can react with moisture to produce **hydrogen cyanide**, a highly **toxic gas**. To **reduce** any **risks**, any person carrying out this experiment should **wear gloves**, **safety goggles** and a **lab coat** and perform the experiment in a **fume cupboard**.

A risk assessment (see page 202) involves assessing the hazards that a reaction poses. You need to take all reasonable precautions to reduce the risk of an accident.

Practice Questions

Q1 What's the difference between the structures of aldehydes and ketones?

Q2 Explain why the reduction of an aldehyde or a ketone is a nucleophilic addition reaction.

Exam Questions

Q1 The compound C_3H_6O can exist as an aldehyde and a ketone.

a) Draw and name the carbonyl isomers of C_3H_6O. [4 marks]

b) Hydrogen cyanide, HCN, reacts with C_3H_6O carbonyl compounds to form a compound with the molecular formula C_4H_6ON.

i) Name the type of mechanism that this reaction proceeds by. [1 mark]

ii) Draw the mechanism for the reaction of hydrogen cyanide with the ketone with the molecular formula C_3H_6O. [5 marks]

c) The aldehyde C_3H_6O can be reduced to an alcohol, C_3H_7OH. Write an equation for the reaction, listing suitable reagents and conditions. [3 marks]

Q2 There are two straight-chain carbonyl compounds with the molecular formula C_4H_8O.

a) Name the two compounds. [2 marks]

b) Describe a test that could distinguish between the isomers and give the expected result for each. [3 marks]

Before I begin baking, I always carry out a whisk assessment...

Make sure you know how aldehydes differ from ketones and what you get when you oxidise or reduce them both. The mechanisms are a pain to learn, but you've just got to do it. Keep trying to write them out from memory till you can.

Carboxylic Acids and Esters

Carboxylic acids are much more interesting than cardboard boxes — as you're about to discover...

Carboxylic Acids contain –COOH

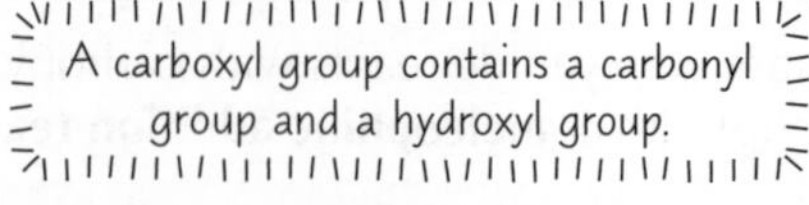

A carboxyl group contains a carbonyl group and a hydroxyl group.

Carboxylic acids contain the **carboxyl** functional group **-COOH**.

To name them, you find and name the longest alkane chain, take off the 'e' and add **'–oic acid'**.

ethanoic acid | 4-hydroxy-2-methylbutanoic acid | benzoic acid

See page 173 for more about naming benzene compounds.

The carboxyl group is always at the **end** of the molecule and when naming, it's more important than other functional groups — so all the other functional groups in the molecule are numbered starting from this carbon.

Carboxylic Acids are *Weak Acids*

Carboxylic acids are **weak acids** — in water they partially dissociate into a **carboxylate ion** and an **H^+ ion**:

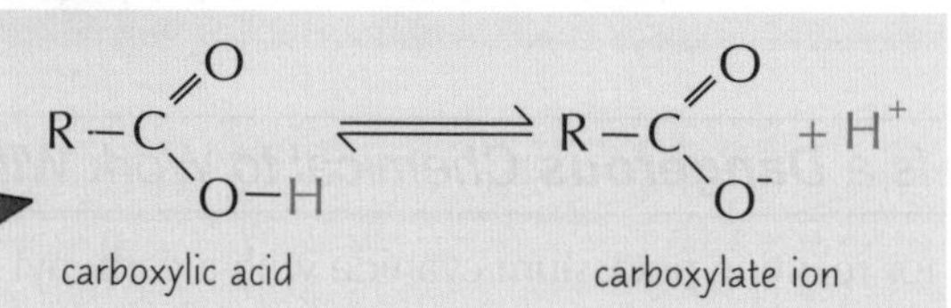

This equilibrium lies to the left because most of the molecules don't dissociate.

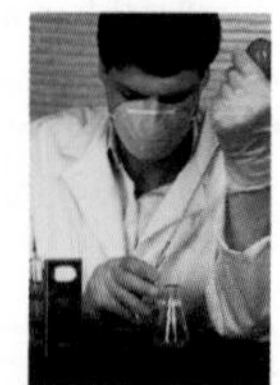

Pete was hoping for an A^- in his carboxylic acid practical, but he ended up with an H^+.

Carboxylic Acids React with **Carbonates** to Form **Carbon Dioxide**

Carboxylic acids react with **carbonates** (which contain the **CO_3^{2-}** ion) to form a **salt**, **carbon dioxide** and **water**.

ethanoic acid → sodium ethanoate

$$2CH_3COOH_{(aq)} + Na_2CO_{3(s)} \rightarrow 2CH_3COONa_{(aq)} + H_2O_{(l)} + CO_{2(g)}$$

$$CH_3COOH_{(aq)} + NaHCO_{3(s)} \rightarrow CH_3COONa_{(aq)} + H_2O_{(l)} + CO_{2(g)}$$

In these reactions, carbon dioxide fizzes out of the solution.

Carboxylic Acids React with **Alcohols** to form **Esters**

1) Esters are organic compounds that contain a -COO- group.
2) They're frequently made by heating a **carboxylic acid** with an **alcohol** in the presence of a **strong acid catalyst**.
3) It's called an **esterification** reaction. Concentrated sulfuric acid is usually used as the acid catalyst.

$$R{-}COOH + R{-}O{-}H \underset{\text{reflux}}{\overset{H^+}{\rightleftharpoons}} R{-}COO{-}R + H_2O$$

carboxylic acid + alcohol ⇌ ester + water

This oxygen comes from the alcohol.

It's also a condensation reaction as it releases water.

Here's how ethanoic acid reacts with ethanol to make the ester, ethyl ethanoate:

$$H_3C{-}COOH + H{-}O{-}CH_2CH_3 \underset{\text{reflux}}{\overset{H^+}{\rightleftharpoons}} H_3C{-}COO{-}CH_2CH_3 + H_2O$$

ethanoic acid + ethanol ⇌ ethyl ethanoate + water

Carboxylic Acids and Esters

Esters have the Functional Group –COO–

You've just seen that an ester is formed by reacting an alcohol with a carboxylic acid. Well, the **name** of an **ester** is made up of **two parts** — the **first** bit comes from the **alcohol**, and the **second** bit from the **carboxylic acid**.

1) Look at the **alkyl group** that came from the **alcohol**. This is the first bit of the ester's name.

H–C(H)(H)–C(=O)–O–C(H)(H)–C(H)(H)–C(H)(H)–H — This is a **propyl** group.

2) Now look at the part that came from the **carboxylic acid**. Swap its '-oic acid' ending for 'oate' to get the second bit of the name.

This came from **ethanoic acid**, so it's an **ethanoate**.

3) Put the two parts together. It's **propyl ethanoate** $CH_3COOCH_2CH_2CH_3$

The name's written the opposite way round from the formula.

Naming esters where either the acid or alcohol chain is branched can be a bit tricker. For an ester, number the carbons from the C atoms in the C–O–C bond.

H–C₄(H)(H)–C₃(H)(H)–C₂(H)(CH₃)–C₁(=O)–O–C₁(H)(H)–C₂(H)(H)–H

This is ethyl 2-methylbutanoate. $CH_3CH_2CH(CH_3)COOCH_2CH_3$

Esters are Used as Food Flavourings, Perfumes, Solvents and Plasticisers

1) Esters have a **sweet smell**, varying from gluey sweet for smaller esters to a fruity 'pear drop' smell for larger ones. This makes them useful in perfumes. The food industry uses esters to **flavour** things like drinks and sweets too.
2) Esters are **polar** liquids so lots of **polar organic compounds** will dissolve in them. They've also got quite **low boiling points**, so they **evaporate easily** from mixtures. This makes them good solvents in **glues** and **printing inks**.
3) Esters are used as **plasticisers** — they're added to plastics during polymerisation to make the plastic more **flexible**. Over time, the plasticiser molecules escape though, and the plastic becomes brittle and stiff.

Practice Questions

Q1 Name the catalyst used in an esterification reaction between an alcohol and a carboxylic acid.

Q2 Name the alcohol and carboxylic acid that can be used to make propyl butanoate.

Q3 What properties of esters make them good solvents?

Exam Questions

Q1 The structures of substances X and Y are shown on the right:

X: H–C(H)(H)–C(=O)–OH

Y: H–C(H)(H)–C(=O)–O–C(H)(H)–H

a) Write a balanced equation for the reaction between substance **X** and sodium carbonate, Na_2CO_3. [3 marks]

b) Substance **Y** can be synthesised from substance **X** in a single step process. Give the name of the other reagent necessary for this synthesis and name the type of reaction. [2 marks]

Q2 Ethyl butanoate is an ester that is made by reacting ethanol and butanoic acid.

a) Draw the structure of ethyl butanoate. [1 mark]

b) Ethyl butanoate is used as a solvent in some adhesives. Name another possible use for ethyl butanoate. [1 mark]

Q3 3-methylbutyl ethanoate is the ester responsible for the odour of pear essence.

a) Write an equation for the formation of this ester from an alcohol and a carboxylic acid. Include any reaction conditions. [3 mark]

b) Name the carboxylic acid used. [1 mark]

Ahh... the sweet smell of success...

You've met a pretty hefty number of organic molecules now, so make sure you don't start to get them muddled up. Esters and carboxylic acids look a little bit similar — they both contain a carbonyl group. But they're not the same — carboxylic acids have an -OH group stuck onto the carbonyl carbon too, whereas esters have an -OR group. Got it?

More on Esters

OK, brace yourself. There's some pure unadulterated lard coming up.

Esters are **Hydrolysed** to Form **Alcohols**

Hydrolysis is when a substance is split up by water — but using just water is often really slow, so an **acid** or an **alkali** is often used to speed it up. There are two types of hydrolysis of esters — **acid hydrolysis** and **base hydrolysis**. With both types you get an **alcohol**, but the second product in each case is different.

ACID HYDROLYSIS Splits the ester into an **acid** and an **alcohol** — it's the reverse of the reaction on page 164. You have to **reflux** the ester with a **dilute acid**, such as hydrochloric or sulfuric.

For example:

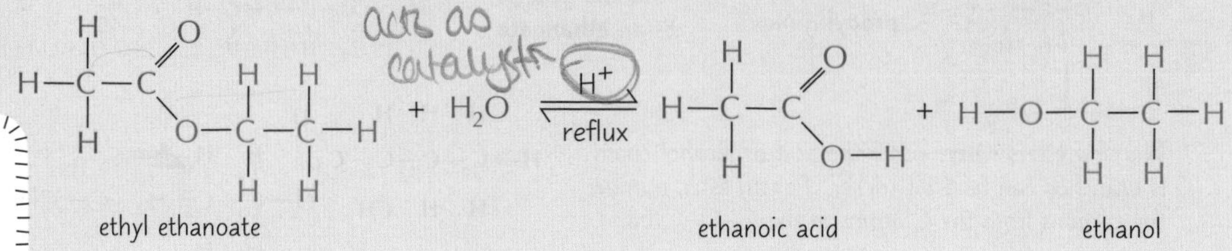

H_2O
H^+
reflux
ethyl ethanoate
ethanoic acid
ethanol

As it's a reversible reaction, you add lots of water to push the equilibrium to the right.

BASE HYDROLYSIS Involves **refluxing** the ester with a **dilute alkali**, such as sodium hydroxide. You get a **carboxylate ion** and an **alcohol**.

For example:

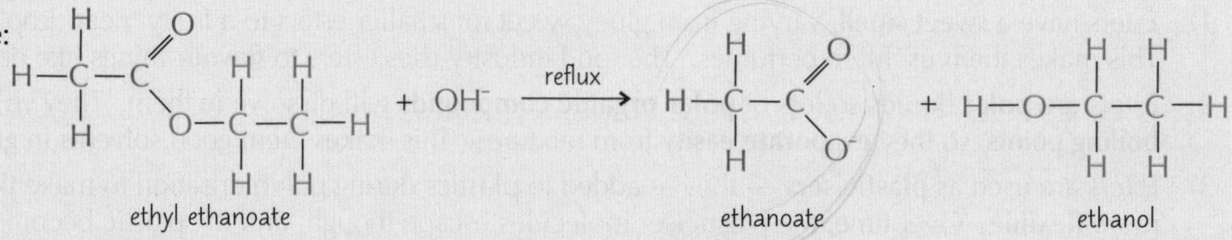

OH^-
reflux
ethyl ethanoate
ethanoate
ethanol

Fats and **Oils** are Esters of **Glycerol** and **Fatty Acids**

Fatty acids are long chain **carboxylic acids**.

They combine with glycerol (propane-1,2,3-triol) to make esters. These esters of glycerol are fats and oils. The fatty acids can be **saturated** (no double bonds) or **unsaturated** (have C=C double bonds).

Most of a fat or oil is made from fatty acid chains — so it's these that give them many of their properties.

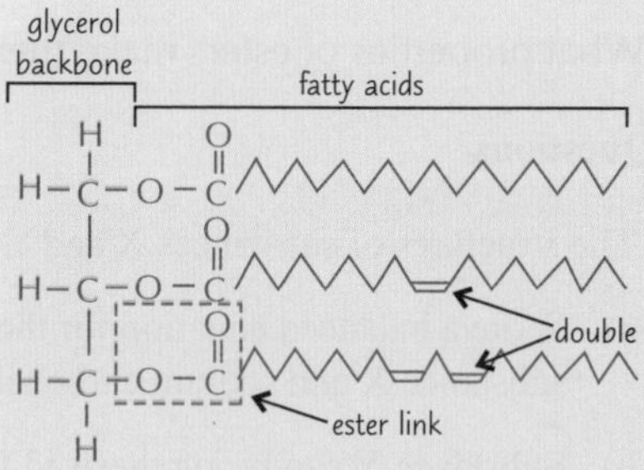

glycerol backbone
fatty acids
double bonds
ester link

- **Animal fats** have mainly **saturated** hydrocarbon chains — they fit neatly together, increasing the van der Waals forces between them. This means you need higher temperatures to melt them, so they're **solid** at room temperature.
- **Vegetable oils** have **unsaturated** hydrocarbon chains — the double bonds mean the chains are bent and don't pack together well, decreasing the van der Waals forces. So they're easier to melt and are **liquids** at room temperature.

Oils and Fats can be **Hydrolysed** to Make **Glycerol**, **Soap** and **Fatty Acids**

Like any ester, you can **hydrolyse** vegetable oils and animal fats by heating them with **sodium hydroxide**. And you'll never guess what the sodium salt produced is — **a soap**.

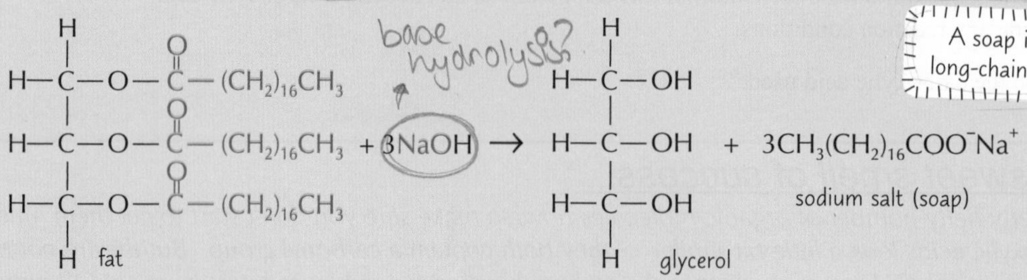

$3NaOH$
$3CH_3(CH_2)_{16}COO^-Na^+$
fat
glycerol
sodium salt (soap)

A soap is the salt of a long-chain carboxylic acid.

More on Esters

Biodiesel is a Mixture of **Methyl Esters of Fatty Acids**

1) Vegetable oils, e.g. rapeseed oil, make good vehicle fuels, but you can't burn them directly in engines.
2) The oils must be converted into **biodiesel** first.
   This involves reacting them with **methanol**, using **potassium hydroxide** as a **catalyst**.
3) You get a mixture of **methyl esters** of fatty acids — this is biodiesel.

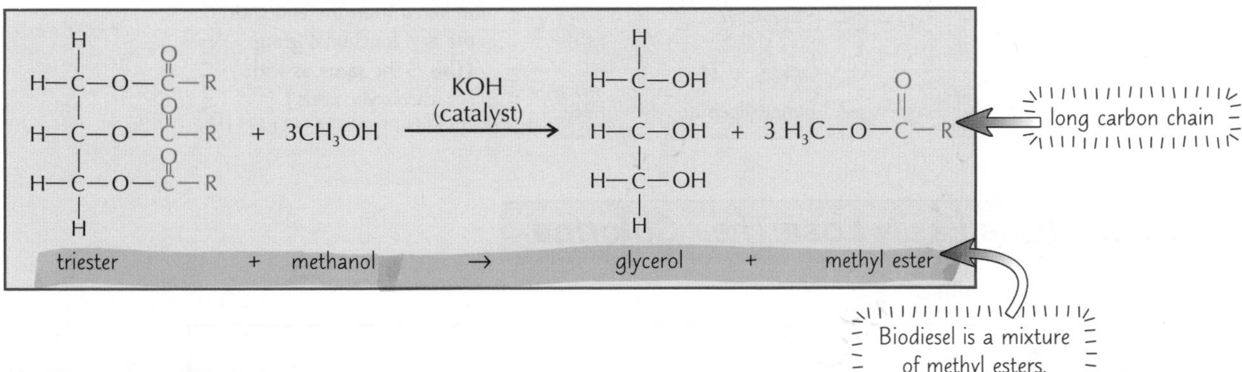

Practice Questions

Q1 Describe two different ways of hydrolysing an ester to make an alcohol.

Q2 Draw the structure of a fat.

Q3 What two products do you get when you hydrolyse an oil or fat with sodium hydroxide?

Q4 What does biodiesel consist of?

Exam Questions

Q1 Compound C, shown below, is found in raspberries.

```
     H    O
     |   //     H  H  H
  H—C—C         |  |  |
     |   \     O—C—C—C—H
     H           |  |  |
                 H CH3 H
```

a) Name compound C. [1 mark]

b) Draw and name the structures of the products formed when compound C is refluxed with dilute sulfuric acid. What kind of reaction is this? [5 marks]

c) If compound C is refluxed with excess sodium hydroxide, a similar reaction occurs. What is the difference between the products formed in this reaction and the products of the reaction described in b)? [1 mark]

Q2 When a vegetable oil was refluxed with concentrated aqueous sodium hydroxide, the products were propane-1,2,3-triol and a salt.

a) Draw the structure of propane-1,2,3-triol. [1 mark]

b) The salt was treated with excess hydrochloric acid and oleic acid, $CH_3(CH_2)_7CH{=}CH(CH_2)_7COOH$, was produced. Write an equation for the formation of this acid from its salt. [1 mark]

c) Describe a simple chemical test that you could use to distinguish oleic acid from stearic acid, $CH_3(CH_2)_{16}COOH$. [2 marks]

Sodium salts — it's all good, clean fun...

I bet you never knew that you could get from something you fry your chips in to something you wash your hands with in one small leap. There are lots of yukky, complicated structures to learn on these pages — you might find it easier if you think through where the ester breaks in each case and where the atoms of the other reactant add on. Keep at it...

Acyl Chlorides

Acyl chlorides are easy to make and are good starting points for making other types of molecule.

Acyl Chlorides have the Functional Group –COCl

Acyl (or acid) chlorides have the functional group **COCl**. All their names end in **–oyl chloride**.

ethanoyl chlorine

4-hydroxy-2,3-dimethylpentanoyl chlorine

The carbon atoms are numbered from the end with the acyl functional group. (This is the same as with carboxylic acids.)

*Acyl Chlorides Easily **Lose** Their **Chlorine***

Acyl chlorides react with...

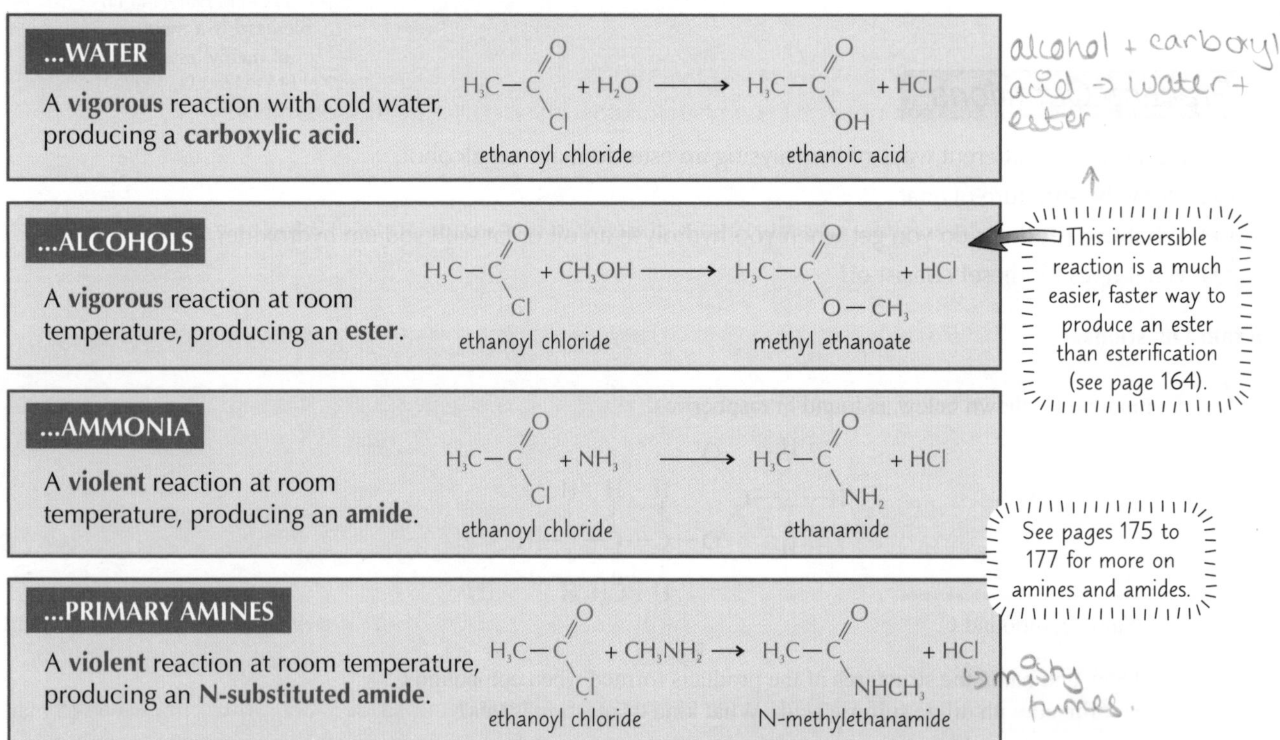

...WATER
A vigorous reaction with cold water, producing a carboxylic acid.
H3C–COCl + H2O → H3C–COOH + HCl
ethanoyl chloride
ethanoic acid
...ALCOHOLS
A vigorous reaction at room temperature, producing an ester.
H3C–COCl + CH3OH → H3C–COO–CH3 + HCl
ethanoyl chloride
methyl ethanoate
...AMMONIA
A violent reaction at room temperature, producing an amide.
H3C–COCl + NH3 → H3C–CONH2 + HCl
ethanoyl chloride
ethanamide
...PRIMARY AMINES
A violent reaction at room temperature, producing an N-substituted amide.
H3C–COCl + CH3NH2 → H3C–CONHCH3 + HCl
ethanoyl chloride
N-methylethanamide

This irreversible reaction is a much easier, faster way to produce an ester than esterification (see page 164).

See pages 175 to 177 for more on amines and amides.

Each time, **Cl** is **substituted** by an oxygen or nitrogen group and misty fumes of **hydrogen chloride** are given off.

*Acyl Chlorides and **Acid Anhydrides** React in the Same Way*

An **acid anhydride** is made from two identical carboxylic acid molecules. If you know the name of the carboxylic acid, they're easy to name — just take away '**acid**' and add '**anhydride**'.

ethanoic acid → ethanoic anhydride + H_2O

You need to know the reactions of **water**, **alcohol**, **ammonia** and **amines** with acid anhydrides. Luckily, they're almost the same as those of acyl chlorides — the reactions are just **less vigorous** and you get a **carboxylic acid** formed instead of HCl.

e.g. $(CH_3CO)_2O_{(l)} + CH_3OH_{(aq)} \rightarrow CH_3COOCH_{3\,(aq)} + CH_3COOH_{(aq)}$

ethanoic anhydride + methanol → methyl ethanoate + ethanoic acid

Acyl Chlorides

Acyl Chloride Reactions are **Nucleophilic Addition-Elimination**

In acyl chlorides, both the chlorine and the oxygen atoms draw electrons **towards** themselves, so the carbon has a slight **positive** charge — meaning it's easily attacked by **nucleophiles**.

Here's the mechanism for a **nucleophilic addition-elimination** reaction between ethanoyl chloride and methanol:

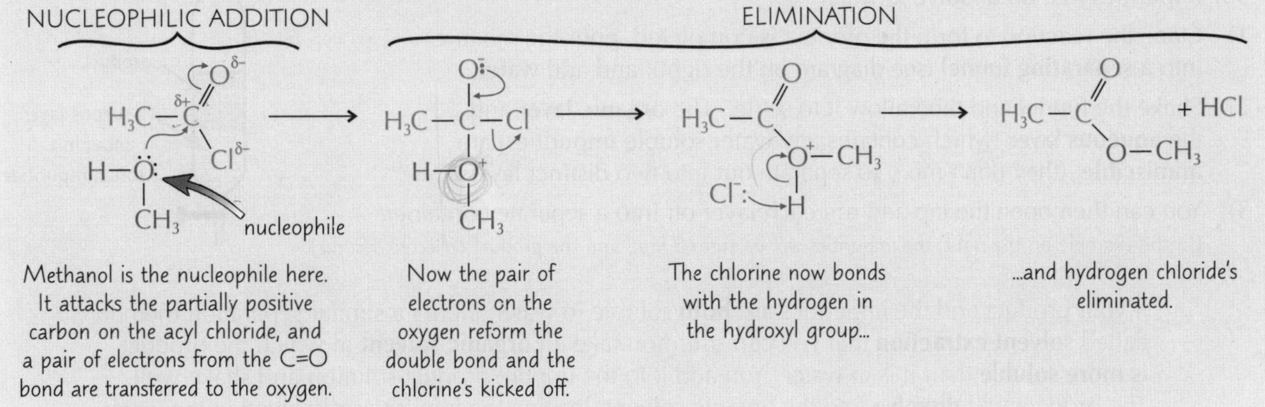

NUCLEOPHILIC ADDITION

ELIMINATION

$$H_3C{-}COCl + CH_3OH \longrightarrow H_3C{-}C(O^-)(Cl){-}O^+H{-}CH_3 \longrightarrow H_3C{-}C(=O){-}O^+(H){-}CH_3 + Cl^- \longrightarrow H_3C{-}C(=O){-}O{-}CH_3 + HCl$$

nucleophile

Methanol is the nucleophile here. It attacks the partially positive carbon on the acyl chloride, and a pair of electrons from the C=O bond are transferred to the oxygen.

Now the pair of electrons on the oxygen reform the double bond and the chlorine's kicked off.

The chlorine now bonds with the hydrogen in the hydroxyl group...

...and hydrogen chloride's eliminated.

The other reactions of acyl chlorides that you need to know all work in exactly the same way. You just need to change the nucleophile to water ($H_2\ddot{O}$:), ammonia ($\ddot{N}H_3$) or an amine (e.g. $CH_3\ddot{N}H_2$).

Ethanoic Anhydride is Used for the **Manufacture** of **Aspirin**

Aspirin is an **ester** — it's made by reacting **salicylic acid** with **ethanoic anhydride** or **ethanoyl chloride**.

Ethanoic anhydride is used in industry because:

- it's **cheaper** than ethanoyl chloride.
- it's **safer** to use than ethanoyl chloride as it's **less corrosive**, reacts **more slowly** with water, and **doesn't** produce dangerous **hydrogen chloride** fumes.

Reacting **salicylic acid** with **ethanoic anhydride** to make **aspirin**:

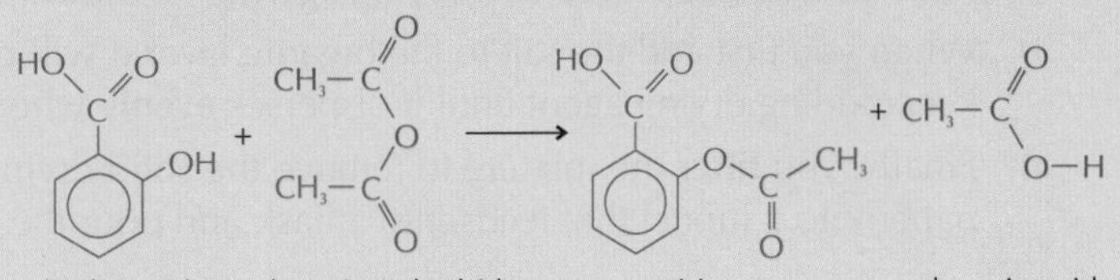

salicylic acid + ethanoic anhydride ⟶ aspirin + ethanoic acid ($CH_3{-}COOH$)

Practice Questions

Q1 Write the equation for the reaction between ethanoyl chloride and ammonia.

Q2 What part of an acyl chloride is attacked by nucleophiles?

Q3 Give TWO reasons why ethanoic anhydride is preferred to ethanoyl chloride when producing aspirin.

Exam Questions

Q1 Ethanoyl chloride and ethanoic anhydride both react with methanol.

a) Write equations for both reactions and name the organic product that is formed in both reactions. [3 marks]

b) Give an observation that could be made for the reaction with ethanoyl chloride that would not occur with ethanoic anhydride. [1 mark]

c) Ethanoic acid can also be used with methanol to prepare the organic product named in (a). Give one advantage of using ethanoyl chloride. [1 mark]

Q2 Ethanoyl chloride and ethylamine react together at room temperature.

a) Write an equation for this reaction and name the organic product. [2 marks]

b) Draw a mechanism for this reaction. [4 marks]

I'll take the low road, you take the hydride...

As if all those acyl chlorides reactions weren't enough, you've got to know about acid anhydrides reactions too. It may sound like a bit of a bore, but acid anhydrides and acyl chlorides do react in the same way — that makes things easier.

Purifying Organic Compounds

Synthesising organic compounds is hardly ever as simple as it sounds. The products of organic reactions are almost always riddled with impurities. It's a good thing there are many magical ways of purifying them, outlined on these pages.

Separation Removes Water Soluble Impurities From a Product

If a product is **insoluble** in water then you can use **separation** to remove any impurities that **do dissolve** in water.

1) Once the reaction to form the product is completed, pour the mixture into a **separating funnel** (see diagram on the right), and add **water**.
2) Shake the funnel and then allow it to settle. The **organic layer** and the **aqueous layer** (which contains any water soluble impurities) are **immiscible**, (they don't mix), so separate out into two distinct layers.
3) You can then open the tap and run each layer off into a separate container. (In the example on the right, the impurities will be run off first, and the product collected second.)

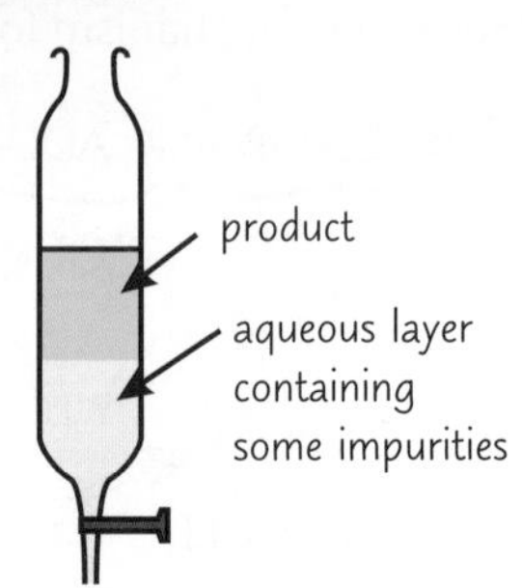

If your product and the impurities are **both** soluble in water, there's a similar separation method called **solvent extraction** that you can use. You take an **organic solvent** in which the product is **more soluble** than it is in water. You add it to the impure product solution and shake well. The product will **dissolve** into the organic solvent, leaving the impurities dissolved in the water. The solvent containing the product can then be run off using a **separating funnel**, as above.

Remove Water from a Purified Product by Drying it

1) If you use separation to purify a product, the organic layer will end up containing **trace amounts** of **water** — so it has to be **dried**.
2) To do this, you add an **anhydrous salt** such as **magnesium sulfate** ($MgSO_4$) or **calcium chloride** ($CaCl_2$). The salt is used as a **drying agent** — it **binds** to any water present to become **hydrated**.
3) When you first add the salt to the organic layer it will **clump** together. You keep adding drying agent until it disperses **evenly** when you swirl the flask.
4) Finally, you **filter** the mixture to remove the solid drying agent — pop a piece of filter paper into a funnel that feeds into a flask and pour the mixture into the filter paper.

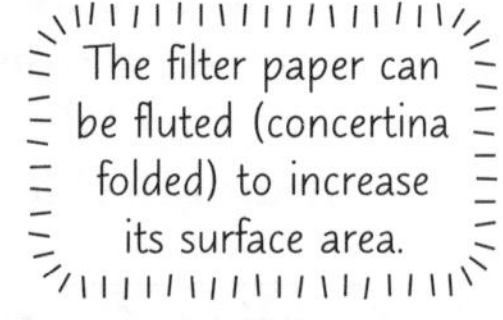

The filter paper can be fluted (concertina folded) to increase its surface area.

Remove Other Impurities by Washing

The product of a reaction can be **contaminated** with leftover reagents or unwanted side products. You can **remove** some of these by **washing** the product (which in this case means adding another liquid and shaking).

For example, aqueous **sodium hydrogencarbonate** can be added to an impure product in solution to remove **acid** from it. The acid reacts with the sodium hydrogencarbonate to give CO_2 gas, and the organic product can be removed using a separating funnel (as above).

Volatile Liquids Can be Purified by Distillation

You've already met distillation on page 149.

Distillation separates out liquids with different boiling points. It works by gently **heating** a mixture in **distillation apparatus**. The substances will evaporate out of the mixture in order of increasing boiling point.

1) Connect a **condenser** to a **round bottomed flask** containing your impure product in solution.
2) Place a **thermometer** in the neck of the flask so that the bulb sits next to the entrance to the condenser. The temperature on the thermometer will show the boiling point of the substance that's evaporating at any given time.
3) **Heat** the impure product. (Many organic chemicals are flammable, so you should use an electric heater.)
4) When the product that you want to collect **boils** (i.e. when the thermometer is showing its boiling point), place a flask at the open end of the condenser to collect your **pure product**.

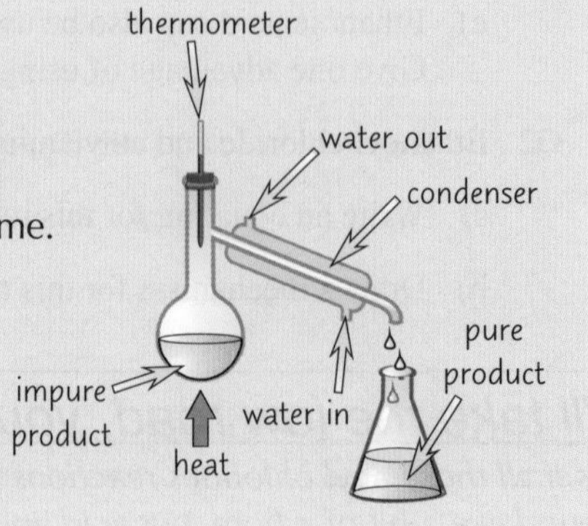

Purifying Organic Compounds

Organic Solids can be *Purified* by *Recrystallisation*

If the product of an organic reaction is a solid, then the simplest way of purifying it is **recrystallisation**.

First, you dissolve your solid in a hot solvent to make a **saturated** solution (that's a solution in which the **maximum possible** amount of solid is dissolved in the solvent). Then you let it cool. As the solution cools, the **solubility** of the product falls. When it reaches the point where it can't stay in solution, it forms pure **crystals**. Here's how it's done:

1) Add **very hot solvent** to the **impure** solid until it **just** dissolves. It's really important not to add too much solvent — this should give a **saturated solution** of the **impure product**.
2) Filter the hot solution through a **heated funnel** to remove any **insoluble impurities**.
3) Leave the solution to **cool** down **slowly**. **Crystals** of the **product** will form as it cools.
4) Remove the liquid containing the **soluble impurities** from the crystals by **filtering** the mixture under **reduced pressure**. (To do this, you pour the mixture into a filter paper lined **Büchner funnel** – a flat-bottomed funnel with holes in the base – that's sitting in a **side-arm flask** attached to a **vacuum line**.)
5) Finally, **wash** the crystals with ice-cold solvent to remove any soluble impurities from their surface. Leave your purified crystals to **dry**.

Melting and *Boiling* Points are Good Indicators of *Purity*

Pure substances have a **specific melting** and **boiling point**. If they're **impure**, the **melting point's lowered** and the **boiling point is raised**. If they're **very impure**, melting and boiling will occur across a wide range of temperatures.

1) You can use **melting point apparatus** to accurately determine the melting point of an **organic solid**.
2) Pack a small sample of the solid into a **glass capillary tube** and place it inside the **heating element**.
3) **Increase the temperature** until the sample turns from solid to **liquid**.
4) You usually measure a **melting range**, which is the range of temperatures from where the solid **begins to melt** to where it has **melted completely**.
5) You can look up the melting point of a substance in **data books** and compare it to your measurements.
6) **Impurities** in the sample will **lower** the **melting point** and **broaden** the **melting range**.

Practice Questions

Q1 Describe the procedure for drying an organic product using a drying agent.

Q2 Describe how you would purify a product using distillation.

Exam Questions

Q1 A student carries out an organic synthesis. The product that he makes is insoluble in water and contains water soluble impurities. Which of the following techniques would be the most appropriate for extracting the product?

A Distillation B Separation C Recrystallisation D Drying [1 mark]

Q2 A scientist has produced some impure solid sodium ethanoate, which she wants to purify using recrystallisation. She begins by dissolving the impure sodium ethanoate in the minimum possible amount of hot solvent.

a) Explain why the scientist used the minimum possible amount of hot solvent. [1 mark]

b) Outline the rest of the procedure that the scientist would need to follow to recrystallise the solid. [5 marks]

c) Describe the melting point range of the impure sodium ethanoate compared to the pure product. [1 mark]

My organic compound isn't volatile — it's just highly strung...

Nobody wants loads of impurities in their reaction products. But now you're well kitted out to get rid of them using these purification techniques. Once you think you've got your product pure, you can check its melting point to make sure.

Aromatic Compounds

We start this section with a fantastical tale about the magical Ring of Benzene. Our story opens in a far away land where our hero is throwing a party for some dwarves... Actually no, that's something else.

Benzene has a **Ring** Of Carbon Atoms

A plane ring

1) Benzene has the formula C_6H_6. It has a **planar cyclic** structure (which is just a complicated way of saying that its six carbon atoms are joined together in a **flat ring**).
2) Each carbon atom forms single covalent bonds to the carbons on either side of it and to one hydrogen atom. The final unpaired electron on each carbon atom is located in a p-orbital that sticks out above and below the plane of the ring. The p-orbitals on each carbon atom combine to form a **ring** of **delocalised electrons**.
('Delocalised' just means that the electrons don't belong to a specific carbon atom, but are shared between them all.)
3) All the carbon-carbon bonds in the ring are the same, so they are the **same length** — 140 pm. This lies in between the length of a single C–C bond (154 pm) and a double C=C bond (135 pm).

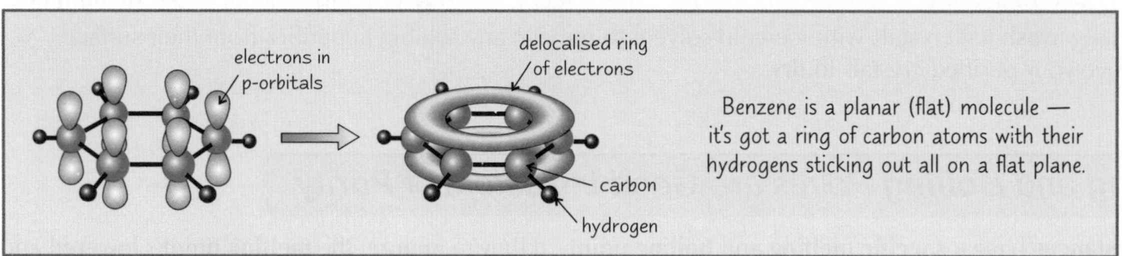
electrons in p-orbitals
delocalised ring of electrons
carbon
hydrogen
Benzene is a planar (flat) molecule — it's got a ring of carbon atoms with their hydrogens sticking out all on a flat plane.

You usually draw benzene like this:

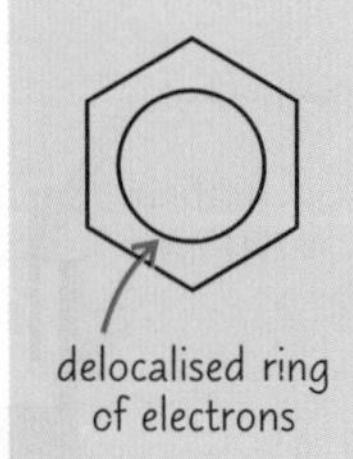
delocalised ring of electrons

Remember, though — there's a hydrogen attached to each of the carbons in the ring, like this

delocalised ring of electrons

You may also see benzene drawn like this:

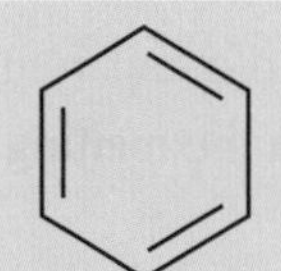

Don't get confused — there aren't really alternating single and double bonds between the carbon atoms. Scientists used to think this was the structure, and their way of drawing it has just stuck around.

The **Delocalised Ring** of Electrons Makes Benzene Very **Stable**

Benzene is far more **stable** than the theoretical compound cyclohexa-1,3,5-triene would be (where the ring would be made up of alternating single and double bonds). You can see this by comparing the enthalpy change of hydrogenation for benzene with the enthalpy change of hydrogenation for cyclohexene:

1) Cyclohexene has **one** double bond. When it's hydrogenated, the enthalpy change is **–120 kJ mol⁻¹**. If benzene had three double bonds, you'd expect it to have an enthalpy of hydrogenation of –360 kJ mol⁻¹.

cyclohexene $+ H_2 \rightarrow$ cyclohexane $\Delta H^{\ominus}_{hydrogenation} = -120\ kJ\ mol^{-1}$

2) But the **experimental** enthalpy of hydrogenation of benzene is **–208 kJ mol⁻¹** — far **less exothermic** than expected.

cyclohexa-1,3,5-triene $+ 3H_2 \rightarrow$ cyclohexane

predicted $\Delta H^{\ominus}_{hydrogenation} = -360\ kJ\ mol^{-1}$

actual $\Delta H^{\ominus}_{hydrogenation} = -208\ kJ\ mol^{-1}$

3) Energy is put in to break bonds and released when bonds are made. So **more energy** must have been put in to break the bonds in benzene than would be needed to break the bonds in a theoretical cyclohexa-1,3,5-triene molecule.
4) This difference indicates that benzene is **more stable** than cyclohexa-1,3,5-triene would be. This is thought to be due to the **delocalised ring of electrons**.

Aromatic Compounds

Aromatic Compounds are Derived from Benzene

Compounds containing a **benzene ring** are called **arenes** or '**aromatic compounds**'.
There are **two** ways of **naming** arenes — here are some examples:

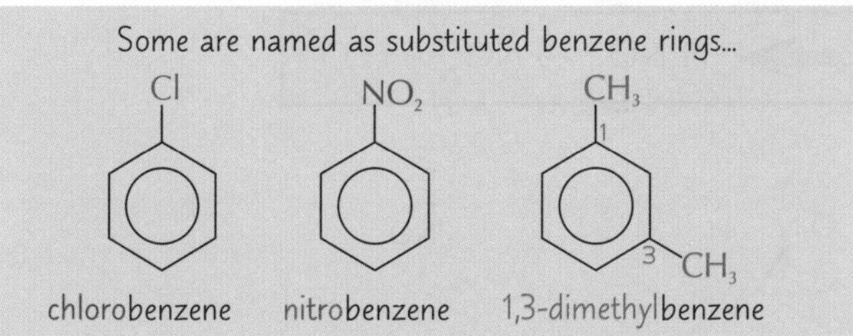

...others are named as compounds with a phenyl group (C_6H_5) attached.

OH
NH_2
phenol
phenylamine

Arenes Undergo Electrophilic Substitution Reactions

The benzene ring is a region of **high electron density**, so it attracts **electrophiles.** As the benzene ring's so stable, it **doesn't** undergo **electrophilic addition** reactions, which would destroy the delocalised ring of electrons. Instead, it undergoes **electrophilic substitution reactions** where one of the hydrogen atoms (or another functional group on the ring) is substituted for the electrophile.

You need to know two **electrophilic substitution mechanisms** for benzene — **Friedel-Crafts acylation** (shown below) and the **nitration reaction** on the next page.

Friedel-Crafts Acylation Reactions Produce Phenylketones

1) Many **useful** chemicals such as **dyes** and **pharmaceuticals** contain benzene rings. But because benzene is so **stable**, it's fairly **unreactive** — so it can be tricky to make chemicals that contain benzene.
2) **Friedel-Crafts acylation** reactions are used to add an **acyl group** (**RCO–**) to the benzene ring. Once an acyl group has been added, the side chains can be **modified** using further reactions to make **useful products**.
3) An electrophile has to have a strong **positive charge** to be able to attack the stable benzene ring — most aren't **polarised** enough. But some can be made into **stronger electrophiles** using a catalyst called a **halogen carrier**.
4) Friedel-Crafts acylation uses an **acyl chloride** (see page 168) as an electrophile and a **halogen carrier**, e.g. $AlCl_3$.

Here's how the $AlCl_3$ makes the acyl chloride electrophile stronger:
$AlCl_3$ accepts a **lone pair of electrons** from the acyl chloride. As the lone pair of electrons is pulled away, the **polarisation** in the acyl chloride **increases** and it forms a **carbocation**. This makes it a much **stronger electrophile**, and gives it a strong enough charge to **react** with the **benzene ring**.

$R-C(=O^{\delta-})-Cl^{\delta-}$ (acyl chloride, $C^{\delta+}$) ---→ $AlCl_3$ (halogen carrier) ⟶ $R-C^{+}=O$ $AlCl_4^-$ (carbocation)

5) Here's how the electrophile is substituted into the benzene ring. The mechanism has **two steps**:

benzene, carbocation, $R-\overset{+}{C}=O$ ⟶ intermediate with H, + charge on ring

$Cl-\overset{-}{Al}(Cl)_2-Cl$ ⟶ phenylketone + HCl + $AlCl_3$

1) **Electrons** in the benzene ring are **attracted** to the positively charged **carbocation**. Two electrons from the benzene **bond** with the carbocation. This **partially breaks** the **delocalised ring** and gives it a **positive charge**.

2) The **negatively charged $AlCl_4^-$** ion is attracted to the **positively charged ring**. One **chloride ion** breaks away from the aluminium chloride ion and **bonds** with the **hydrogen** ion. This **removes the hydrogen** from the ring forming **HCl**. It also allows the catalyst to reform.

6) The reactants need to be **heated under reflux** in a **non-aqueous solvent** (like dry ether) for the reaction to occur.

Aromatic Compounds

Nitration is Used in the *Manufacture* of *Explosives and Dyes*

When you warm **benzene** with **concentrated nitric** and **sulfuric acids**, you get **nitrobenzene**. Sulfuric acid acts as a **catalyst** — it helps to make the nitronium ion, NO_2^+, which is the electrophile.

$$HNO_3 + H_2SO_4 \rightarrow H_2NO_3^+ + HSO_4^- \Longrightarrow H_2NO_3^+ \rightarrow NO_2^+ + H_2O$$

Now here's the electrophilic substitution mechanism:

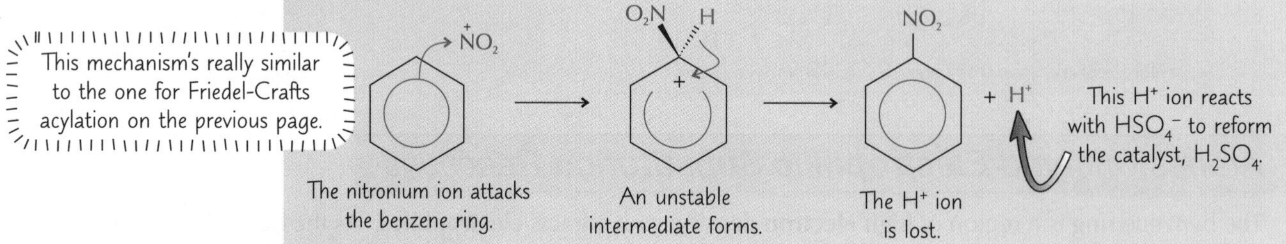
This mechanism's really similar to the one for Friedel-Crafts acylation on the previous page.
NO_2^+
O_2N H
+
NO_2
+ H^+
This H^+ ion reacts with HSO_4^- to reform the catalyst, H_2SO_4.
The nitronium ion attacks the benzene ring.
An unstable intermediate forms.
The H^+ ion is lost.

If you only want one NO_2 group added (**mononitration**), you need to keep the temperature **below 55 °C**. Above this temperature you'll get lots of substitutions.

Nitration reactions are really useful

1) Nitro compounds can be **reduced** to form **aromatic amines** (see page 175). These are used to manufacture **dyes** and **pharmaceuticals**.
2) Some nitro compounds can be used as **explosives**, such as 2,4,6-trinitromethylbenzene (**trinitrotoluene** — TNT).

Practice Questions

Q1 Explain why electrophiles are attracted to benzene.

Q2 Which halogen carrier is used in the Friedel-Crafts acylation reaction?

Q3 What is used as a catalyst in the nitration of benzene?

Q4 What type of compound is this?

Exam Questions

Q1 A halogen carrier, such as $AlCl_3$, is used as a catalyst in the reaction between benzene and ethanoyl chloride.

a) Describe the conditions needed for this reaction. [1 mark]

b) Explain why the halogen carrier is needed as a catalyst for this reaction to occur. [2 marks]

c) Draw the structure of the electrophile that attacks the benzene ring. [1 mark]

Q2 An electrophilic substitution reaction of benzene is summarised in the diagram on the right.

a) Name the product A, and the reagents B and C, and give the conditions D. [3 marks]

b) Write equations to show the formation of the electrophile. [2 marks]

c) Outline a mechanism for this reaction. [2 marks]

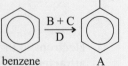

Everyone needs a bit of stability in their life...

The structure of benzene is really odd — even top scientists struggled to find out what its molecular structure was. If you're asked why benzene reacts the way it does, it's bound to be something to do with the ring of delocalised electrons. Remember there's a hydrogen at every point on the benzene ring — it's easy to forget they're there.

Amines and Amides

Two more types of organic compound coming up — amines and amides. Both these families contain nitrogen atoms.

Amines are Organic Derivatives of Ammonia

If one or more of the **hydrogens** in **ammonia** (NH_3) is replaced with an organic group, you get an **amine**. If **one** hydrogen is **replaced** with an organic group, you get a **primary amine**. If **two** are replaced, it's a **secondary amine**, and **three** means it's a **tertiary amine**. The lone pair of electrons on the nitrogen atom in a tertiary amine can also bond with a fourth organic group — that gives you a **quaternary ammonium ion**.

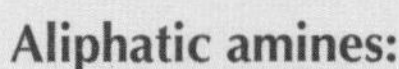

Aliphatic amines:

H_3C-NH_2	$H_3C-NH(CH_3)$	$H_3C-N(CH_3)_2$	$[H_3C-N^+(CH_3)_3]$
methylamine (primary amine)	dimethylamine (secondary amine)	trimethylamine (tertiary amine)	tetramethylamine ion (quaternary ammonium ion)

Aromatic amine:

$C_6H_5-NH_2$

phenylamine (primary amine)

Quaternary Ammonium Salts are Used as Cationic Surfactants

Because quaternary ammonium ions are **positively charged**, they will hang around with any negative ions that are near. The complexes formed are called **quaternary ammonium salts** — like **tetramethylammonium chloride, $(CH_3)_4N^+Cl^-$.**

Quaternary ammonium salts with at least one long hydrocarbon chain are used as **cationic surfactants**. The hydrocarbon tail will bind to nonpolar substances such as **grease**, whilst the **cationic** head will dissolve in water, so they are useful in things like **fabric cleaners** and **hair products**.

CH_3, Cl^-, N^+-CH_3, CH_3

In addition, the **positively charged** part (ammonium ion) will bind to negatively charged surfaces such as hair and fibre. This gets rid of **static**, so they are often used in **fabric conditioners**.

Amines Have A Lone Pair of Electrons

1) Amines act as **weak bases** because they **accept protons**. There's a **lone pair of electrons** on the **nitrogen** atom that can form a **dative covalent (coordinate) bond** with an H^+ ion.

$$RNH_2 + H^+ \longrightarrow [RNH_3]^+$$

2) The **strength** of the **base** depends on how **available** the nitrogen's lone pair of electrons is. The more **available** the **lone pair** is, the more likely the amine is to **accept a proton**, and the **stronger** a base it will be. A **lone pair** of electrons will be **more available** if its **electron density** is **higher**.

3) **Primary aliphatic amines** are **stronger** bases than **ammonia**, which is a **stronger** base than **aromatic amines**. Here's why:

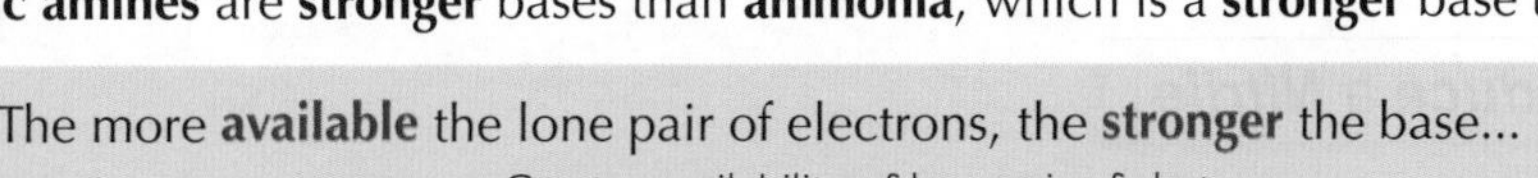

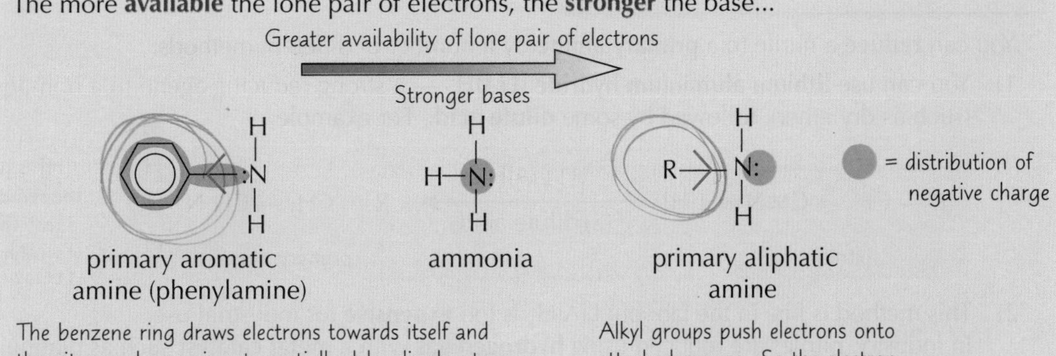

The benzene ring draws electrons towards itself and the nitrogen lone pair gets partially delocalised onto the ring. So the electron density on the nitrogen decreases, making the lone pair much less available.

Alkyl groups push electrons onto attached groups. So the electron density on the nitrogen atom increases. This makes the lone pair more available.

4) The lone pair of electrons also means that amines are **nucleophiles**. They react with **halogenoalkanes** in a **nucleophilic substitution reaction** (see next page), or with **acyl chlorides** and **acid anhydrides** in **nucleophilic addition-elimination** reactions (see pages 168-169).

Amines and Amides

Aliphatic Amines are made from Halogenoalkanes or Nitriles

There are **two** ways to produce aliphatic amines — either from **halogenoalkanes** or by **reducing nitriles**. (The method for producing **aromatic amines** is different again — as you'll see on the next page.)

You Can Heat a Halogenoalkane with Ammonia...

Amines can be made by heating a **halogenoalkane** with **excess ammonia**.

Example: Ethylamine can be made by reacting ammonia with bromoethane:

$$2NH_3 + CH_3CH_2Br \rightarrow CH_3CH_2NH_2 + NH_4Br$$

The mechanism for this reaction is:

ammonia + halogenoalkane ⟶ alkylammonium salt

$$H_3N{:} + CH_3CH_2^{\delta+}{-}Br^{\delta-} \longrightarrow H_3\overset{+}{N}{-}CH_2CH_3 \; Br^-$$

Ammonia attacks the carbon in the halogenoalkane

The halogen is released

...then...

alkylammonium salt ⇌ primary amine + ammonium salt

$$Br^- \; H_3\overset{+}{N}{-}CH_2CH_3 + {:}NH_3 \rightleftharpoons H_2\ddot{N}{-}CH_2CH_3 + NH_4^+ Br^-$$

A second ammonia molecule donates its lone pair of electrons to a hydrogen, which breaks off from the alkylammonium salt.

If you make amines using this method, you end up with a **mixture** of primary, secondary and tertiary amines and quaternary ammonium salts. This is because the primary amine that you produce first has a **lone pair of electrons** — it's a **nucleophile**. This means that it can react with any remaining halogenoalkane in a **nucleophilic substitution reaction**. As long as there's some of the halogenoalkane around, further substitutions can take place. They keep happening until you get a **quaternary ammonium salt**, which can't react any further as it has no lone pair:

$$R{-}\ddot{N}H_2 \xrightarrow{\text{halogenoalkane}} R{-}\ddot{N}HR \xrightarrow{\text{halogenoalkane}} R{-}\ddot{N}R_2 \xrightarrow{\text{halogenoalkane}} R{-}\overset{+}{N}R_3$$

primary amine — secondary amine — tertiary amine — quaternary ammonium ion

The **mechanism** is similar to the reaction of ammonia with a halogenoalkane — two **amine molecules** react with the halogenoalkane in succession to form a **more substituted amine** (e.g. a primary amine forms a secondary amine) and an **ammonium salt** with a similar structure to the original amine. For example:

$$2\,H_3C{-}NH_2 + CH_3CH_2Br \rightarrow H_3C{-}NH{-}CH_2CH_3 + CH_3NH_3^+Br^-$$

...Or You Can Reduce a Nitrile

You can **reduce** a nitrile to a **primary amine** by a number of different methods:

1) You can use **lithium aluminium hydride** ($LiAlH_4$ — a strong reducing agent) in a non-aqueous solvent (such as dry ether), followed by some **dilute acid**. For example:

$$\underset{\text{nitrile}}{R-CH_2-C{\equiv}N} + 4[H] \xrightarrow[\text{(2) dilute acid}]{\text{(1) } LiAlH_4} \underset{\text{primary amine}}{R-CH_2-CH_2NH_2}$$

[H] is just the reducing agent (here it's $LiAlH_4$).

2) This method is fine in the lab, but $LiAlH_4$ is too **expensive** for industrial use. In industry, nitriles are reduced using **hydrogen gas** with a **metal catalyst** such as platinum or nickel at high temperature and pressure. This is called **catalytic hydrogenation**. For example:

$$\underset{\text{nitrile}}{R-CH_2-C{\equiv}N} + 2H_2 \xrightarrow[\text{high temperature and pressure}]{\text{nickel catalyst}} \underset{\text{primary amine}}{R-CH_2-CH_2NH_2}$$

Becky was reduced to tears by lithium aluminium hydride.

Amines and Amides

Aromatic Amines are made by *Reducing* a *Nitro Compound*

Aromatic amines are produced by **reducing** a nitro compound, such as **nitrobenzene**.
There are **two steps** to the method:

1) First you need to heat a mixture of a **nitro compound**, **tin metal** and **concentrated hydrochloric acid** under **reflux** — this makes a salt. For example, if you use nitrobenzene, the salt formed is $C_6H_5NH_3^+Cl^-$.
2) Then to turn the salt into an **aromatic amine**, you need to add an alkali, such as **sodium hydroxide** solution.

$$C_6H_5NO_2 + 6[H] \xrightarrow[\text{(2) NaOH}]{\text{(1) tin, conc. HCl, reflux}} C_6H_5NH_2 + 2H_2O$$

nitrobenzene → phenylamine

3) Aromatic amines are useful compounds in organic synthesis — they're used as the starting molecules for lots of **dyes** and **pharmaceuticals**.

And now for a weeny bit about amides...

Amides are Carboxylic Acid Derivatives

Amides contain the functional group **$-CONH_2$**.
The **carbonyl group** pulls electrons away from the NH_2 group, so amides behave differently from amines.

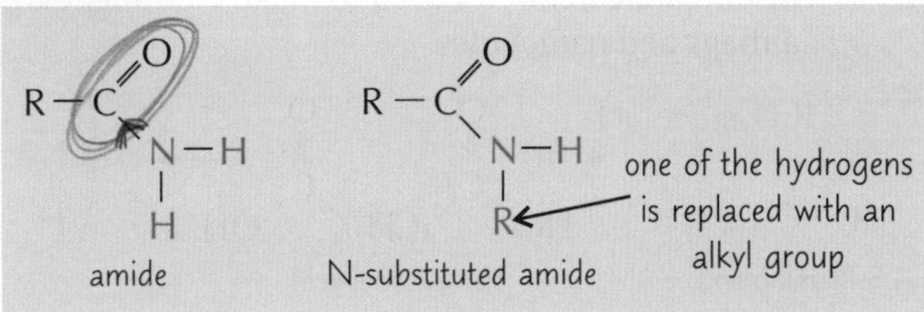

Practice Questions

Q1 Predict, with reasoning, whether ammonia or ethylamine will be a stronger base.

Q2 Explain why amines and ammonia can act as nucleophiles.

Q3 What conditions are needed to reduce nitrobenzene to phenylamine?

Exam Questions

Q1 a) Explain how methylamine, CH_3NH_2, can act as a base. [1 mark]

b) Methylamine is a stronger base than ammonia, NH_3. However, phenylamine, $C_6H_5NH_2$, is a weaker base than ammonia. Explain these differences in base strength. [2 marks]

Q2 a) Propylamine can be synthesised from bromopropane. Suggest a disadvantage of this synthesis route. [1 mark]

b) Propylamine can also be synthesised from propanenitrile.

i) Suggest suitable reagents for its preparation in a laboratory. [1 mark]

ii) Why is this method not suitable for industrial use? [1 mark]

iii) What reagents and conditions are used in industry? [2 marks]

Q3 Ethylamine can react with bromoethane to form a compound of molecular formula $C_4H_{11}N$. Write an equation and outline a mechanism for the reaction [3 marks]

You've got to learn it — amine it might come up in your exam...

Did you know that rotting fish smells so bad because the flesh releases diamines as it decomposes? But the real question is: is it fish that smells of amines or amines that smell of fish — it's one of those chicken or egg things that no one can answer. Well, enough philosophical pondering — we all know the answer to the meaning of life. It's 42.

Condensation Polymers

You met addition polymers back in Unit 3 Section 3. Now it's time to meet their big brothers — condensation polymers...

Condensation Polymers Include **Polyamides**, **Polyesters** and **Polypeptides**

1) **Condensation polymerisation** usually involves two different types of monomer, each with at least **two functional groups**. Each functional group reacts with a group on another monomer to form a link, creating polymer chains.
2) Each time a link is formed, a small molecule is lost (water) — that's why it's called **condensation** polymerisation.
3) Examples of condensation polymers include **polyamides**, **polyesters** and **polypeptides** (or proteins, see page 184).

Reactions Between **Dicarboxylic Acids** and **Diamines** Make **Polyamides**

The **carboxyl** groups of **dicarboxylic acids** react with the **amino** groups of **diamines** to form **amide links**. Dicarboxylic acids and diamines have functional groups at each end of the molecule, so **long chains** can form.

a water molecule is eliminated

HO–C(=O)–R–C(=O)–OH + H–N(H)–R'–N(H)–H ⟶ HO–C(=O)–R–(C(=O)–N(H))–R'–N(H)–H + H_2O

dicarboxylic acid — diamine — amide link

This is the formula of the polymer. The bit inside the brackets is called the repeating unit of the polymer.

Example: Nylon 6,6 is a polyamide made from **1,6-diaminohexane** and **hexanedioic acid**. It's used to make clothing, carpet, rope, airbags and parachutes.

n $H_2N(CH_2)_6NH_2$ + n $HOOC(CH_2)_4COOH$ ⟶ $[\,-NH-(CH_2)_6-NH-CO-(CH_2)_4-CO-\,]_n$ + $2nH_2O$

1,6-diaminohexane — hexanedioic acid — nylon 6,6

Example: Kevlar® is a polyamide made from **1,4-diaminobenzene** and **benzene-1,4-dicarboxylic acid**. It's used in bulletproof vests, boat construction, car tyres and lightweight sports equipment.

n $H_2N-C_6H_4-NH_2$ + n $HOOC-C_6H_4-COOH$ ⟶ $[\,-NH-C_6H_4-NH-CO-C_6H_4-CO-\,]_n$ + $2nH_2O$

1,4-diaminobenzene — benzene-1,4-dicarboxylic acid — Kevlar®

Reactions Between **Dicarboxylic Acids** and **Diols** Make **Polyesters**

The **carboxyl** groups of dicarboxylic acids can also react with the **hydroxyl** groups of **diols** to form **ester links**. Polymers joined by **ester links** are called **polyesters**.

a water molecule is eliminated

HO–C(=O)–R–C(=O)–OH + H–O–R'–O–H ⟶ HO–C(=O)–R–(C(=O)–O)–R'–O–H + H_2O

dicarboxylic acid — diol — ester link

Example: Terylene™ (PET) — formed from **benzene-1,4-dicarboxylic acid** and **ethane-1,2-diol**. It's used in plastic bottles, clothing, sheets and sails.

n $HOOC-C_6H_4-COOH$ + n $HO-CH_2-CH_2-OH$ ⟶ $[\,-O-CO-C_6H_4-CO-O-CH_2-CH_2-\,]_n$ + $2nH_2O$

benzene-1,4-dicarboxylic acid — ethane-1,2-diol — Terylene™

Condensation Polymers

Hydrolysis Produces the Original Monomers

1) Condensation polymerisation can be reversed by **hydrolysis** — water molecules are added back in and the links are broken. For example, this equation shows a polyamide being hydrolysed:

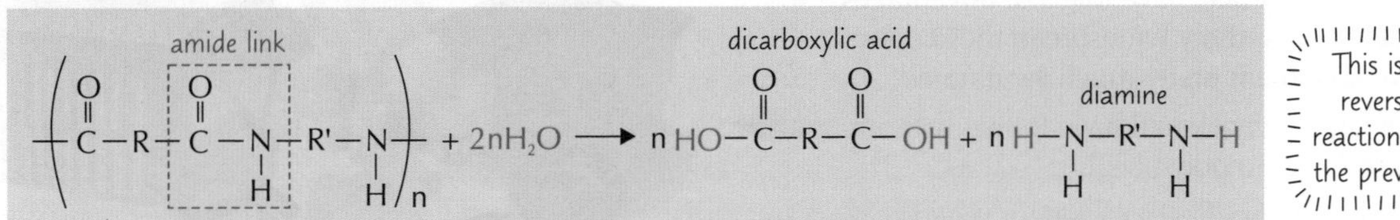

This is just the reverse of the reaction shown on the previous page.

2) To draw the **monomers** from the repeating unit of a condensation polymer, break the chain through the **middle bond** of the amide or ester link. Then just add an **OH** or an **H** to each end of both new molecules.

   If you're starting from a diagram showing a section of the polymer chain and you need to draw the repeating unit first, you just need to find and draw out the longest chunk of the chain that repeats.

3) To work out what you need to add where, just remember — for a **polyamide** you want the monomers to be a **dicarboxylic acid** and a **diamine**, and for a **polyester** you need to end up with a **dicarboxylic acid** and a **diol**.

Condensation Polymers Contain Polar Bonds

1) Condensation polymers are generally stronger and more rigid than addition polymers (see pages 146-147).
2) This is because condensation polymers are made up of chains containing **polar bonds**, e.g. C–N and C–O. So, as well as Van der Waals forces, there are **permanent dipole-dipole forces** and **hydrogen bonds** between the polymer chains.

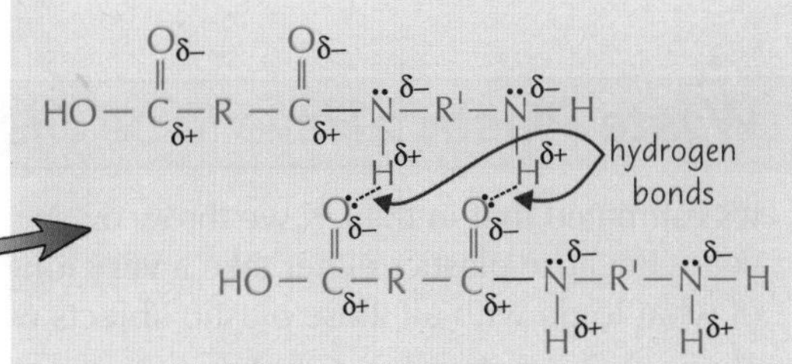

Practice Questions

Q1 What molecule is eliminated when a polyester is made?

Q2 Why are condensation polymers usually stronger than addition polymers?

Exam Questions

Q1 Kevlar® is a polymer used in bulletproof vests. Its repeating unit is shown on the right.

$$-\underset{H}{N}-C_6H_4-\underset{H}{N}-\overset{O}{\overset{\|}{C}}-C_6H_4-\overset{O}{\overset{\|}{C}}-$$

a) What type of polymer is Kevlar®? [1 mark]

b) Kevlar® is made by reacting two different monomers together. What type of compounds are each of these monomers? [1 mark]

c) Which reaction could you use to break up Kevlar® into its constituent monomers? [1 mark]

Q2 Nylon 6,6 is the most commonly produced nylon. A section of the polymer chain is shown on the right.

$$-\underset{H}{N}-(CH_2)_6-\underset{H}{N}-\overset{O}{\overset{\|}{C}}-(CH_2)_4-\overset{O}{\overset{\|}{C}}-\underset{H}{N}-(CH_2)_6-\underset{H}{N}-\overset{O}{\overset{\|}{C}}-(CH_2)_4-\overset{O}{\overset{\|}{C}}-$$

Draw the structural formulas of the monomers from which nylon 6,6 is formed. It is not necessary to draw the carbon chains out in full. [2 marks]

Q3 A polyester is formed by the reaction between the monomers hexanedioic acid and 1,6-hexanediol.

a) Draw the repeating unit for the polyester. [1 mark]

b) Explain why this is an example of condensation polymerisation. [1 mark]

Conversation polymerisation — when someone just goes on and on and on...

Condensation polymers are like people who are in an on-off relationship. They get together, then they hydrolyse apart, only to get back together again. And you have to keep up with it — monomers, polymers, amides, esters and all. You'll also need to know the structures and links involved in nylon 6,6, Kevlar® and Terylene™ for your exams. So get swotting...

Disposing of Polymers

Polymers are amazingly useful. But they have one big drawback...

Polymers — **Useful** but Difficult to **Get Rid Of**

1) Synthetic polymers have loads of **advantages**, so they're incredibly widespread these days — we take them pretty much for granted.
   Just imagine what you'd have to live without if there were no polymers...

(Okay... I could live without the polystyrene head, but the rest of this stuff is pretty useful.)

2) **Polyalkenes** such as poly(ethene) and polystyrene are addition polymers. They are made up of **non-polar** carbon chains, which makes them unreactive and chemically inert.
3) This is an advantage when they are being used — e.g. a polystyrene cup won't react with your coffee, but has the disadvantage of making them **non-biodegradable**.
4) **Condensation polymers** such as polyesters and polyamides do have **polar bonds** in their chains, which makes them open to attack by **nucleophiles**.
5) This means that condensation polymers can be broken down by **hydrolysis** (see page 179). So these polymers are **biodegradable**, although the process is **very slow**.

Waste Plastics Have to be **Disposed Of**

It's estimated that in the UK we throw away over **3 million tons** of plastic (i.e. synthetic polymers) every year. Because plastics either take a **very long time** to biodegrade or are **non-biodegradable**, the question of what to do with all those plastic objects when we've finished using them is an important one.

The options are **burying**, **burning** or sorting for **reusing** or **recycling**. None of these methods is an ideal solution — they all have **advantages** and **disadvantages** associated with them.

Waste plastics → burying in landfill
Waste plastics → burning as fuel
Waste plastics → sorting
sorting → remoulding → new objects
sorting → cracking → processing → other chemicals / new plastics

Waste Plastics can be **Buried**

1) **Landfill** is one option for dealing with waste plastics. It is generally used when the plastic is:
   - difficult to separate from other waste,
   - not in sufficient quantities to make separation financially worthwhile,
   - too difficult technically to recycle.
2) Landfill is a relatively **cheap** and **easy** method of waste disposal, but it requires **areas of land**.
3) As the waste decomposes it can release **methane** — a **greenhouse gas**. **Leaks** from landfill sites can also **contaminate water supplies**.
4) The **amount of waste** we generate is becoming more and more of a problem, so there's a need to **reduce** landfill as much as possible.

Waste Plastics can be **Burned**

1) Waste plastics can be **burned** and the heat used to generate **electricity**.
2) This process needs to be carefully **controlled** to reduce the release of **toxic** gases. For example, polymers that contain **chlorine** (such as **PVC**) produce **HCl** when they're burned — this has to be removed.
3) So, waste gases from the combustion are passed through **scrubbers** which can **neutralise** gases such as HCl by allowing them to react with a **base**.
4) But the waste gases, e.g. carbon dioxide, will still contribute to the **greenhouse effect**.

Rex and Dirk enjoy some waist plastic.

Disposing of Polymers

Waste Plastics can be Recycled

Because many plastics are made from non-renewable **oil-fractions**, it makes sense to recycle plastics as much as possible. There's more than one way to recycle plastics. After **sorting** into different types:

- some plastics (poly(propene), for example) can be **melted** and **remoulded**,
- some plastics can be **cracked** into **monomers**, and these can be used to make more plastics or other chemicals.

Plastic products are usually marked to make sorting easier. The different numbers show different polymers, e.g.

3 = PVC, and 5 = poly(propene)

Like other disposal methods, there are advantages and disadvantages to recycling plastics:

Advantages	Disadvantages
It reduces the amount of waste going into landfill.	It is technically difficult to recycle plastics.
It saves raw materials — which is important because oil is non-renewable.	Collecting, sorting and processing the plastic is more expensive than burning/landfill.
The cost of recycling plastics is lower than making the plastics from scratch.	You often can't remake the plastic you started with — you have to make something else.
It produces less CO_2 emissions than burning the plastic.	The plastic can be easily contaminated during the recycling process.

Practice Questions

Q1 Why aren't polyalkenes biodegradable?

Q2 Name a type of polymer that is biodegradable.

Q3 Explain why burning PVC produces HCl gas, but burning poly(ethene) doesn't.

Q4 Give two advantages and two disadvantages of recycling waste plastics.

Exam Questions

Q1 Waste plastics can be disposed of by burning.

a) Describe one advantage of disposing of waste plastics by burning. [1 mark]

b) Describe a disadvantage of burning waste plastic that contains chlorine, and explain how the negative impact of this disadvantage could be reduced. [2 marks]

Q2 Give one advantage and one disadvantage of landfill as a disposal method for waste plastic. [2 marks]

Q3 The diagram below shows sections of two polymers.

A: $-C(=O)-N(H)-(CH_2)_5-C(=O)-N(H)-(CH_2)_5-$

B: $-CH(CH_3)-CH_2-CH(CH_3)-CH_2-$ (drawn as $-C-C-C-C-$ with substituents CH_3, H, CH_3, H above and H, H, H, H below)

a) State which of these polymers is biodegradable.
Explain why the polymer you have selected is more reactive and can be broken down. [2 marks]

b) Name the type of chemical reaction that occurs when a polymer biodegrades. [1 mark]

Phil's my recycled plastic plane — but I don't know where to land Phil...

You might have noticed that all this recycling business is a hot topic these days. And not just in the usual places, such as chemistry books. No, no, no... recycling even makes it regularly onto the news as well. This suits examiners just fine — they like you to know how useful chemistry is. So learn this stuff, pass your exam, and do some recycling.

Amino Acids

Wouldn't it be nice if you could go to sleep with this book under your pillow and when you woke up you'd know it all.

*Amino Acids have an **Amino Group** and a **Carboxyl** Group*

An amino acid has a **basic amino group** (NH_2) and an **acidic carboxyl group** (COOH). This makes them **amphoteric** — they've got both acidic and basic properties.

They're **chiral molecules** (see page 160) because the carbon has **four** different groups attached. So a solution of a single amino acid enantiomer will **rotate plane polarised light**.

Glycine's the exception to this as its R group is just a hydrogen.

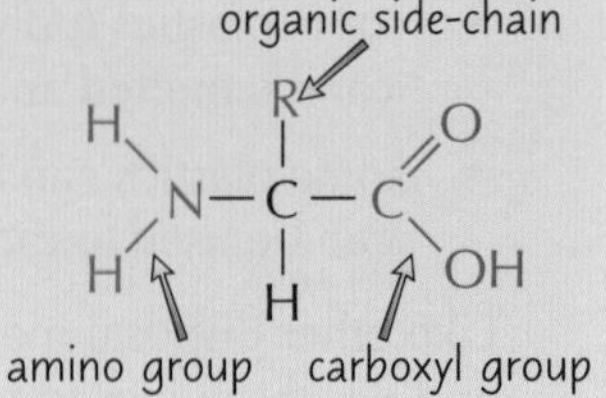

Amino Acids Have Common and Systematic Names

Most amino acids have a **common name** (like glycine or valine), but each one has a **systematic name** too. You should be given the common names if you need any of them, but you might be asked to **work out** their systematic names, using the IUPAC naming system you met in Unit 3 Section 1.

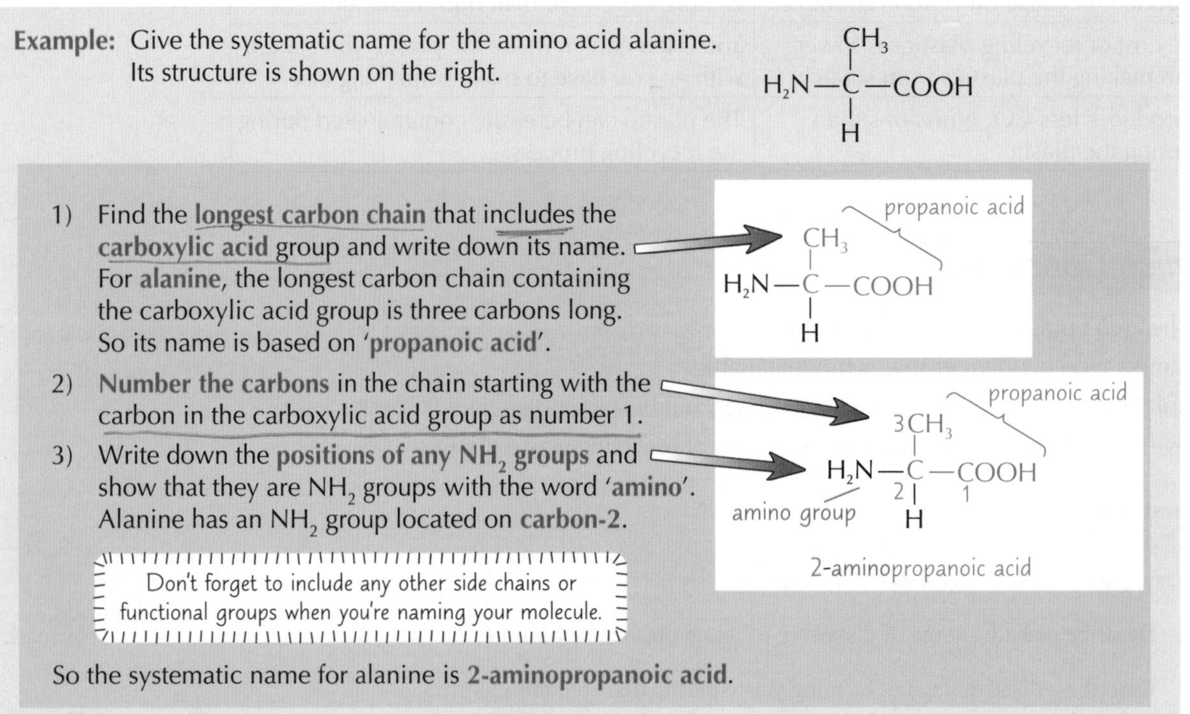

Example: Give the systematic name for the amino acid alanine. Its structure is shown on the right.

1) Find the longest carbon chain that includes the carboxylic acid group and write down its name. For **alanine**, the longest carbon chain containing the carboxylic acid group is three carbons long. So its name is based on '**propanoic acid**'.
2) **Number the carbons** in the chain starting with the carbon in the carboxylic acid group as number 1.
3) Write down the **positions of any NH_2 groups** and show that they are NH_2 groups with the word '**amino**'. Alanine has an NH_2 group located on **carbon-2**.

Don't forget to include any other side chains or functional groups when you're naming your molecule.

So the systematic name for alanine is **2-aminopropanoic acid**.

Amino Acids Can Exist As Zwitterions

A zwitterion is a **dipolar ion** — it has both a **positive** and a **negative charge** in different parts of the molecule. Zwitterions only exist near an amino acid's **isoelectric point**. This is the **pH** where the **average overall charge** on the amino acid is zero. It's different for different amino acids — it depends on their R-group.

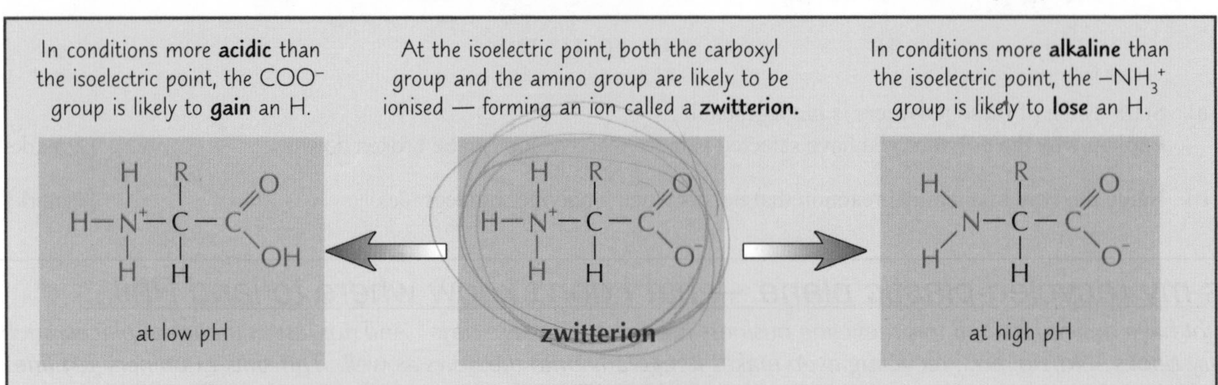

In conditions more **acidic** than the isoelectric point, the COO^- group is likely to **gain** an H.
At the isoelectric point, both the carboxyl group and the amino group are likely to be ionised — forming an ion called a **zwitterion.**
In conditions more **alkaline** than the isoelectric point, the $-NH_3^+$ group is likely to **lose** an H.
at low pH
zwitterion
at high pH

Amino Acids

Thin-Layer Chromatography is used to *Identify Unknown* Amino Acids

Since different amino acids have different 'R' groups, they will all have different **solubilities** in the same solvent. This means you can easily separate and identify the different amino acids in a mixture using **thin-layer chromatography**. Each amino acid will move up the thin-layer chromatography plate at a different rate depending on how soluble it is in the solvent you've used. Here's what you do:

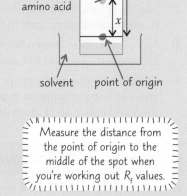

This is a piece of plastic or glass covered with a thin layer of silica gel or alumina powder.

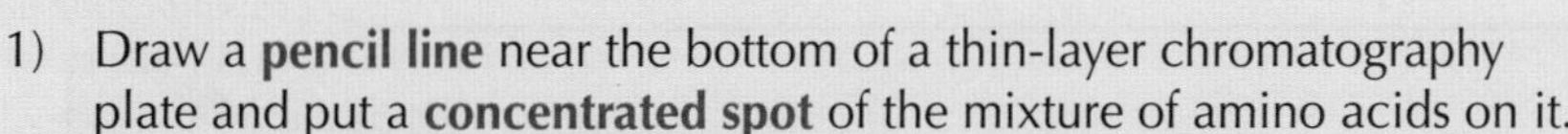

1) Draw a **pencil line** near the bottom of a thin-layer chromatography plate and put a **concentrated spot** of the mixture of amino acids on it.
2) Dip the bottom of the plate (not the spot) into a solvent.
3) As the solvent spreads up the plate, the different amino acids move with it, but at **different rates**, so they separate out.
4) When the solvent's **nearly** reached the top, take the plate out and **mark** the **solvent front** with pencil. Then leave the plate to dry.
5) Amino acids aren't coloured so you will need to make the spots **visible**.
   - You can do this by spraying **ninhydrin solution** on the plate, which will turn the spots purple.
   - Alternatively, you can use a special plate that has a **fluorescent dye** added to it. The dye glows when **UV light** shines on it. Where there are spots of chemical on the plate, they cover the fluorescent dye — so the spots appear dark. You can put the plate under a **UV lamp** and draw around the dark patches to show where the spots are.

thin-layer chromatography plate
distance moved by solvent ('solvent front')
spot of amino acid
y
x
solvent
point of origin

6) You can work out the **R_f value** of each amino acid spot using this formula:

$$R_f \text{ value of amino acid} = \frac{x}{y} = \frac{\text{distance travelled by spot}}{\text{distance travelled by solvent}}$$

Measure the distance from the point of origin to the middle of the spot when you're working out R_f values.

7) Then you can use a **table of known amino acid R_f values** to identify the amino acids in the mixture.

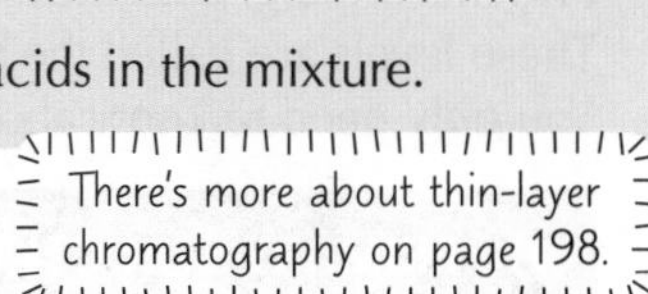

There's more about thin-layer chromatography on page 198.

Practice Questions

Q1 Draw the structure of a typical amino acid.

Q2 Draw the zwitterion that would be formed from this typical amino acid.

Q3 Describe how you could use thin-layer chromatography to identify the amino acids present in a mixture.

Exam Questions

Q1 Valine is an amino acid with the molecular formula $C_5H_{11}NO_2$.
The longest carbon chain in valine is four carbons long.

a) Draw the displayed formula of valine. Label any chiral carbons with an asterisk*. [2 marks]

b) Give the systematic name for valine. [1 mark]

Q2 Leucine is an amino acid with the systematic name 2-amino-4-methylpentanoic acid.

a) Draw the displayed formula of leucine. [1 mark]

b) Draw the displayed formula of the zwitterion formed by leucine when it is at its isoelectric point. [1 mark]

c) Draw the displayed formula of the species formed by leucine when it is dissolved in an alkaline solution. [1 mark]

Everybody run — the Zwitterions are coming...

'The Zwitterions' do sound a bit like a bunch of aliens from a sci-fi show. Or a band. Zwitterion is a lovely word though — it flutters off your tongue like a butterfly. Well, these pages aren't too bad. A few structures, a bit on how to name them and an experiment. Make sure you know how to do thin-layer chromatography and how to work out R_f values.

Proteins and Enzymes

Amino acids are often called the building blocks of life. They're like little plastic building bricks, but for chemistry. Instead of putting them together to make houses and rockets though, they're used to make all the proteins in your body.

***Proteins** are **Condensation Polymers** of Amino Acids*

1) Proteins are made up of **lots** of amino acids joined together by **peptide links**. The chain is put together by **condensation** reactions and broken apart by **hydrolysis** reactions.
2) Here's how two amino acids join together to make a **dipeptide**:

> Proteins are really polyamides — the monomers are joined by amide groups. In proteins these are called peptide links.

$$\underbrace{H_2N-CHR-COOH}_{\text{amino acid 1}} + \underbrace{H_2N-CHR'-COOH}_{\text{amino acid 2}} \underset{\text{Hydrolysis}}{\overset{\text{Condensation}}{\rightleftharpoons}} \underbrace{H_2N-CHR-CO-NH-CHR'-COOH}_{\text{dipeptide}} + H_2O$$

amino acid 1 · amino acid 2 · Condensation · Hydrolysis · dipeptide · Peptide link

3) The dipeptide still has an NH_2 group at one end and a COOH group at the other. So if you want to **add** more amino acids to the chain, you can just keep repeating the **condensation reaction**.
4) A protein can be broken back down into its individual amino acids (**hydrolysed**), but pretty harsh conditions are needed. You add hot aqueous 6 M hydrochloric acid and heat the mixture under reflux for 24 hours.
5) Proteins are condensation polymers, so to work out which **amino acids** a protein chain was made from, you can use the same method as on page 179 — just **break** each of the peptide links down the middle, then add either an **H atom** or an **OH group** to each of the broken ends to get the amino acids back.

*Proteins have **Different** Levels of **Structure***

Proteins are **big**, **complicated** molecules. They're easier to explain if you describe their structure in four 'levels'. These levels are called the **primary**, **secondary**, **tertiary** and **quaternary** structures.
You only need to know about the first three though.

1) PRIMARY STRUCTURE

The **primary structure** is the **sequence of amino acids** in the long chain that makes up the protein (the **polypeptide chain**).

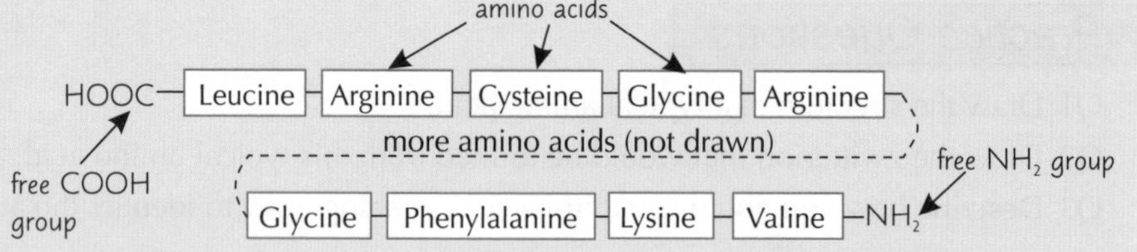

2) SECONDARY STRUCTURE

The **peptide links** can form **hydrogen bonds** with each other (see next page), meaning the chain isn't a straight line. The shape of the chain is called its **secondary structure**. The most common secondary structure is a **spiral** called an **alpha (α) helix**. Another common type of secondary structure is a **β–pleated sheet**. This is a layer of protein folded like a concertina.

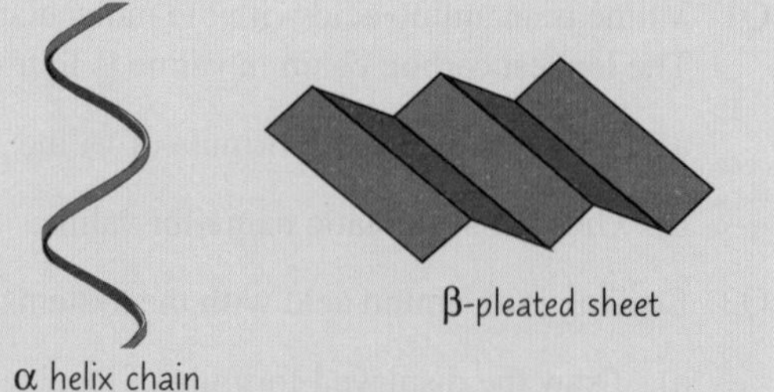

3) TERTIARY STRUCTURE

The chain of amino acids is itself often coiled and folded in a characteristic way that identifies the protein. **Extra bonds** can form between different parts of the polypeptide chain, which gives the protein a kind of **three-dimensional shape**. This is its **tertiary structure**.

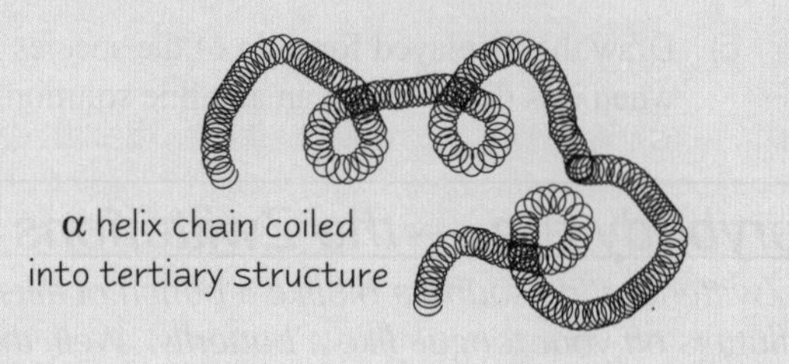

Proteins and Enzymes

Hydrogen Bonds and Disulfide Bonds Help Keep Proteins in Shape

The secondary and tertiary structures of proteins are formed by **intermolecular forces** causing the amino acid chains to **fold** or **twist**. These intermolecular forces are really important, because the three-dimensional shape of a protein is **vital** to how it **functions**. For example, changing the shape of an **enzyme** (see below) can stop it working.

There are two main types of bond that hold proteins in shape:

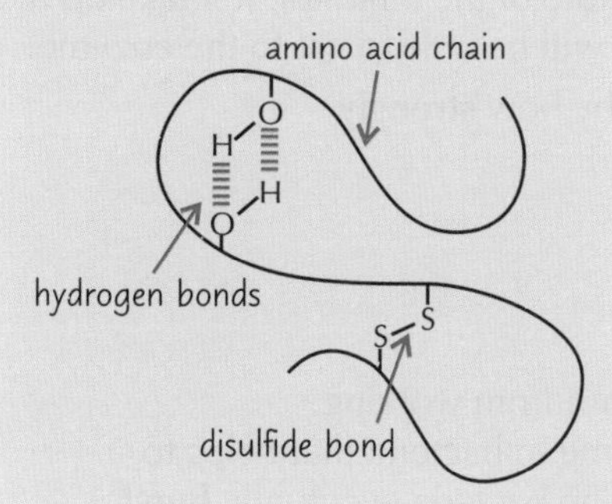

1) **Hydrogen bonding** is one type of force that holds proteins in shape. Hydrogen bonds exist between polar groups — e.g. –OH and $-NH_2$. They stabilise both the secondary and the tertiary structure of the protein.
2) The amino acid **cysteine** contains a **thiol group** (-SH). Thiol groups on different cysteine residues can lose their H atoms and **join together** by forming a **disulfide** (or **sulfur-sulfur**) **bond** (-S–S-). These disulfide bonds link together different parts of the protein chain, and help to stabilise the tertiary structure.

Factors such as **temperature** and **pH** can affect hydrogen bonding and the formation of disulfide bonds and so can **change the shape** of proteins.

Enzymes are Proteins that Act as Biological Catalysts

Enzymes speed up chemical reactions by acting as biological catalysts.

1) They catalyse every **metabolic reaction** in the bodies of living organisms.
2) Enzymes are **proteins**. Some also have **non-protein components**.
3) Every enzyme has an area called its **active site**. This is the part that the **substrate** fits into so that it can interact with the enzyme.
4) The active site is three-dimensional — it's part of the **tertiary structure** of the enzyme protein (see page 184).

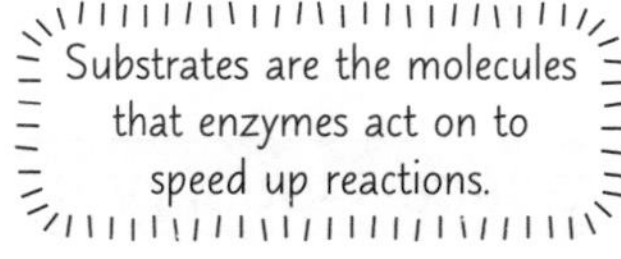

Substrates are the molecules that enzymes act on to speed up reactions.

Enzymes have High Specificity

1) Enzymes are a bit picky. They only work with **specific substrates** — usually only one.
2) This is because, for the enzyme to work, the substrate has to **fit** into the **active site**. If the substrate's shape doesn't match the active site's shape, the reaction won't be catalysed. This is called the '**lock and key**' model.

The substrate fits into the enzyme the same way a key fits into a lock.

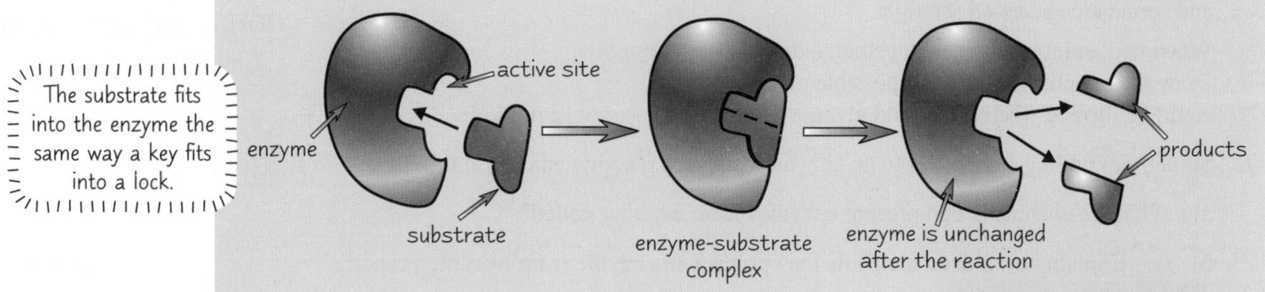

3) Enzymes are made up of **amino acids**, so they contain **chiral centres**.
4) This makes their active sites **stereospecific** — they'll only work on **one enantiomer** of a substrate. The other enantiomer won't **fit properly** in the active site, so the enzyme can't work on it — it's a bit like how your left shoe doesn't fit on your right foot properly.

Dougal had found the perfect loch, but he was in completely the wrong key.

Proteins and Enzymes

Inhibitors Slow Down the Rate of Reaction

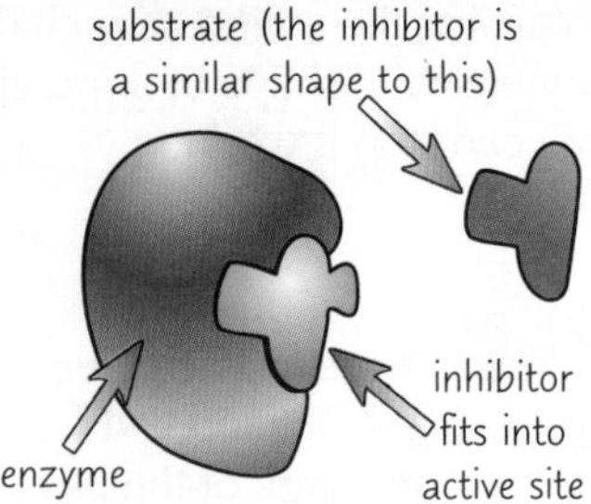

1) Molecules that have a **similar shape** to the **substrate** act as enzyme **inhibitors**.
2) They compete with the substrate to bond to the active site, but no reaction follows. Instead they **block** the active site, so **no substrate** can **fit** in it.
3) How much inhibition happens depends on the **relative concentrations** of inhibitor and substrate — if there's a lot more of the inhibitor, it'll take up most of the active sites and very little substrate will be able to get to the enzyme.
4) The amount of inhibition is also affected by how **strongly** the inhibitor bonds to the active site.

Some Drugs Work as Inhibitors

1) Some **drugs** are **inhibitors** that block the active site of an enzyme and **stop it** from working. For example, some **antibiotics** work by blocking the active site of an enzyme in bacteria that helps to make their cell walls. This causes their cell walls to **weaken** over time, so the bacteria eventually **burst**.
2) The active site of an enzyme is very **specific**, so it takes a lot of effort to find a drug molecule that will **fit** into the active site. It's even trickier if the drug molecule is **chiral** — then only one **enantiomer** will fit into the active site, because the active sites of enzymes are **stereospecific** (see page 185).
3) Often, new drug molecules are found by **trial and error**. Scientists will carry out experiments using lots of compounds to see if they work as inhibitors for a particular enzyme. They'll then **adapt** any that work to try and improve them. This process takes a long time.
4) One way that scientists are speeding this process up is by using **computers** to **model** the shape of an enzyme's active site and **predict** how well potential drug molecules will interact with it. They can **quickly** examine hundreds of molecules to look for ones that might be the right shape **before** they start synthesising and testing things in the laboratory.

Practice Questions

Q1 What type of link joins the two amino acids in a dipeptide?

Q2 Name two types of secondary structure that proteins can form.

Q3 Name two types of bond that can help to hold together the tertiary structure of a protein.

Q4 What is an enzyme?

Q5 Explain the advantage of using computer-aided drug design instead of trial and error.

Exam Questions

Q1 The structures of the amino acids glycine and serine are shown on the right.
When two amino acids react together, a dipeptide is formed. Draw the structures of the two possible dipeptides that could be formed when serine and glycine react together. [2 marks]

glycine: $NH_2-C(H)(H)-COOH$

serine: $NH_2-C(CH_2-OH)(H)-COOH$

Q2 Sometimes adding a chemical to an enzyme-catalysed reaction can cause the enzyme to stop working properly.

a) What are chemicals that prevent enzymes from working called? [1 mark]

b) i) Explain how a drug molecule may prevent an enzyme from working properly. [2 marks]

ii) A scientist has developed a drug that stops a bacterial enzyme from working. The drug molecule is chiral. Explain why only one enantiomer of the drug will be effective against the enzyme. [2 marks]

Procrastination — the ultimate revision inhibitor...

There's almost as many proteins on these pages as in a big steak topped with cheese and a peanut sauce. Have another look at the reaction that links amino acids together — it's the same as the condensation polymerisation reactions on page 178, just using amino acids rather than diamines and dicarboxylic acids — so you're already halfway to knowing it.

DNA

DNA — the molecule of life. And unfortunately, just like life, it's complicated.

DNA is a Polymer of Nucleotides

DNA (**d**eoxyribo**n**ucleic **a**cid) contains all the genetic information of an organism.
DNA is made up from lots of **monomers** called **nucleotides**.
Nucleotides are made from the following:

1) **A phosphate group:** $^{-}O-P(=O)(OH)_2$

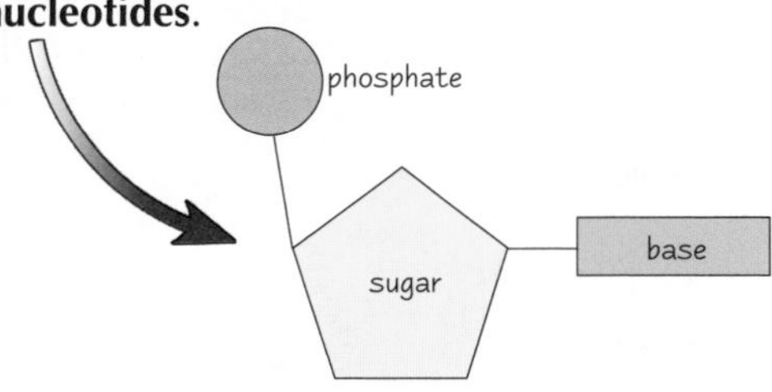

Chris and Rita had forgotten about the rising nucleotide

2) **A pentose sugar** — a five-carbon sugar. In DNA, the sugar is 2-deoxyribose.

You don't need to learn the structures of phosphate and 2-deoxyribose, or of the bases — they'll all be in your data booklet.

3) **A base** — one of four different bases. In DNA they are **adenine** (A), **cytosine** (C), **guanine** (G) and **thymine** (T).

adenine guanine

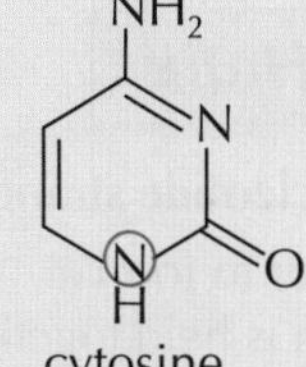

cytosine thymine

The circled nitrogens are the atoms that bond with the deoxyribose molecule (see below).

Here are the **structures** of all four DNA **nucleotides**:

adenine nucleotide

guanine nucleotide

cytosine nucleotide

thymine nucleotide

You should be able to work out these structures from the info in the exam data booklet if you need to, so you don't need to learn them. But you do need to know where the phosphate group and the bases join to the sugar.

The nucleotides join together to form a **polynucleotide chain**. Covalent bonds form between the phosphate group of one nucleotide and the sugar of another — this makes what's called the **sugar-phosphate backbone** of the chain.

cytosine thymine adenine guanine

sugar-phosphate backbone

DNA

DNA Forms by Condensation Polymerisation

The **sugar-phosphate backbone** of DNA is formed by condensation polymerisation, like this:

The sugar group of another nucleotide can attach here

Phosphodiester bond

The phosphate group of another nucleotide can attach here

A molecule of H_2O is lost

1) A molecule of water is lost and a covalent **phosphodiester bond** is formed.
2) There are still OH groups at the top and bottom of the chain, so further links can be made. This allows the nucleotides to form a **polymer** made up of an alternating sugar-phosphate-sugar-phosphate chain.

DNA Forms a Double Helix

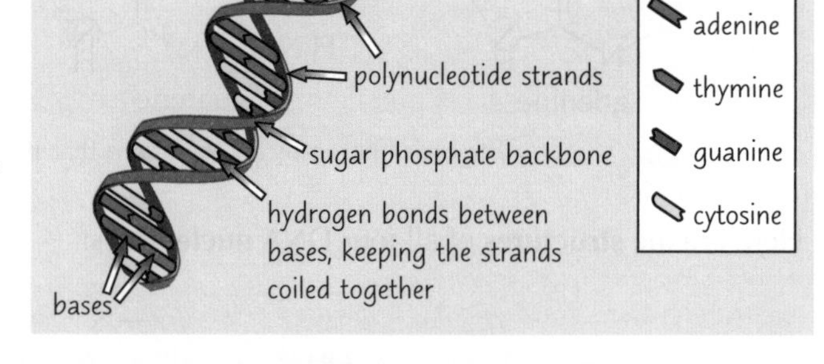

1) **DNA** is made of **two polynucleotide strands**.
2) The two strands spiral together to form a **double helix** structure, which is held together by **hydrogen bonds** between the bases.
3) Each base can only join with one particular partner — **adenine** always pairs with **thymine** (**A – T**) and **guanine** always pairs with **cytosine** (**G – C**).
4) This causes the two strands of DNA to be **complementary** — this means that they match up so that whenever there is an **adenine** base on one strand, there will be a **thymine** base on the other, and whenever there is a **guanine** base on one strand, there will be a **cytosine** base on the other, and vice versa.

Hydrogen Bonding Causes the Bases to Form Specific Pairs

As you saw above, **complementary base pairing** exists in the DNA helix, where **adenine (A)** always pairs to **thymine (T)** and **guanine (G)** always pairs to **cytosine (C)**. It happens because of the arrangement and number of atoms in the base molecules that are capable of forming **hydrogen bonds**.

A **hydrogen bond** forms between a polar positive **H atom** (an H attached to anything highly electronegative like N) and a lone pair of electrons on a nearby **O** or **N atom**. To bond, the two atoms have to be the **right distance apart**.

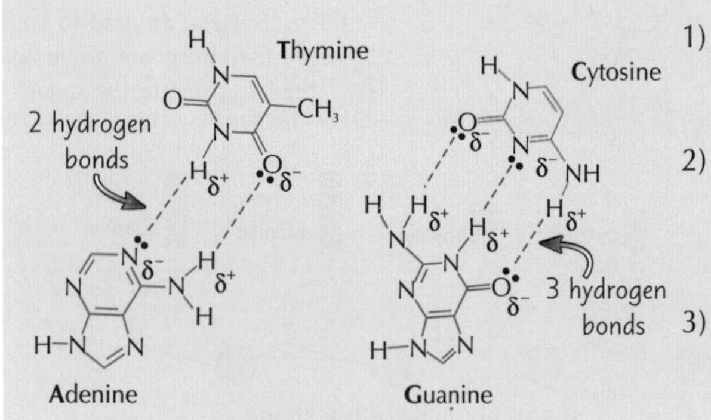

1) A and T have the right atoms in the right places to each form **2 hydrogen bonds**, so they can pair up. G and C can each form **3 hydrogen bonds**, so they can pair up too.
2) These are the **only** possible base combinations. Other base pairings would put the partially charged atoms too close together (they'd repel each other), or too far apart, or the bonding atoms just wouldn't line up properly.
3) The DNA helix has to twist so that the bases are in the **right alignment** and at the **right distance** apart for the complementary base pairs to form.

DNA

Cisplatin Can Bind to DNA in Cancer Cells

1) **Cisplatin** is a complex of platinum(II) with two chloride ion ligands and two ammonia ligands in a **square planar shape** (see page 110). It is used as an **anti-cancer drug**.

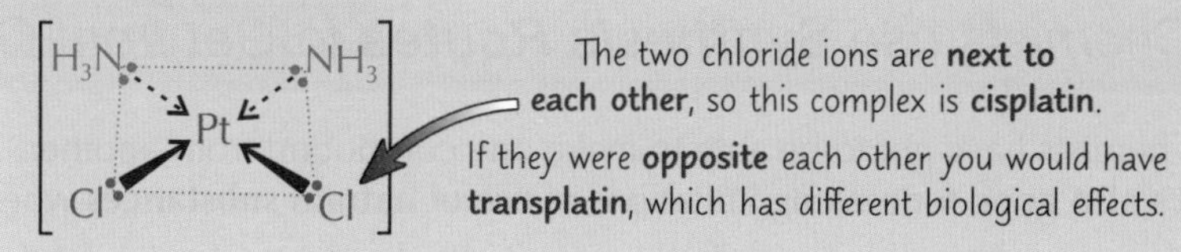

H_3N
NH_3
Pt
Cl
Cl
The two chloride ions are **next to each other**, so this complex is **cisplatin**.
If they were **opposite** each other you would have **transplatin**, which has different biological effects.

2) **Cancer** is caused by cells in the body **dividing uncontrollably** and forming **tumours**.
3) In order for a cell to divide it has to **replicate** its DNA.
4) Cisplatin binds to DNA, causing **kinks** in the DNA helix which **stop** the proteins that replicate the DNA from copying it properly. This **stops** tumour cells reproducing. Here's how it works:

- A nitrogen atom on a **guanine base** in DNA forms a **co-ordinate bond** with cisplatin's platinum ion, replacing one of the chloride ion ligands. (This is a **ligand substitution** reaction, see page 116).
- A **second** nitrogen atom from a nearby guanine (either on the same strand of the DNA or the opposite strand) can bond to the platinum and replace the **second** chloride ion.
- The presence of the cisplatin complex bound to the DNA strands causes the strands to **kink**. This means that the DNA strands **can't unwind** and be **copied** properly — so the cell can't **replicate**.

This damage to the DNA also triggers mechanisms that lead to the death of the cell.

5) Unfortunately, cisplatin can bind to DNA in **normal cells** as well as cancer cells. This is a particular problem for any healthy cells that **replicate frequently**, such as **hair cells** and **blood cells**, because cisplatin stops them from replicating in the same way as it does the cancer cells. This means that cisplatin can cause **hair loss** and suppress the **immune system** (which is controlled by white blood cells). It can also cause kidney damage.
6) These side effects can be lessened by giving patients very **low dosages** of cisplatin.
7) Another way to reduce the side effects of cisplatin is to **target** it to the tumour — this means using a method that delivers the drug only to the cancer cells, so it doesn't get the chance to attack healthy cells.
8) Despite the side effects of cisplatin, it is still used as a chemotherapy drug. This is because the **balance** of the long-term positive effects (curing cancer) **outweigh** the negative short-term effects.

Practice Questions

Q1 What are the three components of a DNA nucleotide?

Q2 Briefly describe how the sugar-phosphate backbone of the DNA polymer is formed.

Q3 What is the complementary base of adenine?

Q4 Draw the structure of cisplatin. Explain how it stops DNA from replicating.

Exam Question

Q1 The structures of a phosphate group, the base guanine and the sugar 2-deoxyribose are shown below.

phosphate group OH, $^-O-P=O$, OH

guanine O, HN, N, H_2N, N, N, H

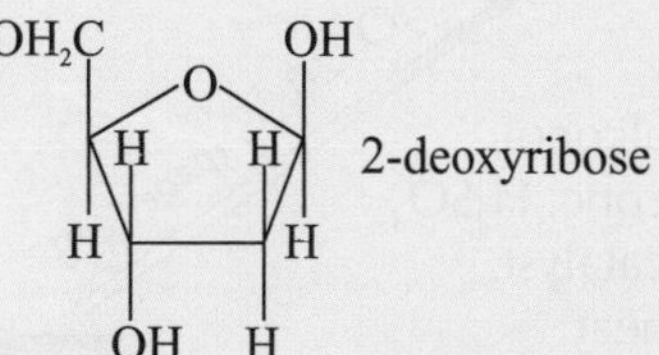

2-deoxyribose

a) Use these structures to draw a diagram showing a DNA nucleotide containing the base guanine. [2 marks]

b) The complementary base of guanine is cytosine, shown on the right. Draw a diagram to show how guanine bonds to cytosine in a DNA double helix. What type of bonds are formed between the bases? [4 marks]

NH_2, N, N, H, O

cytosine

Sissy-platin: scared of spiders, loud noises and sandwiches...

Let's face it, I'm a science geek, but this DNA stuff never ceases to amaze me. It's just so flippin' clever. Sadly, even if you don't share my enthusiasm for genetics, you do have to know it. If you happen to be skipping the extremely useful exam questions (naughty, naughty), then have a go at this one. Seriously, try it. You can have a biscuit afterwards.

Organic Synthesis

In your exam you may be asked to suggest a pathway for the synthesis of a particular molecule. These pages contain a summary of some of the reactions you should know.

Chemists use **Synthesis Routes** to Get from One Compound to Another

Chemists have got to be able to make one compound from another. It's vital for things like **designing medicines**. It's also good for making imitations of **useful natural substances** when the real things are hard to extract.

These reactions are covered elsewhere in the book, so check back for extra details.

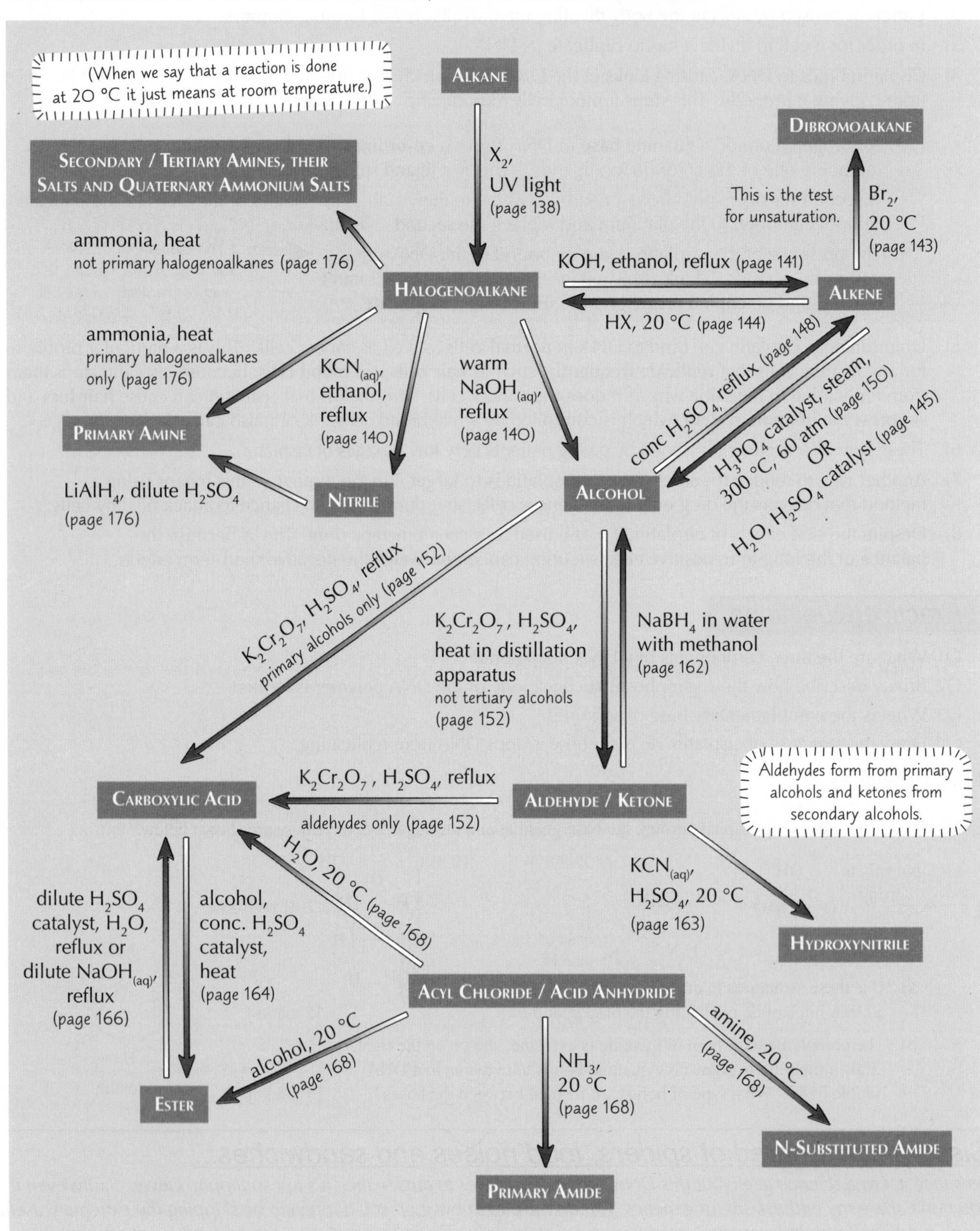

Organic Synthesis

Synthesis Routes for Making **Aromatic Compounds**

There are not so many of these reactions to learn — but make sure you still know all the itty-bitty details.

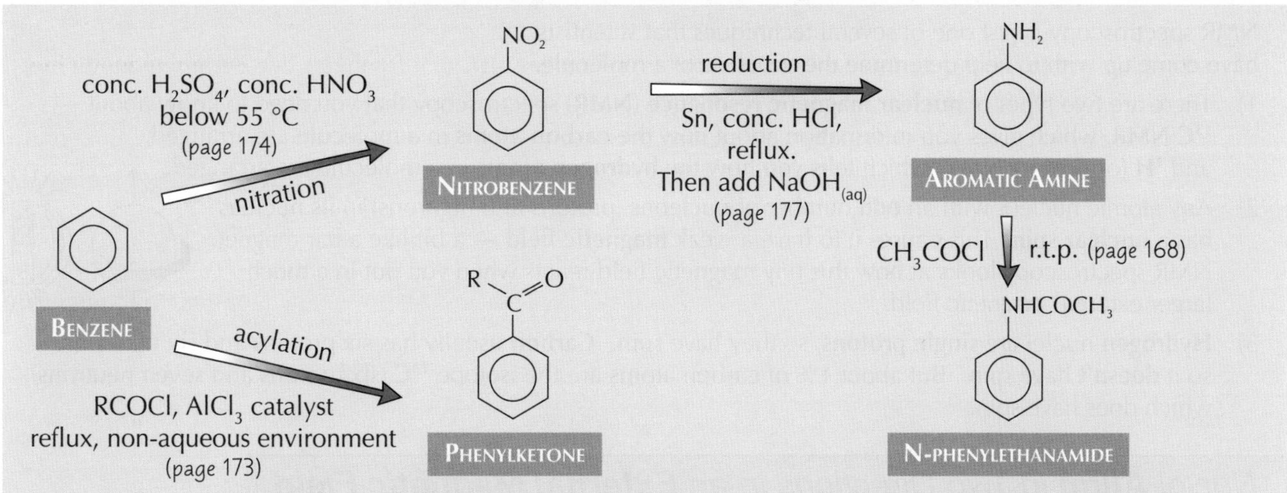

Chemists Aim to Design Synthesis Routes Which are **Safe** and **Efficient**

1) Chemists try to design synthesis routes that use **non-hazardous starting materials** to limit the potential for accidents and environmental damage.
2) Chemists are also concerned with designing processes that are **not too wasteful**. Processes with **high atom economies** and **high percentage yields** are preferred, because they convert more of the starting materials into useful products. Waste can also be reduced by designing synthesis routes that have as **few steps** as possible.

Avoiding using **solvents** wherever possible is one way of **reducing** both the **hazards** associated with a process and the amount of **waste** created by a synthesis route. Solvents are often **flammable** and **toxic** so can pose safety risks. If the solvent has to be disposed of after the reaction is complete that can create a lot of **waste** too.

Practice Questions

Q1 What type of organic product is formed by the reaction between a primary halogenoalkane and ammonia?

Q2 Give two reagents that when combined with an alcohol will give an ester.

Q3 What reagents and conditions are needed to synthesise nitrobenzene from phenylamine?

Exam Questions

Q1 The diagram on the right shows a possible reaction pathway for the two-step synthesis of a ketone from a halogenoalkane.

$$\underset{\text{Halogenoalkane P}}{CH_3CHXCH_3} \xrightarrow[\text{NaOH}]{\text{Step 1}} \underset{\text{Alcohol Q}}{CH_3CH(OH)CH_3} \xrightarrow{\text{Step 2}} \underset{\text{Ketone R}}{CH_3COCH_3}$$

a) Give the conditions needed to carry out Step 1. [1 mark]

b) Give the reagents and the conditions needed to carry out Step 2. [2 marks]

Q2 Ethyl methanoate is one of the compounds responsible for the smell of raspberries.
Outline, with reaction conditions, how it could be synthesised in the laboratory from methanol. [7 marks]

Q3 How would you synthesise propanol starting with propane? State the reaction conditions and reagents needed for each step and any particular safety considerations. [8 marks]

I saw a farmer turn a tractor into a field once — now that's impressive...

There's loads of information here, but you need to know it all. If you're asked in an exam how you would make one compound from another, make sure you include any procedures needed (e.g. refluxing), any reaction conditions and any safety precautions that should be taken. Why? Because that's the only way you can be sure you'll get all the marks.

NMR Spectroscopy

NMR isn't the easiest of things, so ingest this information one piece at a time — a bit like eating a bar of chocolate...

NMR Gives You Information About a Molecule's Structure

NMR spectroscopy is just one of several techniques that scientists have come up with to help determine the structure of a molecule.

1) There are two types of **nuclear magnetic resonance** (**NMR**) spectroscopy that you need to know about — ^{13}**C NMR**, which gives you information about how the **carbon atoms** in a molecule are arranged, and ^{1}**H** (or **proton**) **NMR**, which tells you how the **hydrogen atoms** in a molecule are arranged.
2) Any atomic nucleus with an **odd** number of nucleons (protons and neutrons) in its nucleus has a **nuclear spin**. This causes it to have a weak **magnetic field** — a bit like a bar magnet. NMR spectroscopy looks at how this tiny magnetic field reacts when you put in a much larger external magnetic field.
3) **Hydrogen** nuclei are **single protons**, so they have spin. **Carbon** usually has six protons and six neutrons, so it **doesn't** have spin. But about 1% of carbon atoms are the isotope ^{13}**C** (six protons and seven neutrons), which does have spin.

Nuclei Align in Two Directions in an External Magnetic Field

1) Normally the nuclei are spinning in **random directions** — so their magnetic fields **cancel out**.
2) But when a strong **external** magnetic field is applied the nuclei will all align either **with the field** or **opposed to it**.
3) The nuclei aligned with the external field are at a **slightly lower energy level** than the opposed nuclei.

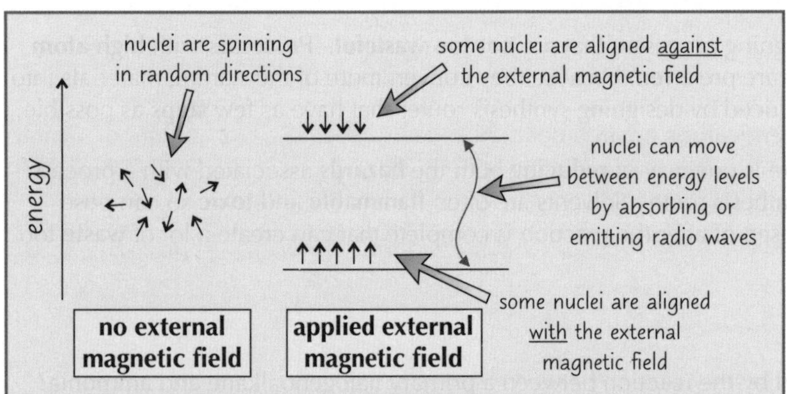

4) **Radio waves** of the right frequency can give the nuclei that are aligned with the external magnetic field enough energy to flip up to the higher energy level. The nuclei opposed to the external field can **emit** radio waves and flip down to the lower energy level.
5) To start with, there are more nuclei **aligned** with the external field, so there will be an **overall absorption** of energy. NMR spectroscopy **measures** this **absorption**.

Nuclei in Different Environments Absorb Different Amounts of Energy

1) A nucleus is partly **shielded** from the effects of external magnetic fields by its **surrounding electrons**.
2) Any **other atoms** and **groups of atoms** that are around a nucleus will also affect its amount of electron shielding. E.g. If a carbon atom bonds to a more electronegative atom (like oxygen) the amount of electron shielding around its nucleus will decrease.
3) This means that the nuclei in a molecule feel different magnetic fields depending on their **environments**. Nuclei in different environments will absorb **different amounts** of energy at **different frequencies**.
4) It's these **differences in energy absorption** between environments that you're looking for in **NMR spectroscopy**.
5) An atom's **environment** depends on **all** the groups that it's connected to, going **right along the molecule** — not just the atoms it's actually bonded to. To be in the **same environment**, two atoms must be joined to **exactly the same things**.

```
   H H
   | |
 H-C-C-Cl
   | |
   H H
```

Chloroethane has **2** hydrogen environments — **3Hs** in a CH_3 group bonded to CH_2Cl and **2Hs** in a CH_2Cl group bonded to CH_3.

```
   H Cl H
   |  | |
 H-C--C-C-H
   |  | |
   H  H H
```

2-chloropropane has **2** carbon environments:
- **1 C** in a CHCl group, bonded to $(CH_3)_2$
- **2 Cs** in CH_3 groups, bonded to $CHCl(CH_3)$

```
   H H H H
   | | | |
 H-C-C-C-C-Cl
   | | | |
   H H H H
```

1-chlorobutane has **4** carbon environments. (The two carbons in CH_2 groups are **different distances** from the **electronegative** Cl atom — so their **environments** are **different**.)

NMR Spectroscopy

Chemical Shift is Measured Relative to Tetramethylsilane

1) Nuclei in different environments absorb energy of **different frequencies**. NMR spectroscopy measures these differences relative to a **standard substance** — the difference is called the **chemical shift (δ)**.
2) The standard substance is **tetramethylsilane (TMS)**, $Si(CH_3)_4$. This molecule has 12 hydrogen atoms all in **identical environments**, so it produces a **single** absorption peak, well away from most other absorption peaks.
3) Chemical shift is measured in **parts per million** (or **ppm**) relative to TMS. So the single peak produced by TMS is given a **chemical shift value of 0**.
4) You'll often see a peak at δ = 0 on spectra because TMS is added to the test compound for calibration purposes.

Tetramethylsilane is also inert (so it doesn't react with the sample), non-toxic, and volatile (so it's easy to remove from the sample).

^{13}C NMR Tells You How Many Different Carbon Environments a Molecule Has

The number of peaks on a ^{13}C NMR spectrum tells you how many different carbon environments are present in a particular molecule. The spectrum will have **one peak** on it for each **carbon environment** in the molecule.

There are **two carbon atoms** in a molecule of **ethanol**:

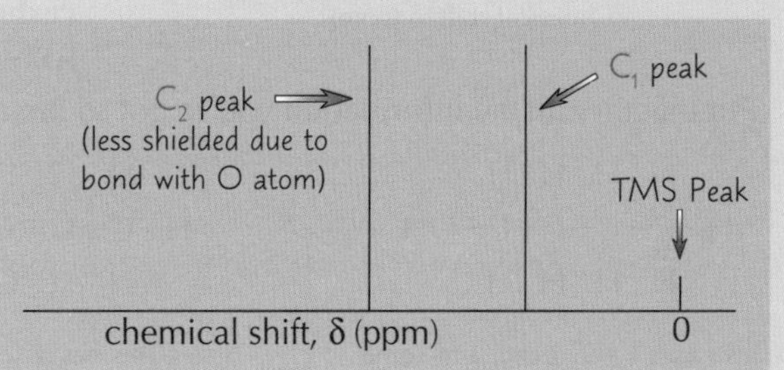

Because they are bonded to **different** atoms, each has a **different** amount of **electron shielding** — so there are **two carbon environments** in the ethanol molecule and **two peaks** on its ^{13}C NMR spectrum.

Molecules containing an **aromatic ring** look a bit more complicated, but you can still **predict** what their spectra will look like by looking at the number of different carbon environments. Just keep a keen eye out for **lines of symmetry**.

In cyclohexane-1,3-diol there are **four different carbon environments**. If you think about the symmetry of the molecule you can see why this is. So cyclohexane-1,3-diol's ^{13}C NMR spectrum will have **4 peaks**.

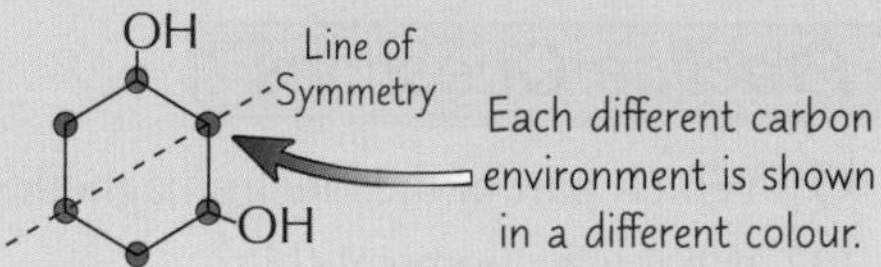

You Can Look Up Chemical Shifts in a Data Table

In your exam you'll get a **data sheet** that will include a **table** like this one. The table shows the **chemical shifts** experienced by **carbon–13 nuclei** in **different environments**.

You need to **match up** the **peaks** in the spectrum with the **chemical shifts** in the table to work out which **carbon environments** they could represent.

Matching peaks to the groups that cause them isn't always straightforward, because the chemical shifts can **overlap**. For example, a peak at **δ ≈ 30** might be caused by **C–C**, **C–Cl** or **C–Br**. A peak at **δ ≈ 210**, is due to a **C=O** group in an **aldehyde** or a **ketone** — but you **don't** know which.

^{13}C NMR Chemical Shifts Relative to TMS

Chemical Shift, δ (ppm)	Type of Carbon
5 – 40	C – C
10 – 70	R – C – Cl or Br
20 – 50	R – C(=O) – C
25 – 60	R – C – N (amines)
50 – 90	C – O (alcohols, ethers or esters)
90 – 150	C = C (alkenes)
110 – 125	R – C ≡ N
110 – 160	(benzene ring) aromatic
160 – 185	R – C(=O) – carbonyl (ester or carboxylic acid)
190 – 220	R – C(=O) – carbonyl (ketone or aldehyde)

Ralph was trying to get a better look at his table.

NMR Spectroscopy

Interpreting NMR Spectra Gets Easier with *Practice*

^{13}C NMR spectra are usually much **simpler** than ^{1}H NMR spectra (see page 195) — they have fewer, sharper peaks. So, interpreting ^{13}C spectra really isn't all that bad — it's just a case of using your data sheet to work out what **carbon environment** is responsible for **each peak**.

Example: The diagram shows the carbon-13 NMR spectrum of an alcohol with the molecular formula $C_4H_{10}O$. Analyse and interpret the spectrum to identify the structure of the alcohol.

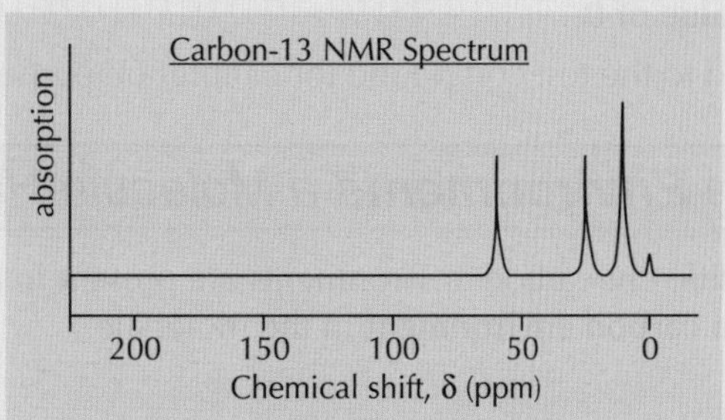

1) Looking at the **table** on the **previous page**, the peak with a **chemical shift** of **δ ≈ 65** is likely to be due to a **C–O** bond.
2) The two peaks around **δ ≈ 20** probably both represent carbons in **C–C** bonds, but with slightly different environments. Remember the alcohol doesn't contain any **chlorine, bromine** or **nitrogen** so you can **ignore** those entries in the table.
3) The spectrum has **three peaks**, so the alcohol must have three **carbon environments**. There are **four carbons** in the alcohol, so two of the carbons must be in the **same environment**.
4) Put together all the **information** you've got so far, and try out some **structures**:

```
    H  H  H  H
    |  |  |  |
 H–C–C–C–C–OH
    |  |  |  |
    H  H  H  H
```

This has a C–O bond, and some C–C bonds, which is right. But all four carbons are in different environments.

```
    H  H  H  H
    |  |  |  |
 H–C–C–C–C–H
    |  |  |  |
    H  H  OH H
```

Again, this has a C–O bond, and some C–C bonds. But the carbons are still all in different environments.

```
      OH
      |
    H–C–H
   H  |  H
   |  |  |
 H–C––C––C–H
   |  |  |
   H  H  H
```

This molecule has a C–O bond and C–C bonds and two of the carbons are in the same environment. So this must be the correct structure.

Practice Questions

Q1 Which part of the electromagnetic spectrum is absorbed in NMR spectroscopy?

Q2 What is a chemical shift?

Q3 Explain what the number of peaks in a ^{13}C NMR shows.

Exam Questions

Q1 Draw the molecular structures of each of the following compounds and predict the number of peaks in the ^{13}C NMR spectrum of each compound.

a) ethyl ethanoate [2 marks]

b) 1-chloro-2-methylpropane [2 marks]

c) 1,3,5-trichlorocyclohexane [2 marks]

Q2 The molecule shown on the right is methoxyethane. It has the molecular formula C_3H_8O.

```
   H  H     H
   |  |     |
 H–C––C–O–C–H
   |  |     |
   H  H     H
```

a) Draw the displayed formulas of the two other possible isomers of C_3H_8O. [2 marks]

b) The ^{13}C NMR spectrum of one of the three isomers is on the right. Deduce which of the three isomers it represents. Explain your answer. [2 marks]

chemical shift, δ (ppm) 0

c) What is responsible for the peak at 0? [1 mark]

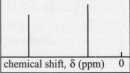

NMR, TMS — IMO this page goes a bit OTT on the TLAs...

The ideas behind NMR are difficult, so don't worry if you have to read these pages quite a few times before they make sense. You've got to make sure you really understand the stuff on these three pages as there's loads more about NMR on the next few pages — and it isn't any easier. Keep bashing away at it though — you'll eventually go "aaah... I get it."

^{1}H NMR

And now that you know the basics, here's the really crunchy bit for you to get your teeth stuck in.

*^{1}H NMR — How Many **Environments** and How Many **Hydrogens** Are In Each*

1) ^{1}H NMR is all about how **hydrogen nuclei** react to a magnetic field. The nucleus of a hydrogen atom is a **single proton**. So ^{1}H NMR is also known as '**proton NMR**' — and you might see the hydrogen atoms involved being called 'protons'.
2) **Each peak** on a ^{1}H NMR spectrum is due to one or more hydrogen nuclei (protons) in a **particular environment** — this is similar to a ^{13}C NMR spectrum (which tells you the number of different carbon environments).

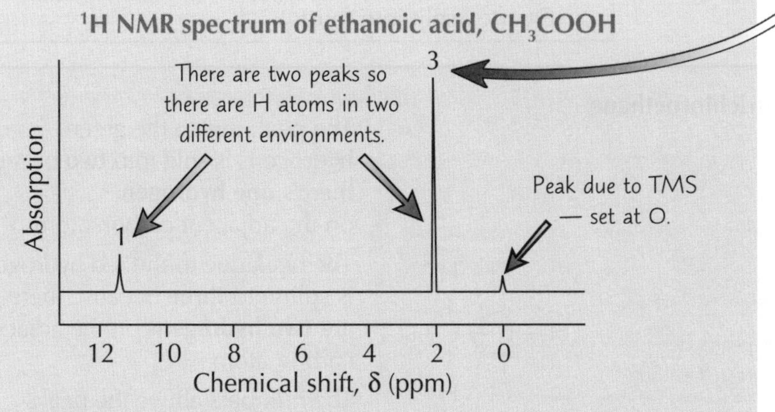

1H NMR spectrum of ethanoic acid, CH3COOH
There are two peaks so there are H atoms in two different environments.
Peak due to TMS — set at O.
Absorption
1
3
12 10 8 6 4 2 0
Chemical shift, δ (ppm)

1) There are **two peaks** — so there are **two environments.**
2) The area ratio is **1:3** — so there's 1 H atom in the environment at $\delta \approx 11.5$ ppm to every 3 H atoms in the other environment.
3) If you look at the structure of ethanoic acid, this makes sense:

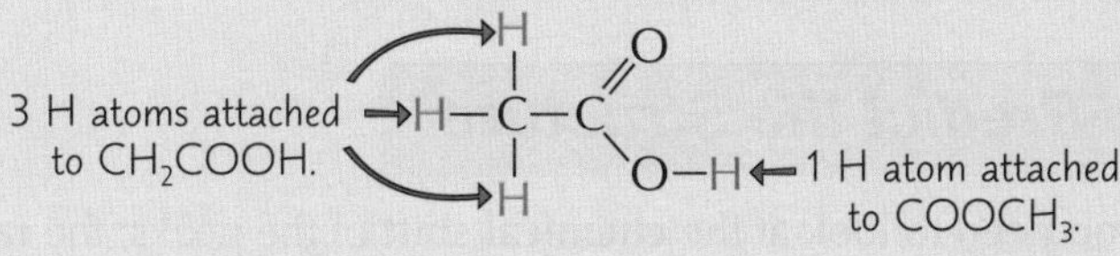

3 H atoms attached to CH2COOH.
1 H atom attached to COOCH3.

3) The **numbers above the peaks** on a ^{1}H NMR spectrum tell you the **ratio** of the areas under the peaks. This **relative area** under each peak also tells you the **relative number** of H atoms in each environment. Don't worry if they're not always whole numbers — they are ratios and not exact numbers.

Sometimes an integration trace is drawn on ^{1}H NMR spectrums, to show more clearly the ratio of the areas under the peaks (see page 197 for more).

Doug took great pride in his peak counting ability.

*Use a **Table** to Identify the **Hydrogen Atom** Causing the **Chemical Shift***

You use a table like this to **identify** which functional group each peak is due to.

Don't worry — **you don't need to learn it**. You'll be given one in your exam, so you just need to learn how to use it. The copy you get in your exam may look a little different, and have different values — they depend on the solvent, temperature and concentration.

The hydrogen atoms that cause the shift are highlighted in red. R stands for any alkyl group.

^{1}H NMR Chemical Shifts Relative to TMS

Chemical Shift, δ (ppm)	Type of H atom
0.5 – 5.0	ROH
0.7 – 1.2	RCH_3
1.0 – 4.5	RNH_2
1.2 – 1.4	R_2CH_2
1.4 – 1.6	R_3CH
2.1 – 2.6	R–C(=O)–C(H)–
3.1 – 3.9	R–O–C(H)–
3.1 – 4.2	RCH_2Br or Cl
3.7 – 4.1	R–C(=O)–O–C(H)–
4.5 – 6.0	R(C=C)H
9.0 – 10.0	R–C(=O)H
10.0 – 12.0	R–C(=O)O–H

According to the table, ethanoic acid (CH_3COOH) should have a peak at **10.0 – 12.0** ppm due to R-COO**H**, and a peak at **2.1 – 2.6 ppm** due to R-COC**H$_3$**.
You can see these peaks on ethanoic acid's spectrum above.

^{1}H NMR

Splitting Patterns Provide More Detail About Structure

The peaks may be **split** into smaller peaks.
Peaks always split into the number of hydrogens on the neighbouring carbon, **plus one**.
It's called the **n+1 rule**.

Type of Peak	Number of Hydrogens on Adjacent Carbon
Singlet (not split)	0
Doublet (split into two)	1
Triplet (split into three)	2
Quartet (split into four)	3

Here's the ^{1}H NMR spectrum for **1,1,2-trichloroethane**:

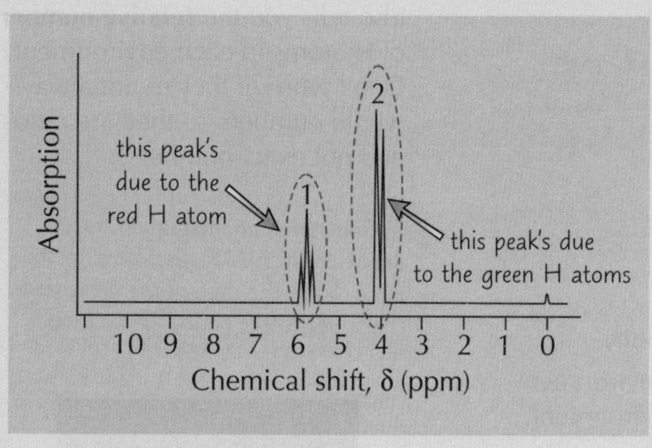

Cl H
H—C—C—Cl
Cl H

The peak due to the green hydrogens is split into **two** because there's **one hydrogen** on the adjacent carbon atom.

The peak due to the red hydrogen is split into **three** because there are **two hydrogens** on the adjacent carbon atom.

The numbers above the peaks confirm that the **ratio** of **hydrogens** in the red environment to those in the green environment is **1 : 2**.

Put All the Information Together to Predict the Structure

To successfully interpret a proton NMR spectrum, you need to look at the **chemical shift** of the peaks, the **ratio of the areas under the peaks** and the **splitting patterns**. It sounds like an awful lot to think about, but if you go through each step carefully, you can predict the structure of an entire molecule just from the ^{1}H NMR spectra.

Example: Using the spectrum below, and the table of chemical shift data on page 195, predict the structure of the compound.

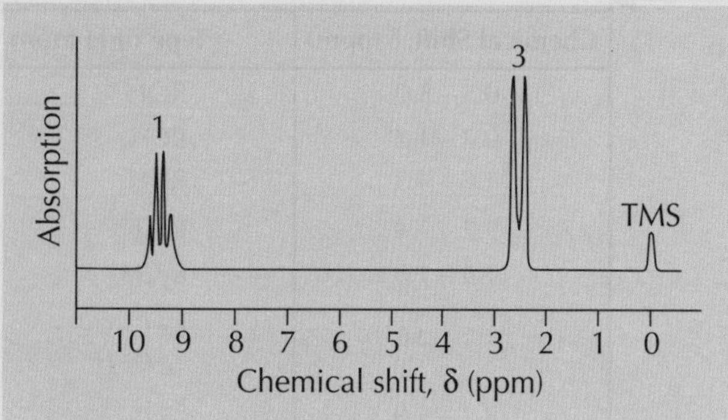

1) The peak at **δ = 2.5 ppm** is likely to be due to an **R–COCH$_3$** group,and the peak at **δ = 9.5 ppm** is likely to be due to an **R–CHO** group.
2) From the areas, there's one proton in the peak at **δ = 9.5 ppm**, for every three in the peak at **δ = 2.5 ppm**. This fits nicely with the first bit — so far so good.
3) The quartet's got **three** neighbouring hydrogens, and the doublet's got **one** — so it's likely these two groups are next to each other.

Now you know the molecule has to contain: H–C(=O)– and –C(=O)–CH$_3$

All you have to do is fit them together: H–C(=O)–CH$_3$

^{1}H NMR

Integration Traces Show Areas More Clearly

When the peaks are split, it's not as easy to see the ratio of the **areas** under the peaks. So, sometimes an **integration trace** is shown instead of peak ratios. The increases in height are proportional to the areas under each peak.

You can use a ruler to measure the height of each vertical bit of the trace and then use the heights to work out the ratio of the peak areas.

The **integration ratio** for this spectrum is **1:2** — this means that there's 1 H atom in the first environment for every 2 H atoms in the second environment.

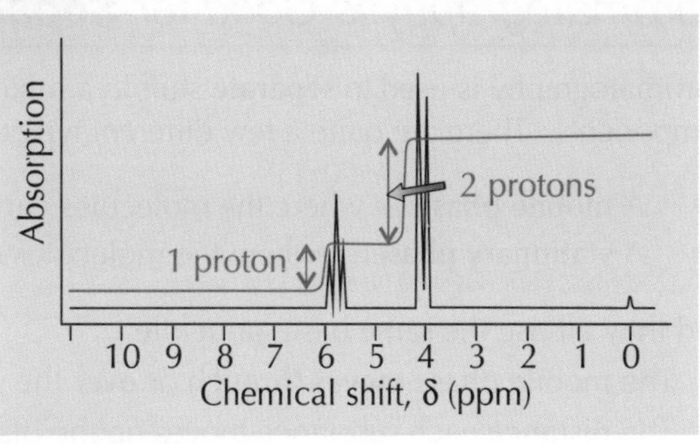

Samples are Dissolved in Hydrogen-Free Solvents

1) If a sample has to be dissolved, then a solvent is needed that doesn't contain any **^{1}H atoms** — because these would show up on the spectrum and confuse things.
2) **Deuterated solvents** are often used — their hydrogen atoms have been replaced by **deuterium** (D or ^{2}H). Deuterium's an isotope of hydrogen that's got two nucleons (a proton and a neutron).
3) Because deuterium has an **even number** of nucleons, it doesn't have a spin (so it doesn't create a magnetic field).
4) **CCl_4** can also be used as a solvent — it doesn't contain any ^{1}H atoms either.

Practice Questions

Q1 What causes peaks to split in proton NMR?

Q2 What causes a triplet of peaks?

Q3 Why are deuterated solvents used when carrying out ^{1}H NMR spectroscopy?

Exam Questions

Q1 The ^{1}H NMR spectrum on the right is that of an haloalkane.
Use the table of chemical shifts on page 195 to do the following:

a) Predict the environment of the two H atoms with a shift of 3.6 ppm. [1 mark]

b) Predict the environment of the three H atoms with a shift of 1.0 ppm. [1 mark]

c) The relative molecular mass of the molecule is 64.5. Suggest a possible structure and explain your suggestion. [2 marks]

d) Explain the shapes of the two peaks. [2 marks]

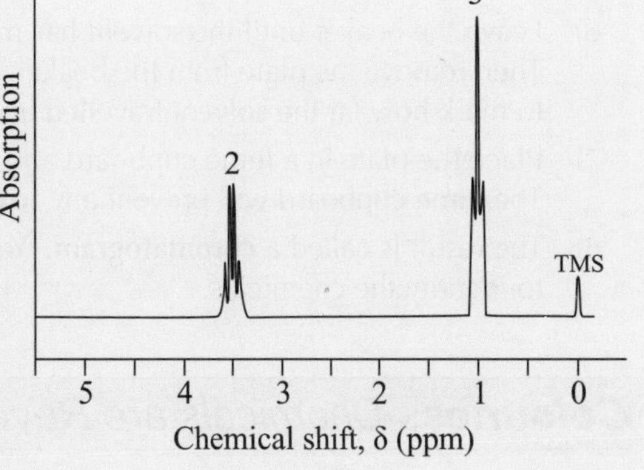

Q2 How many hydrogen environments are present in the molecule pentan-3-one, $CH_3CH_2COCH_2CH_3$? [1 mark]

Q3 The molecule ethyl ethanoate has three hydrogen environments, as shown in the diagram on the right. For the ^{1}H NMR spectrum of ethyl ethanoate, state:

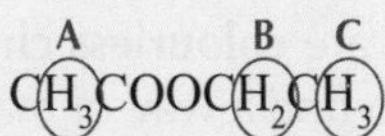

a) the integration ratio of the peaks, in the form A : B : C. [1 mark]

b) the type of peak that will be caused by each of the hydrogen environments. [3 marks]

Never mind splitting peaks — this stuff's likely to cause splitting headaches...

Is your head spinning yet? I know mine is. Round and round like a merry-go-round. It's a hard life when you're tied to a desk trying to get NMR spectroscopy firmly fixed in your head. You must be looking quite peaky yourself by now... so go on, learn this stuff, take the dog around the block, then come back and see if you can still remember it all.

Chromatography

You've probably tried chromatography with a spot of ink on a piece of filter paper — it's a classic experiment.

Chromatography is Good for **Separating** and **Identifying** Things

Chromatography is used to **separate** stuff in a mixture — once it's separated out, you can often **identify** the components. There are quite a few different types of chromatography — but they all have the same basic set up:

- A **mobile phase** — where the molecules can move. This is always a liquid or a gas.
- A **stationary phase** — where the molecules can't move. This must be a solid, or a liquid on a solid support.

And they all use the same basic principle:

1) The mobile phase **moves through** or **over** the stationary phase.
2) The **distance** each substance moves up the plate depends on its **solubility** in the mobile phase and its **retention** by the stationary phase.
3) Components that are **more soluble** in the mobile phase will **travel further** up the plate.
4) It's these **differences** in solubility and retention by the stationary phase that **separate** out the different substances.

Thin-Layer Chromatography is a Simple Way of **Separating Mixtures**

1) In thin-layer chromatography (TLC), the **stationary phase** is a thin layer of **silica** (**silicon dioxide**) or **alumina** (**aluminium oxide**) fixed to a glass or metal plate.
2) Draw a line **in pencil** near the bottom of the TLC plate (the baseline) and put a very small drop of each mixture to be separated on the line.
3) Allow the spots on the plate to **dry**.
4) Place the plate in a beaker with a small volume of solvent (this is the **mobile phase**). The solvent level must be **below** the baseline, so it doesn't dissolve your samples away.
5) The solvent will start to move up the plate. As it moves, the solvent carries the substances in the mixture with it — some chemicals will be carried **faster** than others and so travel further up the plate.

It's a good idea to wear gloves when handling the plate, to avoid any contamination by substances on your hands.

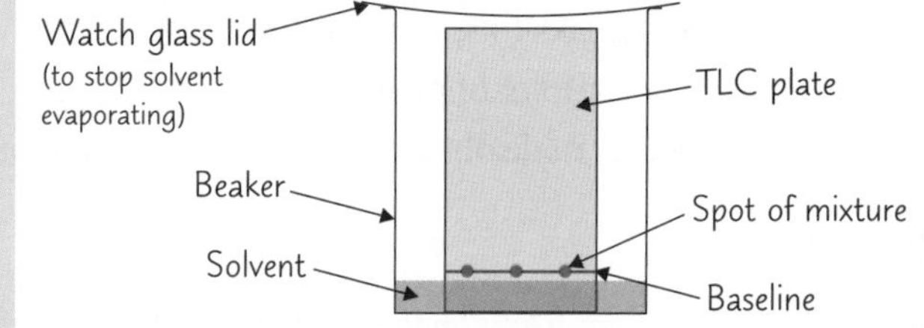

6) Leave the beaker until the solvent has moved almost to the top of the plate. Then remove the plate from the beaker. Before it evaporates, use a pencil to mark how far the solvent travelled up the plate (this line is called the **solvent front**).
7) Place the plate in a fume cupboard and leave it to **dry**. The fume cupboard will prevent any **toxic** or **flammable fumes** from escaping into the room.
8) The result is called a **chromatogram**. You can use the **positions of the spots** on the chromatogram to identify the chemicals.

Colourless Chemicals are Revealed Using **UV Light** or **Iodine**

1) If the chemicals in the mixture are **coloured** (such as the dyes that make up an ink) then you'll see them as a **set of coloured dots** at different heights on the TLC plate...
2) But if there are **colourless chemicals**, such as amino acids, in the mixture, you need to find a way of making them **visible**. Here are two ways:

Many TLC plates have a special **fluorescent dye** added to the silica or alumina layer that glows when **UV light** shines on it. You can put the plate under a **UV lamp** and draw around the dark patches to show where the spots of chemical are.

Expose the chromatogram to **iodine vapour** (leaving the plate in a sealed jar with a couple of iodine crystals does the trick). Iodine vapour is a **locating agent** — it sticks to the chemicals on the plate and they'll show up as **brown/purple spots**.

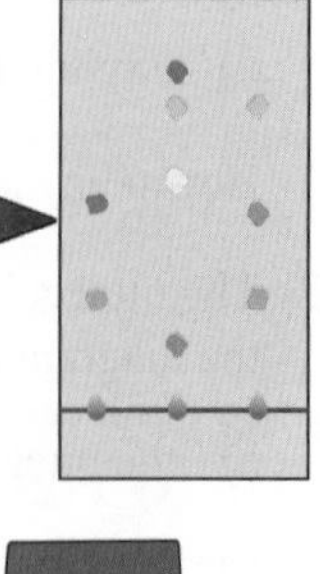

Chromatography

The **Position** of the Spots on a Plate Can Help to **Identify Substances**

1) If you just want to know **how many** chemicals are present in a mixture, all you have to do is **count the number of spots** that form on the plate.
2) But if you want to find out what each chemical **is**, you can calculate something called an **R_f value**. The formula for this is:

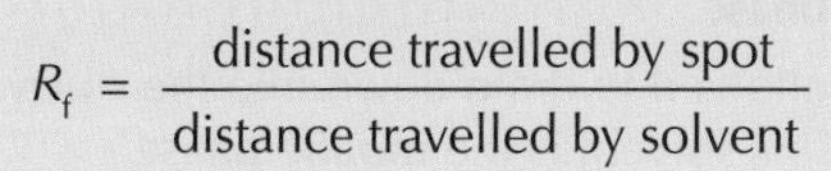

$$R_f = \frac{\text{distance travelled by spot}}{\text{distance travelled by solvent}}$$

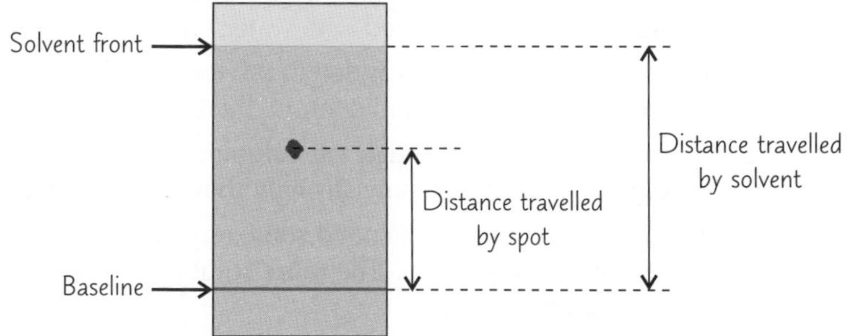

3) R_f values aren't dependent on how big the plate is or how far the solvent travels — they're properties of the chemicals in the mixture and so can be used to identify those chemicals.
4) This means you can look your R_f value up in a table of **standard R_f values** to identify what that substance is.
5) BUT — if the composition of the TLC plate, the solvent, or the temperature change even slightly, you'll get **different R_f values**.
6) It's hard to keep the conditions identical. So, if you suspect that a mixture contains, say, chlorophyll, it's best to put a spot of chlorophyll on the baseline of the **same plate** as the mixture and run them both at the **same time**.

Practice Questions

Q1 Explain the terms 'stationary phase' and 'mobile phase' in the context of chromatography.
Q2 What is the stationary phase in TLC?
Q3 Describe how you would calculate the R_f value of a substance on a TLC plate.

Exam Questions

Q1 A student is carrying out a thin-layer chromatography experiment.

a) The student uses a pair of gloves to hold the TLC plate. Give one reason for taking this precaution. [1 mark]

b) Once the plate has been placed inside the beaker of solvent, a lid is placed on top. Why is it necessary to seal the plate inside the beaker? [1 mark]

c) Once the solvent is nearing the top of the plate, the student removes it from the beaker and marks how far the solvent has travelled. What is the next step that the student needs to do? [1 mark]

Q2 The diagram below shows a chromatogram of four known substances (1 to 4) and two unknowns, labelled X and Y. One of the unknowns is pure and the other is a mixture.

a) Which of the unknowns, X or Y is a pure substance? [1 mark]

b) Suggest which of the known substances (1 to 4) are present in the unknown that is a mixture. Explain your answer. [2 marks]

c) The solvent front on the chromatogram was measured at 8 cm from the baseline, and Substance 1 travelled 5.6 cm. Calculate the R_f value of Substance 1. [1 mark]

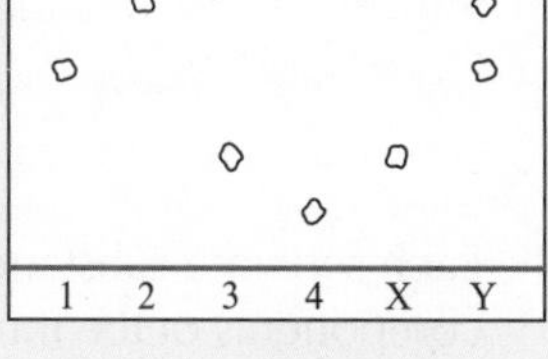

A little bit of TLC is what you need...

There's nothing better than watching small dots racing up a plate. Hours of fun. Working out R_f values comes close though and it's easy marks if you remember that it's the distance travelled by the spot divided by distance travelled by the solvent, and not the other way around — it might help to think of it as a fraction (and it'll always be less than 1).

More on Chromatography

As well as the thin-layer chromatography on the previous pages, there are also some weirder, wackier versions. The good thing is the principle is still the same — you've still got a mobile phase and a stationary phase. Phew.

Column Chromatography *is Used To Separate Out* ***Solutions***

Column chromatography is mostly used for **purifying an organic product.**

This is done to separate the product from unreacted chemicals and by-products.

1) It involves packing a glass column with a slurry of an absorbent material such as aluminium oxide, coated with water. This is the **stationary phase.**
2) The mixture to be separated is added to the top of the column and allowed to drain down into the slurry. A **solvent** is then run slowly and continually through the column. This solvent is the **mobile phase.**
3) As the mixture is washed through the column, its components **separate out** according to **how soluble** they are in the mobile phase and **how strongly they are adsorbed** onto the stationary phase (**retention**).
4) Each different component will spend some time adsorbed onto the stationary phase and some time dissolved in the mobile phase. The **more** soluble each component is in the mobile phase, the **quicker** it'll pass through the column.

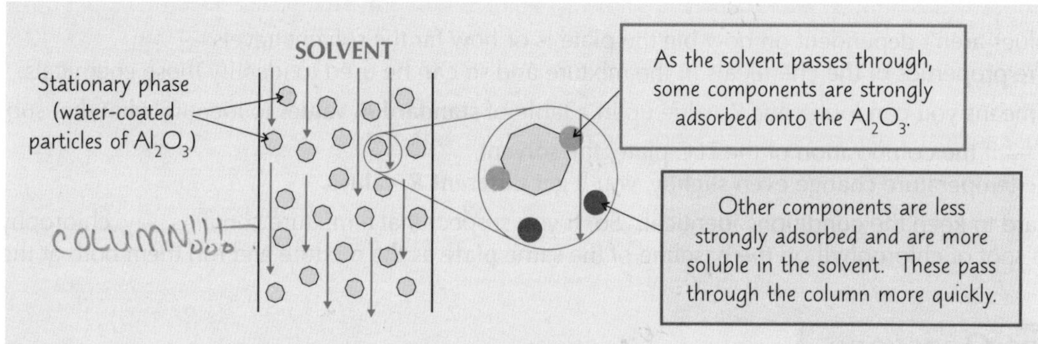

SOLVENT
Stationary phase (water-coated particles of Al_2O_3)
As the solvent passes through, some components are strongly adsorbed onto the Al_2O_3.
Other components are less strongly adsorbed and are more soluble in the solvent. These pass through the column more quickly.

Gas Chromatography *is Used To Separate* ***Mixtures of Volatile Liquids***

1) If you've got a mixture of **volatile liquids** (ones that turn into gases easily), then **gas chromatography** (GC) is the way to separate them out so that you can **identify** them.
2) The stationary phase is a **solid** or a solid coated by a **viscous liquid**, such as an oil, packed into a long tube. The tube is coiled to save space and built into an oven. The mobile phase is an **unreactive carrier gas** such as nitrogen.
3) Each component takes a different amount of time from being **injected** into the tube to being **recorded** at the other end. This is the **retention time**.

The retention time depends on how much time the component spends moving along with the carrier gas, and how much time it spends stuck to the viscous liquid.

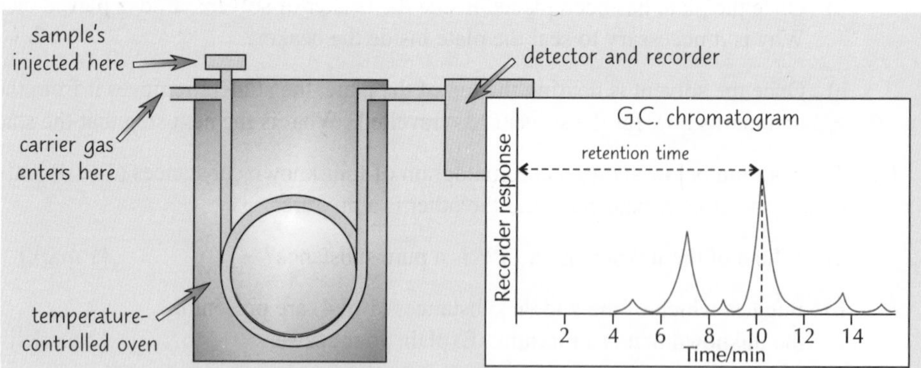

sample's injected here
carrier gas enters here
temperature-controlled oven
detector and recorder
G.C. chromatogram
retention time
Recorder response
2 4 6 8 10 12 14
Time/min

4) Each separate substance will have a unique retention time — so you can use the retention time to **identify** the components of the mixture. (You have to run a known sample under the same conditions for comparison.)
   For example, if you wanted to know if a mixture contained **octane**, you could run a sample of the **mixture** through the system, then run a sample of **pure octane** through, and see if there's a peak at the **same retention time** on both spectra.
5) The **area** under each peak tells you the relative **amount** of each component that's present in the mixture.
6) GC can be used to find the **level of alcohol** in **blood** or **urine** — the results are **accurate** enough to be used as evidence in court. It's also used to find the **proportions** of various **esters in oils** used in **paints** — this lets picture restorers know exactly what paint was originally used.

More on Chromatography

Mass Spectrometry can be Combined with Gas Chromatography

1) **Mass spectrometry** is a technique used to identify substances from their mass/charge ratio (see page 7). It is very good at **identifying** unknown compounds, but would give confusing results from a mixture of substances.
2) **Gas chromatography** (see previous page), on the other hand, is very good at **separating** a mixture into its individual components, but not so good at identifying those components.
3) If you put these **two techniques together**, you get an **extremely useful** analytical tool.

> **Gas chromatography-mass spectrometry** (or GC-MS for short) **combines the benefits** of gas chromatography and mass spectrometry to make a super analysis tool.
> The sample is **separated** using **gas chromatography**, but instead of going to a detector, the separated components are fed into a **mass spectrometer**.
> The spectrometer produces a **mass spectrum** for each **component**, which can be used to **identify** each one and show what the original **sample** consisted of.

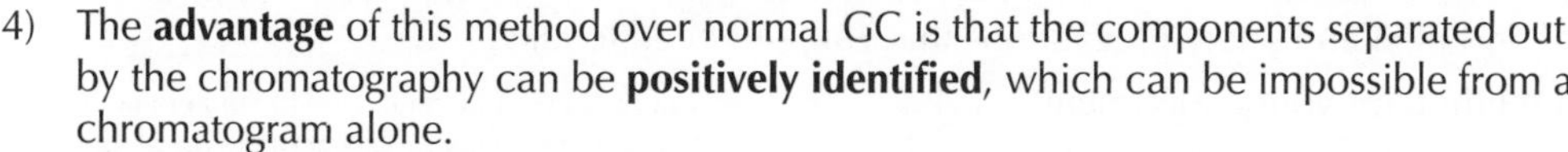

4) The **advantage** of this method over normal GC is that the components separated out by the chromatography can be **positively identified**, which can be impossible from a chromatogram alone.

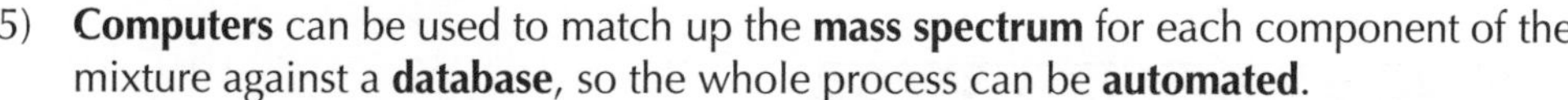

5) **Computers** can be used to match up the **mass spectrum** for each component of the mixture against a **database**, so the whole process can be **automated**.

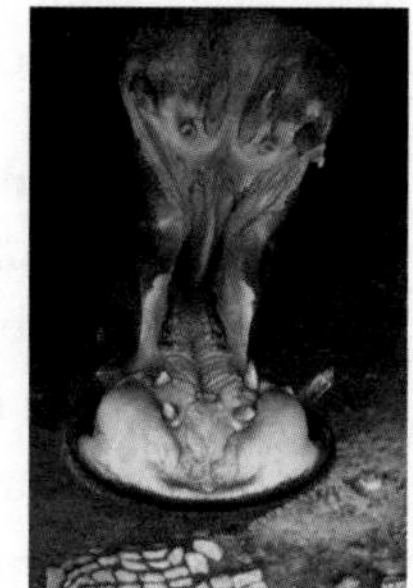

Oh, excuse me — chemistry always sends me to sleep, I'm afraid.

Practice Questions

Q1 In column chromatography, explain why the components pass through the column at different rates.

Q2 What is the mobile phase in GC?

Q3 GC can identify which substances are present and their relative amounts. How is the relative amount determined?

Exam Questions

Q1 A mixture of 25% ethanol and 75% benzene is run through gas chromatography apparatus.

a) Describe what happens to the mixture in the apparatus. [4 marks]

b) Explain why the substances separate. [2 marks]

c) How will the resulting chromatogram show the proportions of ethanol and benzene present in the mixture? [1 mark]

Q2 GC can be used to detect the presence and quantity of alcohol in the blood or urine samples of suspected drink-drivers.

a) What do the letters GC stand for? [1 mark]

b) Explain how 'retention time' is used to identify ethanol in a sample of blood or urine. [2 marks]

c) Why is nitrogen used as the carrier gas? [1 mark]

Q3 Which of these statements about column chromatography are correct?

1. It can be used as a purification technique.
2. The stationary phase is a slurry of an adsorbent material coated in water.
3. The more soluble each component is in the mobile phase, the slower it will pass through the column.

A Only 1　　B 1 and 2　　C 2 and 3　　D 1, 2 and 3 [1 mark]

Cromer-tography — pictures from my holiday in Norfolk...

The fun need not stop with thin-layer chromatography. Column chromatography and gas chromatography are potentially even more exciting. It may seem like a lot of techniques to learn, but the theory behind all these different types of chromatography is the same. You've got a mobile phase, a stationary phase and a mixture that wants separating.

Planning Experiments

As well as doing practical work in class, you can get asked about it in your exams too. Harsh I know, but I'm afraid that's how it goes. You need to be able to plan experiments and to spot the good and bad points of plans that you're shown.

*Experiments Need to be **Carefully Planned***

Scientists solve problems by **suggesting answers** and then doing **experiments** that **test** their ideas to see if the evidence supports them. Being able to plan experiments that will give you **accurate** and **precise results** is an important part of this process. Here's how you go about it:

There's more about what accurate and precise results are on page 208.

1) State the **aim** of your experiment — what question are you trying to answer?
2) Make a **prediction** — a specific testable statement about what will happen in the experiment, based on observation, experience or a **hypothesis** (a suggested explanation for a fact or observation).
3) Identify the **independent**, **dependent** and other **variables** (see below) in your experiment.
4) Decide what **data** you need to collect.
5) Select **appropriate equipment** for your experiment.
6) Do a **risk assessment** and plan any safety precautions that you will need to take.
7) Write out a **detailed method** for your experiment.

After step 7), you can actually go ahead and do your experiment. Lucky you...

*Make it a **Fair Test** — Control your **Variables***

You probably already know what the different kinds of **variable** are, but they're easy to mix up, so here's a recap:

Variable — A variable is a **quantity** that has the **potential to change**, e.g. temperature, mass, or volume. There are two types of variable commonly referred to in experiments:

- **Independent variable** — the thing that you **change** in an experiment.
- **Dependent variable** — the thing that you **measure** in an experiment.

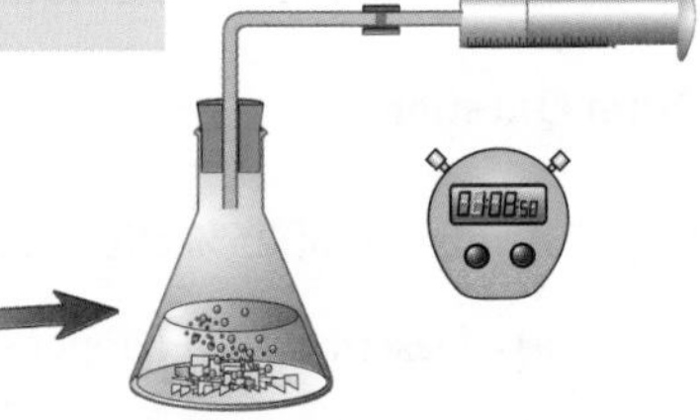

As well as the independent and dependent variables, you need to think of all the **other variables** (sometimes called **control variables**) that could affect the result of the experiment and plan ways to keep each of them **the same**.

So, if you're investigating the effect of temperature on the rate of a reaction using this equipment, then the variables will be:

Independent variable	Temperature.
Dependent variable	Volume of gas produced in a set period of time.
Other variables — you MUST keep these the same	Concentration and volume of solutions, mass of solids, pressure, whether or not you use a catalyst, the surface area of any solid reactants, time over which the gas is collected.

*Work **Safely** and **Ethically** — Don't Blow Up the Lab or Harm Small Animals*

1) When you plan an experiment, you need to think about how you're going to make sure that you work **safely**.
2) The first step is to identify all the **hazards** that might be involved in your experiment (e.g. dangerous chemicals or naked flames). Then you need to come up with ways to reduce the **risks** that these hazards pose.

   This means things like wearing **goggles** and a **lab coat** when handling any **hazardous chemicals** (e.g. those that are **irritants**, **toxic** or **corrosive**), using a **fume cupboard** to do any reactions that produce nasty gases, or heating anything flammable with a **water bath**, **sand bath** or **electric heater** (rather than over a flame).

3) Doing this procedure is sometimes referred to as doing a '**risk assessment**'.
4) You need to make sure you're working **ethically** too. This is most important if there are other people or animals involved. You have to put their welfare first.

Planning Experiments

Choose Appropriate Equipment — Think about Size and Sensitivity

Selecting the right equipment may sound easy but it's something you need to think carefully about.

1) The equipment has to be **appropriate** for the experiment.

   E.g. if you want to measure the amount of gas produced in a reaction, you need to make sure you use apparatus which will collect the gas, without letting any escape.

2) The equipment needs to be the right **size**.

   E.g. if you're using a gas syringe to measure the volume of gas produced by a reaction, it needs to be big enough to collect all the gas, or the plunger will be pushed out of the end. You might need to do some rough calculations to work out what size of equipment to use.

3) The equipment needs to have the right level of **sensitivity**.

   E.g. if you want to measure out 3.8 g of a substance, you need a balance that measures to the nearest tenth of a gram, not the nearest gram. If you want to measure out 6 cm^3 of a solution, you need to use a measuring cylinder that has a scale marked off in steps of 1 cm^3, not one that only has markings every 10 cm^3.

Moira was very sensitive to comments about the inappropriate size of her hat.

If you want to measure out a solution really accurately (e.g. 20.0 cm^3 of solution) you'll need to use a burette or a pipette.

Know Your Different Sorts of Data

Experiments always involve some sort of measurement to provide **data** and you need to decide what data to collect. There are different types of data — so it helps to know what they are.

Discrete — a discrete variable can only have **certain values** on a scale. For example the number of bubbles formed in a reaction is discrete (you can't have 1.77 bubbles). You usually get discrete data by **counting** things.

Continuous — a continuous variable can have **any value** on a scale. For example, the volume of gas produced or the voltage of an electrochemical cell. You can never measure the exact value of a continuous variable.

Categoric — a categoric variable has values that can be sorted into **categories**. For example, the colours of solutions might be blue, red and green, or types of material might be wood, steel and glass.

Ordered (ordinal) — ordered data is similar to categoric, but the categories can be **put in order**. For example, if you classified reactions as 'slow', 'fairly fast' and 'very fast' you'd have ordered data.

Methods Must be Clear and Detailed

When **writing** or **evaluating** a method, you need to think about all of the things on these two pages. The method must be **clear** and **detailed** enough for **anyone** to follow — it's important that other people can recreate your experiment and get the **same** results. Make sure your method includes:

1) All the **substances** needed and what **quantity** of each to use.
2) How to **control** variables.
3) The exact **apparatus** needed (a **diagram** is often helpful to show the set-up).
4) Any **safety precautions** that should be taken.
5) What **data** to collect and **how** to collect it.

Presenting Results

Once you've collected the data from your experiment, it's not time to stop, put your feet up and have a cup of tea — you've got to present your results too. That might well mean putting them in a table or turning them into a graph.

*Organise Your Results in a **Table***

It's a good idea to set up a table to **record** the **results** of your experiment. Make sure that you **include** enough **rows** and **columns** to **record all of the data** you need. You might also need to include a column for **processing** your data (e.g. working out an average).

Make sure each **column** has a **heading** so you know what's going to be recorded where.

The **units** should be in the **column heading**, not the table itself.

Temperature (°C)	Time (s)	Volume of gas evolved (cm^3)			Average volume of gas evolved (cm^3)
		Run 1	Run 2	Run 3	
20	**10**	8.1	8.4	8.1	(8.1 + 8.4 + 8.1) ÷ 3 = 8.2
	20	19.8	19.6	19.4	(19.8 + 19.6 + 19.4) ÷ 3 = 19.6
	30	29.8	29.9	30.0	(29.8 + 29.9 + 30.0) ÷ 3 = 29.9

You can find the **mean result** by **adding up** the data from each repeat and **dividing** by the number of repeats.

*Graphs: **Line, Bar or Scatter** — Use the **Best Type***

When drawing graphs, the dependent variable should go on the y-axis, and the independent on the x-axis.

You'll often need to make a **graph** of your results. Not only are graphs **pretty**, they make your data **easier to understand** — so long as you choose the right type.

Scatter plots are great for showing how two sets of continuous data are related (or **correlated** — see page 206). Don't try to join all the points on a scatter plot — draw a straight or curved **line of best fit** to show the **trend.**

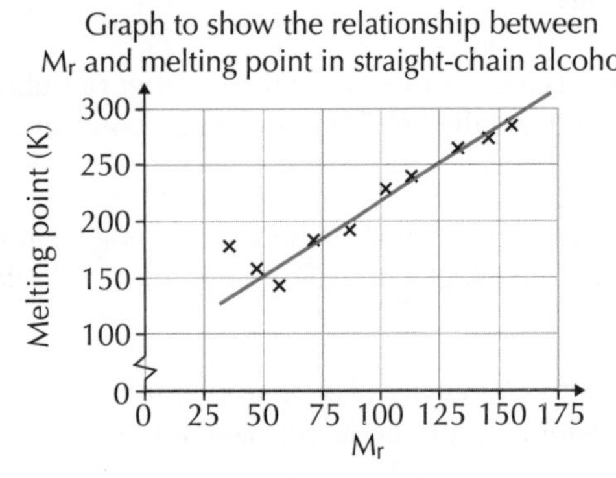

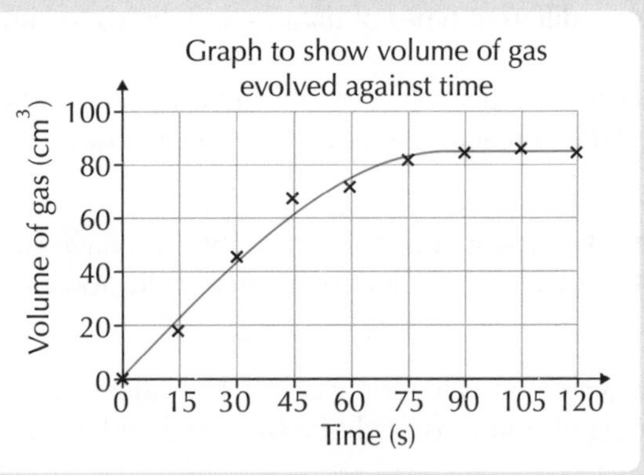

You should use a bar chart when one of your data sets is **categoric or ordered data**. For example:

Graph to show chlorine concentration in water samples

Chlorine concentration (ppm)

0.5
0.4
0.3
0.2
0.1
0

A B C

Water samples

Pie charts can also be used to display categoric data.

Whatever type of graph you draw, you'll ONLY get full marks if you:

- Choose a sensible scale — don't draw a tiny graph in the corner of the paper.
- Label both axes — including units.
- Plot your points accurately — using a sharp pencil.

Presenting Results

Don't Forget About Units

Units are really important — 10 g is very different from 10 kg — so make sure you don't forget to add them to your **tables** and **graphs**. They're also important in **calculations**, particularly if you need to **convert** between two different units.

Here are some useful examples:

Volume can be measured in $\mathbf{m^3}$, $\mathbf{dm^3}$ and $\mathbf{cm^3}$.

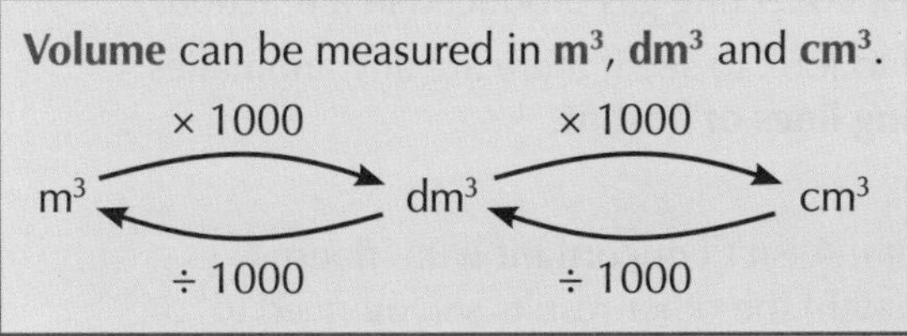

Example: Write 6 dm^3 in m^3 and cm^3.

First, to convert 6 dm^3 into m^3 you divide by 1000.

$6\ dm^3 \div 1000 = 0.006\ m^3 = \mathbf{6 \times 10^{-3}\ m^3}$

Then, to convert 6 dm^3 into cm^3 you multiply by 1000.

$6\ dm^3 \times 1000 = 6000\ cm^3 = \mathbf{6 \times 10^{3}\ cm^3}$

This is written in standard form. Standard form is a useful way to write very big or very small numbers neatly.

Temperature can be measured in **°C** and **K**.

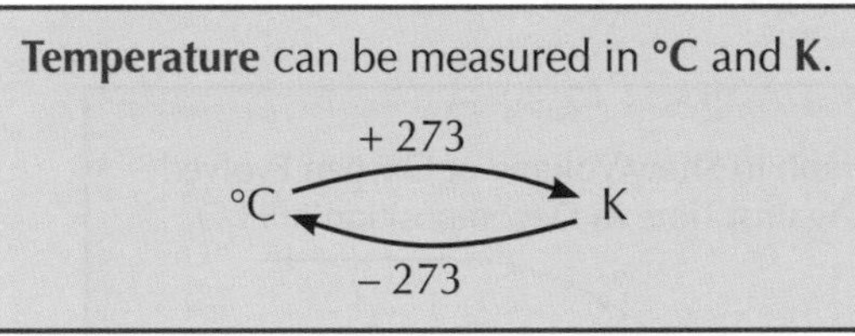

Example: Write 33 °C in K.

To convert 33 °C into K you add 273.

33 °C + 273 = **306 K**

Round to the *Lowest Number* of *Significant Figures*

The first **significant figure** (or **s.f.**) of a number is the **first digit that isn't a zero**. The second, third and fourth significant figures follow on immediately after the first (even if they're zeros).
For example, the number **0.01072** is **0.01** to 1 s.f., **0.011** to 2 s.f., and **0.0107** to 3 s.f.

1) When you're doing a calculation, use the number of significant figures given in the data as a guide for how many you need to give in your answer.
2) Whether you're doing calculations with the results from an experiment or doing calculations in an exam, the rule is the same — round your answer to the **lowest number of significant figures** that's in your data.

Example: 13.5 cm^3 of a 0.51 mol dm^{-3} solution of sodium hydroxide reacts with 1.5 mol dm^{-3} hydrochloric acid. Calculate the volume of hydrochloric acid, in cm^3, required to neutralise the sodium hydroxide.

Moles of NaOH = 0.51 mol dm^{-3} × (13.5 cm^3 ÷ 1000) = 6.885 × 10^{-3} mol

(0.51 — 2 s.f.; 13.5 — 3 s.f.)

Volume of HCl = (6.885 × 10^{-3}) mol ÷ 1.5 mol dm^{-3} = 0.00459 dm^3

(1.5 — 2 s.f.)

= 0.00459 dm^3 × 1000 = 4.59 cm^3

= **4.6 cm^3 (2 s.f.)** ← Final answer should be rounded to 2 s.f.

Don't round any intermediate answers. Rounding too early will make your final answer less accurate.

3) You should always **write down** the number of significant figures you've rounded to after your answer (as in the example above), so that other people can see what rounding you've done.
4) If you get told in an exam question **how many** significant figures you should give your answer to, make sure you follow those instructions — you'll **lose marks** if you don't.
5) If you're converting an answer into **standard form**, keep the same number of significant figures, e.g. 0.0060 mol dm^{-3} has the same number of significant figures as 6.0 × 10^{-3} mol dm^{-3}.

If you're ever asked to give an answer to "an appropriate degree of precision", this just means "to a sensible number of significant figures".

Analysing Results

You're not quite finished yet... there's still time to look at your results and try and make sense of them. Graphs are really useful for helping you to spot patterns in your data. There are lots of examples on these pages. Ooh, pretty...

Watch Out For **Anomalous** Results

1) **Anomalous results** are ones that **don't fit** in with the other values — this means they are likely to be wrong.
2) They're often caused by mistakes or problems with apparatus, e.g. if a drop in a titration is too big and puts you past the end point, or if a syringe plunger gets stuck whilst collecting gas produced in a reaction.
3) When looking at results in tables or graphs, you always need to look to see if there are any anomalies — you **ignore** these results when **calculating means** or **drawing lines of best fit**.

Example: Calculate the mean titre volume from the results in the table below.

Titration Number	1	2	3	4
Titre Volume (cm³)	15.20	15.30	15.25	(15.70)

Titre **4** isn't **concordant** with (doesn't match) the other results so you need to ignore that one and just use the other three:

$$\frac{15.20 + 15.30 + 15.25}{3} = \mathbf{15.25\ cm^3}$$

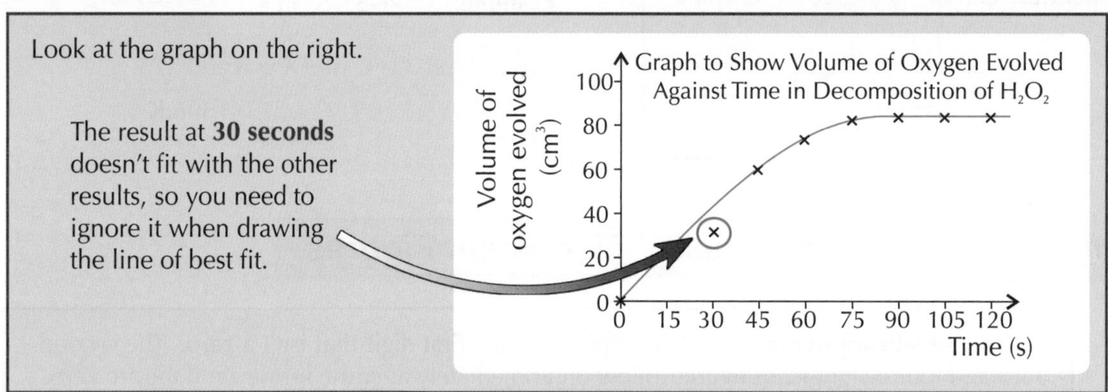

Look at the graph on the right.
The result at 30 seconds doesn't fit with the other results, so you need to ignore it when drawing the line of best fit.
Graph to Show Volume of Oxygen Evolved Against Time in Decomposition of H₂O₂
Volume of oxygen evolved (cm³)
0
20
40
60
80
100
0 15 30 45 60 75 90 105 120
Time (s)

Scatter Graphs Show How Two Variables are **Correlated**

Correlation describes the **relationship** between the independent variable and dependent variable. Data can show:

Positive correlation
As one variable **increases**, the other **increases**.

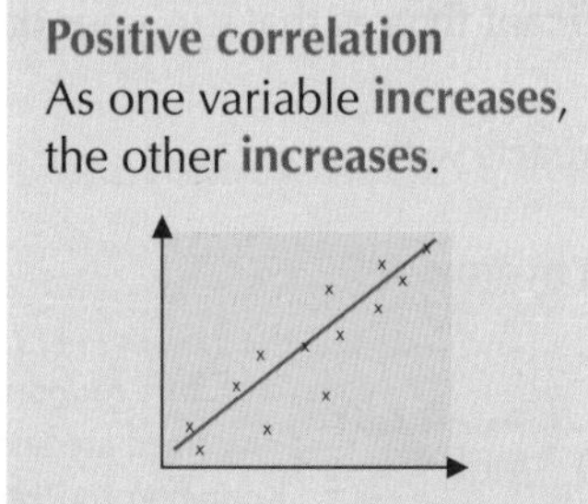

Negative correlation
As one variable **increases**, the other **decreases**.

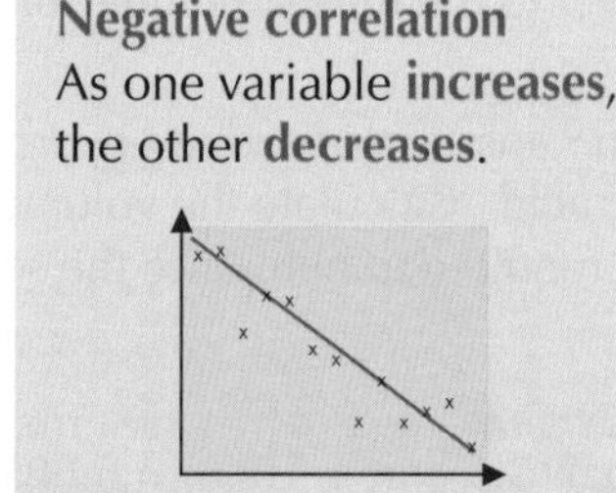

No correlation
There is **no relationship** between the two variables.

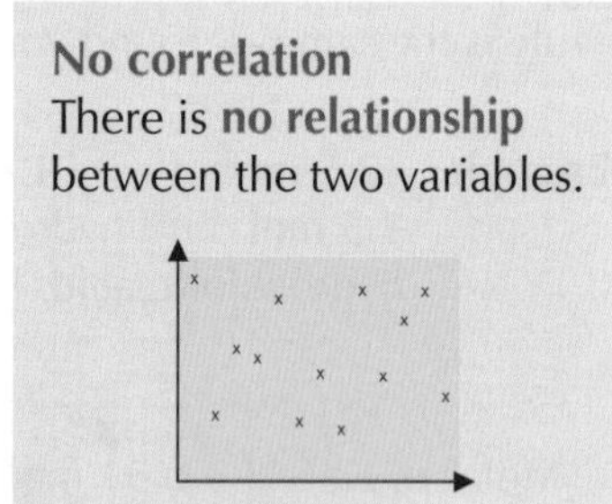

Correlation **Doesn't** Mean **Cause** — Don't Jump to Conclusions

1) Ideally, only **two** quantities would **ever** change in any experiment — everything else would remain **constant**.
2) But in experiments or studies outside the lab, you **can't** usually control all the variables. So even if two variables are correlated, the change in one may **not** be causing the change in the other. Both changes might be caused by a **third variable**.

For example, some studies have found a correlation between drinking **chlorinated tap water** and the risk of developing certain cancers. Some people argue that this means water shouldn't have chlorine added.
BUT it's hard to design a study that **controls all the variables** between people who drink tap water and people who don't. The risk of getting different cancers is affected by many lifestyle factors. Or there could be some other risk factor present in the tap water (or in whatever the non tap water drinkers drink instead).

Analysing Results

Don't Get **Carried Away** When Drawing Conclusions

1) The **data** should always **support** your conclusion. This may sound obvious but it's easy to **jump** to conclusions.
2) Also, conclusions have to be **specific** — you can't make sweeping generalisations.

For example: the **rate** of an enzyme-controlled reaction was measured at **10 °C, 20 °C, 30 °C, 40 °C, 50 °C** and **60 °C**. All other variables were kept constant. The results of this experiment are shown in the graph below.

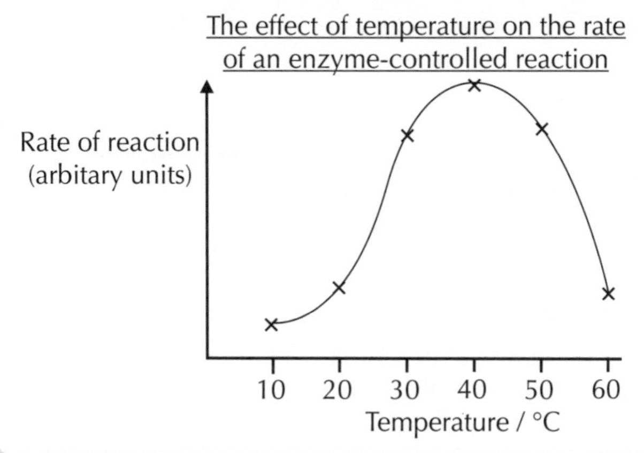

1) A science magazine **concluded** from this data that the enzyme used in the experiment works best at **40 °C**.
2) The data **doesn't** support this exact claim. The enzyme **could** work best at 42 °C or 47 °C, but you can't tell from the data because **increases** of **10 °C** at a time were used. The rate of reaction at in-between temperatures **wasn't** measured.
3) All you can say for certain is that this particular reaction was faster at **40 °C** than at any of the other temperatures tested.
4) Also you can't be sure that if you did the experiment under **different conditions**, e.g. at a **different pressure**, you wouldn't get a **different optimum temperature**.
5) It's also worth remembering that this experiment **ONLY** gives you information about this particular reaction. You can't conclude that **all** enzyme-controlled reactions happen fastest at this temperature — only this one.

You Can Find **Rate** By Finding the **Gradient** of a Graph

Rate is a **measure** of how much something is **changing over time**.
Calculating a rate can be useful when analysing data, e.g. you might want to the find the **rate of a reaction**.

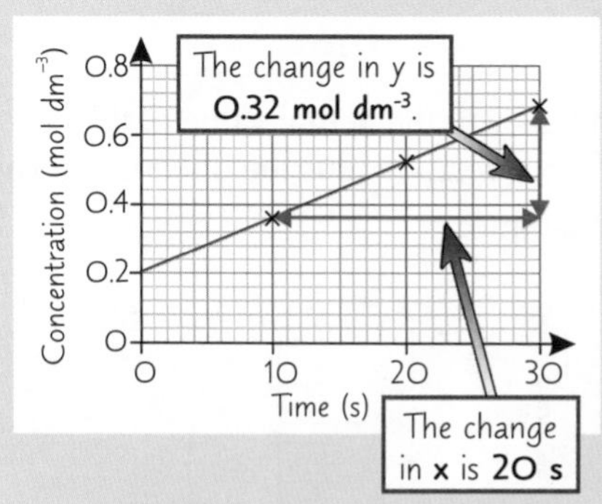

For a **linear** graph you can calculate the **rate** by finding the **gradient of the line**:

$$\text{Gradient} = \frac{\text{Change in } y}{\text{Change in } x}$$

So in this **example**:

$$\text{rate} = \frac{0.32 \text{ mol dm}^{-3}}{20 \text{ seconds}} = \mathbf{0.016 \text{ mol dm}^{-3} \text{ s}^{-1}}$$

The **equation** of a **straight line** can always be written in the form $y = mx + c$, where **m** is the **gradient** and **c** is the ***y*-intercept** (this is the **value of** *y* when the line crosses the ***y*-axis**).
In this example, the equation of the line is $y = 0.016x + 0.2$.

For a **curved** graph you find the **gradient** by drawing a **tangent**:

1) Position a ruler on the graph at the **point** where you want to know the **rate**.
2) **Angle** the **ruler** so there is **equal space** between the **ruler** and the **curve** on **either** side of the point.
3) **Draw** a **line** along the ruler to make the tangent.
   Extend the line right across the graph — it'll help to make your **gradient calculation easier** as you'll have **more points** to choose from.
4) **Calculate** the **gradient** of the **tangent** to find the **rate**:

rate = 55 °C ÷ 44 s = **1.25 °C s^{-1}**

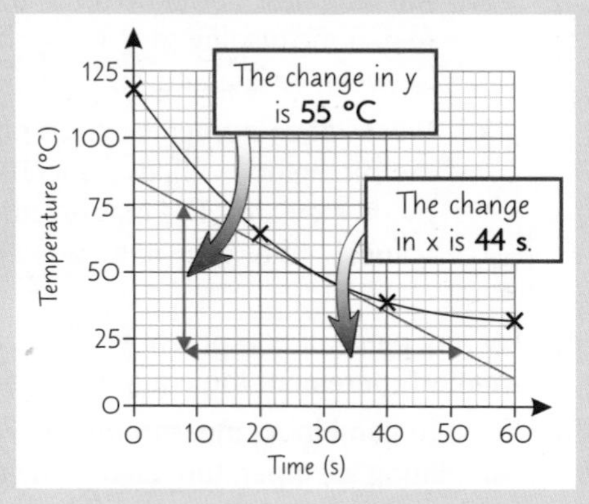

Evaluating Experiments

So you've planned an experiment, done the practical work, collected lots of data and plotted it all on a beautiful graph. Now it's time to sit back, relax and... work out everything you did wrong. That's science, I'm afraid.

You Need to Look Critically at Your Results

Here are a few terms that will come in handy when you're evaluating how convincing your results are...

Valid

Valid results are results that answer the **original question**. For example, if you haven't **controlled all the variables** your results won't be valid, because you won't be testing just the thing you wanted to.

Accurate

Accurate results are results that are **really close** to the **true** answer.

You might see results that fulfil all of these being called reliable.

Precise

The smaller the amount of **spread** of your data around the **mean** (see page 204), the more **precise** it is. Calculating a **mean** (average) result from your repeats will **increase** the **precision** of your result, because it helps to reduce the effect of **random errors** on the answer (see page 209).

Repeatable

Your results are **repeatable** if **you** get the same results when you repeat the experiment using the same method and the same equipment. You really need to repeat your readings **at least three times** to demonstrate that your results really are repeatable.

Reproducible

Your results are **reproducible** if **other people** get the same results when they repeat your experiment.

You Need to Think About the Uncertainty Your Measurements Might Have

1) Any measurements you make will have **uncertainties** (or **errors**) in them, due to the limits of the **sensitivity** of the equipment.
2) Uncertainties are usually written with a ± sign, e.g ±0.05 cm. The ± sign tells you the **actual value** of the measurement lies between your reading **minus** the uncertainty value and your reading **plus** the uncertainty value.
3) The uncertainty will be different for different pieces of equipment. For example:

- The scale on a **50 cm³ burette** usually has marks every **0.1 cm³**. You should be able to tell which mark the level's closest to, so any reading you take won't be more than **0.05 cm³** out. So the **uncertainty** on each burette reading is **±0.05 cm³**.
- If a **mass balance** measures masses to the **nearest 0.1 g**, the real mass could be up to **0.05 g smaller or larger** (e.g. if the display says 1.7 g, the real mass could be anywhere between 1.65 g and 1.75 g.) So the **uncertainty** is **±0.05 g**.
- Pieces of equipment such as pipettes, volumetric flasks and thermometers will have uncertainties that depend on how well made they are. The manufacturers provide these **uncertainty values** — they're usually written on the equipment somewhere.

45
46

The level in this burette is between the 44.9 cm³ and 45.0 cm³ marks. It's closer to 45.0 — so the level is between 44.95 and 45.0. So a reading of 45.0 cm³ can't have an uncertainty of more than 0.05 cm³.

4) For any piece of equipment you use, the uncertainty will be **half** the **smallest increment** the equipment can measure, in either direction.
5) If you're **combining measurements**, you'll need to combine their **uncertainties**. For example, if you're calculating a temperature change by measuring an initial and a final temperature, the **total** uncertainty for the temperature change will be the uncertainties for both measurements added together.

Evaluating Experiments

You Can Calculate The **Percentage Uncertainty** in a Result

If you know the **uncertainty** (error) in a reading that you've taken using a certain piece of equipment, you can use it to calculate the **percentage uncertainty** in your measurement.

$$\text{percentage uncertainty} = \frac{\text{uncertainty}}{\text{reading}} \times 100$$

Example: 2.00 g of solid $CaCl_2$ is measured out on a balance with an uncertainty of ±0.005 g. Calculate the percentage uncertainty on this reading.

$$\text{percentage uncertainty} = \frac{0.005}{2.00} \times 100 = \mathbf{0.25\%}$$

If you're finding the uncertainty on a mean value, you'll need these formulas instead:

uncertainty on the mean = range ÷ 2

$$\% \text{ uncertainty} = \frac{\text{uncertainty on the mean}}{\text{mean}} \times 100$$

Example: In a titration a burette with an uncertainty of ±0.05 cm^3 is used. The initial reading on the burette is 50.0 cm^3. The final reading is 28.8 cm^3. Calculate the percentage uncertainty on the titre value.

Titre value = 50.0 cm^3 – 28.8 cm^3 = 21.2 cm^3

Uncertainty on each burette reading = ±0.05 cm^3

Two burette readings have been combined to find the titre value.
So the total uncertainty on the measurement = 0.05 + 0.05 = ±0.1 cm^3

Percentage uncertainty on titre value = $\frac{0.1}{21.2} \times 100 = 0.472\% = \mathbf{0.5\%\ (1\ d.p.)}$

When you combine uncertainties, give the total uncertainty to the same number of decimal places as the measurement. (That's the titre value here.)

Percentage uncertainty is useful because it tells you how **significant** the uncertainty in a reading is compared to its **size**, e.g. ±0.1 g uncertainty is more significant when weighing out 0.2 g of a solid than when weighing out 100.0 g.

You Can **Minimise** the Percentage Uncertainty

1) One obvious way to **reduce uncertainty** in your measurements is to use the most **sensitive equipment** available. There's not much you can do about this at school or college though — you're stuck with whatever's available.
2) But there are other ways to **lower the uncertainty** in experiments. The **larger the reading** you take with a piece of equipment, the **smaller the percentage uncertainty** on that reading will be. Here's a quick example:
   - If you measure out **5.0 cm^3** of liquid in a burette with an uncertainty of **±0.05 ml** then the percentage uncertainty is (0.05 ÷ 5.0) × 100 = **1%**.
   - But if you measure **10.0 cm^3** of liquid in the same burette, the percentage uncertainty is (0.05 ÷ 10.0) × 100 = **0.5%**. Hey presto — you've just halved the percentage uncertainty.
   - So you can reduce the percentage uncertainty of this experiment by using a **larger volume** of liquid.
3) You can apply the same principle to other measurements too. For example, if you weigh out a small mass of a solid, the **percentage uncertainty** will be larger than if you weighed out a larger mass using the same balance.

Errors Can Be **Systematic** or **Random**

1) **Systematic errors** cause each reading to be different to the true value by the same amount, i.e. they shift all of your results. They may be caused by the **set-up** or the **equipment** you're using. If the 10.00 cm^3 pipette you're using to measure out a sample for titration actually only measures 9.95 cm^3, your sample will be 0.05 cm^3 too small **every time** you repeat the experiment.
2) **Random errors** cause readings to be spread about the true value due to the results varying in an **unpredictable** way. You get random error in all measurements and no matter how hard you try, you can't correct them. The tiny errors you make when you read a burette are random — you have to estimate the level when it's between two marks, so sometimes your figure will be **above** the real one and sometimes **below**.
3) **Repeating an experiment** and finding the mean of your results helps to deal with **random errors**. The results that are a bit high will be **cancelled out** by the ones that are a bit low, so your results will be more **precise**. But repeating your results won't get rid of **systematic errors**, so your results won't get more **accurate**.

This should be a photo of a scientist. I don't know what happened — it's a random error...

Exam Structure and Technique

Passing exams isn't all about revision — it really helps if you know how the exam is structured and have got your exam technique nailed so that you pick up every mark you can.

Make Sure You Know the **Structure** of Your **Exams**

If you're sitting AS-level Chemistry, you'll be doing different papers so this info isn't relevant to you.

The AQA A-Level Chemistry course is split into three units: **Physical Chemistry** (Unit 1), **Inorganic Chemistry** (Unit 2) and **Organic Chemistry** (Unit 3).

For AQA A-Level Chemistry you're going to have to sit **three exams** — Paper 1, Paper 2 and Paper 3.

- **Paper 1** will test you on most of Physical Chemistry and all of Inorganic Chemistry.
- **Paper 2** will test you on some of Physical Chemistry and all of Organic Chemistry.
- **Paper 3** can test you on any of the material covered in the course.
- All three papers could include questions on **practical skills**.

If you want more detail on what could come up in each paper, look at the table below:

If you haven't got a copy of the AQA specification, you can download it from the AQA website (or ask your teacher).

Paper	Content Assessed
1	• Unit 1: Physical chemistry. (Covered in Unit 1 of this book.) AQA specification references 3.1.1.1 to 3.1.4.4 (Year 1), 3.1.6.1 to 3.1.7 (Year 1), 3.1.8.1 to 3.1.8.2 (Year 2) and 3.1.10 to 3.1.12.6 (Year 2). • Unit 2: Inorganic chemistry. (Covered in Unit 2 of this book.) AQA specification references 3.2.1.1 to 3.2.3.2 (Year 1) and 3.2.4 to 3.2.6 (Year 2). • Relevant practical skills. (Covered in the Practical Skills section of this book.)
2	• Unit 1: Physical chemistry. (Covered in Unit 1 of this book.) AQA specification references 3.1.2.1 to 3.1.6.2 (Year 1) and 3.1.9.1 to 3.1.9.2 (Year 2). • Unit 3: Organic chemistry. (Covered in Unit 3 of this book.) AQA specification references 3.3.1.1 to 3.3.6.3 (Year 1) and 3.3.7 to 3.3.16 (Year 2). • Relevant practical skills. (Covered in the Practical Skills section of this book.)
3	• Any content from Units 1, 2 and 3. • Relevant practical skills. (Covered in the Practical Skills section of this book.)

All the exams are **2 hours** long. Papers 1 and 2 are each worth **105 marks**, which is **35%** of your total mark. Paper 3 is worth **90 marks**, which is **30%** of your total mark.

- Paper 1 and Paper 2 are made up of **short and long answer questions**.
- Paper 3 will be made up of 40 marks of questions relating to **practical techniques** and **data analysis**, 20 marks of **general questions** from all parts of the specification, and 30 marks of **multiple choice** questions.

***Manage** Your **Time** Sensibly*

1) **How long** you spend on each question is important in an exam — it could make all the difference to your grade.
2) The **number of marks** tells you roughly how long to spend on a question. Some questions will require lots of work for only a few marks but other questions will be much quicker.

Example:

Q1 Define the term 'standard enthalpy change of neutralisation'. [2 marks]

Q2 Draw the structures of the two monomers that react to form the condensation polymer shown above. [2 marks]

Question 1 only asks you to write down a **definition** — if you can remember it this shouldn't take too long.

Question 2 asks you to **work out and draw** the structures of two monomers used to make a condensation polymer — this may take longer than writing down a definition.

If you're running out of time, it makes sense to do Q1 first and come back to Q2 if you have time at the end.

3) Don't spend ages struggling with questions that are only worth a couple of marks — **move on**. You can come back to them later when you've bagged loads of other marks elsewhere.
4) If you get really stuck on a question, it's also probably best to move on and come back to it later.

Exam Structure and Technique

Make Sure You **Read the Question**

Not all of the questions will have a command word — they may just be a which / what / how type of question.

1) **Command words** are just the bit of the question that tell you what to do.
2) You'll find answering exam questions much easier if you understand exactly what they mean, so here's a summary of the most common command words.

Command word:	What to do:
Give / Name / State	Write a concise answer, from fact recall or from information that you've been given in the question.
Identify	Say what something is.
Describe	Write about what something is like or how it happens.
Explain	Give reasons for something.
Suggest / Predict	Use your scientific knowledge to work out what the answer might be.
Outline	Give a brief description of the main characteristics of something.
Calculate	Work out the solution to a mathematical problem.
Draw	Produce a diagram or graph.
Sketch	Draw something approximately — for example, draw a rough line graph to show the main trend of some data.

Elsie worked hard on making perfect points in her Chemistry exams.

Some Questions Will Test Your Knowledge of **Practical Skills**

At least 15% of the marks in your A-Level Chemistry exams will focus on practical skills.

This means you'll get questions where you're asked to do things like comment on the design of experiments, make predictions, draw graphs, calculate percentage uncertainties — basically, anything related to planning experiments or analysing results. These skills are covered in the Practical Skills section of this book on pages 202 to 209.

Be **Careful** With **Calculations**

At least 20% of the marks up for grabs in A-level Chemistry will require maths skills, so make sure you know your stuff.

1) In calculation questions you should always **show your working** — you may get some marks for your **method** even if you get the answer wrong.
2) Don't **round** your answer until the **very end**. Some of the calculations in A-level Chemistry can be quite **long**, and if you round too early you could introduce errors to your final answer.

Remember to Use the **Exam Data Booklet**

When you sit your exams, you'll be given a data booklet. It will contain lots of useful information, including:

- the characteristic infrared absorptions, ^{13}C NMR shifts and ^{1}H NMR shifts of some common functional groups,
- the structures of some biologically important molecules, such as the DNA bases, some amino acids, common phosphates and sugars and Heme B.
- a copy of the periodic table.

I'd put a joke here, but there's just nothing funny about exams...

The real key to preparing for your exams is to get as much practice as possible. Get hold of some practice papers and try doing them in two hours so you can be sure you've got the timing right. It'll help you get used to the different types of questions that might pop up too. And it'll flag up any topics that you're a bit shaky on, so you can go back and revise.

Answers

Unit 1: Section 1 — Atomic Structure

Page 3 — The Atom

1 a) Similarity — They've all got the same number of protons/electrons ***[1 mark]***.
Difference — They all have different numbers of neutrons ***[1 mark]***.
b) 1 proton, 1 neutron (2 – 1), 1 electron ***[1 mark]***.
c) $^{3}_{1}H$ ***[1 mark]***

2 a) i) Same number of electrons. $^{32}S^{2-}$ has 16 + 2 = 18 electrons. ^{40}Ar has 18 electrons too ***[1 mark]***.
ii) Same number of protons. Each has 16 protons (the atomic number of S must always be the same) ***[1 mark]***.
iii) Same number of neutrons. ^{40}Ar has 40 – 18 = 22 neutrons. ^{42}Ca has 42 – 20 = 22 neutrons ***[1 mark]***.
b) **A** and **C** ***[1 mark]***. They have the same number of protons but different numbers of neutrons ***[1 mark]***.
It doesn't matter that they have a different number of electrons because they are still the same element.

Page 7 — Relative Mass and the Mass Spectrometer

1 a) First multiply each relative abundance by the relative mass:
120.8 × 63 = 7610.4
54.0 × 65 = 3510.0
Next add up the products:
7610.4 + 3510.0 = 11 120.4 ***[1 mark]***
Total abundance = 120.8 + 54.0 = 174.8
Divide the sum of the products by the total abundance:
A_r(Cu) = 11 120.4 ÷ 174.8 ≈ **63.6** ***[1 mark]***
You can check your answer by seeing if A_r(Cu) is in between 63 and 65 (the lowest and highest relative isotopic masses).
b) A sample of copper is a mixture of 2 isotopes in different abundances ***[1 mark]***. The average mass of these isotopes isn't a whole number ***[1 mark]***.

2 a) Mass spectrometry ***[1 mark]***
b) Multiply each % relative abundance by the relative mass:
93.11 × 39 = 3631.29,
0.12 × 40 = 4.8,
6.77 × 41 = 277.57
Next add up the products:
3631.29 + 4.8 + 277.57 = 3913.66 ***[1 mark]***
This time you divide by 100 because they're percentages.
A_r(K) = 3913.66 ÷ 100 = **39.1** (3 s.f.) ***[1 mark]***
Again check your answer's between the lowest and highest relative isotopic masses, 39 and 41. A_r(K) is closer to 39 because most of the sample (93.11%) is made up of this isotope.

3 a) So that they can be accelerated ***[1 mark]*** and detected ***[1 mark]***.
b) Positive ions are accelerated by an electric field ***[1 mark]*** to a constant kinetic energy ***[1 mark]***. Lighter ions will end up moving faster ***[1 mark]***. The faster ions arrive first at the detector ***[1 mark]***.
c) B Gallium ***[1 mark]***.
The percentage isotope abundance is around 60-40.

Page 9 — Electronic Structure

1 a) K atom: $1s^2\ 2s^2\ 2p^6\ 3s^2\ 3p^6\ 4s^1$ ***[1 mark]***
K^+ ion: $1s^2\ 2s^2\ 2p^6\ 3s^2\ 3p^6$ ***[1 mark]***
b) Oxygen atom:

1s	2s	2p
↑↓	↑↓	↑↓ ↑ ↑

Correct number of electrons in each sub-shell ***[1 mark]***.
Having spin-pairing in one of the p orbitals and parallel spins in the other two p orbitals ***[1 mark]***.
A box filled with 2 arrows is spin pairing — 1 up and 1 down. If you've put the four p electrons into just 2 orbitals, it's wrong.

2 a) $1s^2\ 2s^2\ 2p^6\ 3s^2\ 3p^6\ 3d^5\ 4s^2$ ***[1 mark]***
b) Al^{3+} ion:

1s	2s	2p
↑↓	↑↓	↑↓ ↑↓ ↑↓

Correct number of electrons in each sub-shell ***[1 mark]***.
One arrow in each box pointing up, and one pointing down ***[1 mark]***.
c) Germanium ($1s^2\ 2s^2\ 2p^6\ 3s^2\ 3p^6\ 3d^{10}\ 4s^2\ 4p^2$) ***[1 mark]***
d) E.g. Ar (atom) ***[1 mark]***, K^+ (positive ion) ***[1 mark]***, Cl^- (negative ion) ***[1 mark]***.
You also could have suggested Ca^{2+}, S^{2-} or P^{3-}.

Page 11 — Ionisation Energy

1 a) $C_{(g)} \rightarrow C^+_{(g)} + e^-$
Correct equation ***[1 mark]***. Both state symbols showing gaseous state ***[1 mark]***.
b) First ionisation energy increases as nuclear charge increases ***[1 mark]***.
c) As the nuclear charge increases there is a stronger force of attraction between the nucleus and the electron ***[1 mark]*** and so more energy is required to remove the electron ***[1 mark]***.

2 a) Group 3 ***[1 mark]***
There are three electrons removed before the first big jump in energy.
b) The electrons are being removed from an increasingly positive ion ***[1 mark]*** so the force of attraction that has to be broken is greater ***[1 mark]***.
c) When an electron is removed from a different shell there is a big increase in the energy required (since that shell is closer to the nucleus) ***[1 mark]***.
d) There are 3 shells ***[1 mark]***.
You can tell there are 3 shells because there are 2 big jumps in energy. There is always one more shell than big jumps.

Page 13 — Trends in First Ionisation Energy

1 a) The shielding from the electrons is the same in these atoms ***[1 mark]*** but there is an increase in the number of protons in the nucleus / nuclear charge ***[1 mark]***. So it takes more energy to remove an electron from the outer shell ***[1 mark]***.
b) i) Boron has the configuration $1s^2\ 2s^2\ 2p^1$ compared to $1s^2\ 2s^2$ for beryllium ***[1 mark]***. The 2p orbital is at a slightly higher energy level than the 2s orbital. The extra distance and partial shielding by the 2s electrons make it easier to remove the outer electron ***[1 mark]***.
ii) Oxygen has the configuration $1s^2\ 2s^2\ 2p^4$ compared to $1s^2\ 2s^2\ 2p^3$ for nitrogen ***[1 mark]***. Electron repulsion in the shared 2p sub-shell in oxygen makes it easier to remove an electron ***[1 mark]***.

2 As you go down Group 2, it takes less energy to remove an electron ***[1 mark]***. This is evidence that the outer electrons are increasingly distant from the nucleus ***[1 mark]*** and additional inner shells of electrons exist to shield the outer shell ***[1 mark]***.

3 a) D Neon ***[1 mark]***
b) Ionisation energy increases across a period ***[1 mark]*** so the Group 0 elements have the highest first ionisation energies of the elements in their period. Ionisation energy decreases down a group ***[1 mark]*** so neon has a higher ionisation energy than krypton.

4 a) A Sodium ***[1 mark]***
b) E.g. in sodium, the first electron is removed from the third shell but the second is removed from the second shell ***[1 mark]***. The second shell electron is nearer to the nucleus and experiences less shielding, which means it is much more strongly attracted to the nucleus ***[1 mark]***. In all of the other elements given, the second electron is being removed from the same shell as the first one ***[1 mark]***.

Answers

Unit 1: Section 2 — Amount of Substance

Page 15 — The Mole

PQ1 6.02×10^{23}
This is just Avogadro's constant.

PQ2 Number of atoms = $0.500 \times (6.02 \times 10^{23})$
$= \mathbf{3.01 \times 10^{23}}$

PQ3 concentration = 0.100 moles ÷ 0.500 dm^3
$= \mathbf{0.200\ mol\ dm^{-3}}$

1 M_r of $CaSO_4$ = 40.1 + 32.1 + (4 × 16.0) = 136.2
number of moles = $\frac{34.05}{136.2}$ = **0.2500 moles** ***[1 mark]***

2 M_r of CH_3COOH = (2 × 12.0) + (4 × 1.0) + (2 × 16.0) = 60.0
Mass = 60.0 × 0.360 = **21.6 g** ***[1 mark]***

3 number of moles = $0.250 \times \frac{50.0}{1000}$ = **0.0125 moles** ***[1 mark]***

4 number of moles = $0.250 \times \frac{60.0}{1000}$ = 0.0150 moles ***[1 mark]***
M_r of H_2SO_4 = (2 × 1.0) + (1 × 32.1) + (4 × 16.0) = 98.1
Mass = 0.0150 × 98.1 = **1.48 g** ***[1 mark]***

5 M_r of HCl = 1.0 + 35.5 = 36.5
number of moles = $\frac{3.65}{36.5}$ = 0.100 ***[1 mark]***
Volume of water in dm^3 = 100 ÷ 1000 = 0.100 dm^3
Concentration = $\frac{\text{moles}}{\text{volume}} = \frac{0.100}{0.100}$ = **1.00 mol dm^{-3}** ***[1 mark]***

6 M_r of C_3H_8= (3 × 12.0) + (8 × 1.0) = 44.0
number of moles of C_3H_8 = $\frac{88.0}{44.0}$ = 2.00 moles ***[1 mark]***
p = 100 × 10^3 Pa,
T = (25 + 273) = 298 K ***[1 mark]***
V = nRT ÷ p = (2 × 8.31 × 298) ÷ (100 × 10^3)
= **0.0495 m^3 (3 s.f.)** ***[1 mark]***
You get the mark if you wrote the answer in standard form here — that would be 4.95×10^{-2} m^3.

7 B 100 kPa ***[1 mark]***
M_r of CO_2 = 12.0 + (2 × 16.0) = 44.0
number of moles of CO_2 = (35.2 ÷ 44.0) = 0.800
p = nRT ÷ V
n = 0.800 moles, R = 8.31 J K^{-1} mol^{-1}, T = 301 K,
V = 20.0 × (1 × 10^{-3}) = 0.0200 m^3
p = (0.800 × 8.31 × 301) ÷ 0.0200
= 100052.4 Pa = 100 kPa (to 3 s.f.)

Page 17 — Equations and Calculations

PQ2 a) $Cl_2 + 2KBr \rightarrow 2KCl + Br_2$
b) $Cl_2 + 2Br^- \rightarrow 2Cl^- + Br_2$

1 M_r of C_2H_5Cl = (2 × 12.0) + (5 × 1.0) + (1 × 35.5) = 64.5
Number of moles of C_2H_5Cl = $\frac{258}{64.5}$ = 4.00 moles ***[1 mark]***
From the equation, 1 mole C_2H_5Cl is made from 1 mole C_2H_4, so 4 moles C_2H_5Cl is made from 4 moles C_2H_4 ***[1 mark]***.
M_r of C_2H_4 = (2 × 12.0) + (4 × 1.0) = 28.0
mass of 4 moles C_2H_4 = 4 × 28.0 = **112 g** ***[1 mark]***

2 a) M_r of $CaCO_3$ = 40.1 + 12.0 + (3 × 16.0) = 100.1
Number of moles of $CaCO_3$ = $\frac{15.0}{100.1}$ = 0.150 moles ***[1 mark]***
From the equation, 1 mole $CaCO_3$ produces 1 mole CaO, so 0.150 moles of $CaCO_3$ produces 0.150 moles of CaO ***[1 mark]***
M_r of CaO = 40.1 + 16.0 = 56.1
mass of 0.15 moles of CaO = 56.1 × 0.150 = **8.42 g** ***[1 mark]***

b) From the equation, 1 mole $CaCO_3$ produces 1 mole CO_2, so 0.150 moles of $CaCO_3$ produces 0.150 moles of CO_2 ***[1 mark]***
T = 25.0 + 273 = 298 K and p = 100 × 10^3 Pa ***[1 mark]***
V = nRT ÷ p
Volume of CO_2 = $\frac{0.150 \times 8.31 \times 298}{100 \times 10^3}$ = 0.00371 m^3
= **3.71 × 10^{-3} m^3** ***[1 mark]***

3 **$2KI + Pb(NO_3)_2 \rightarrow PbI_2 + 2KNO_3$** ***[1 mark]***
The LHS needs 2 Is, so pop a 2 in front of KI. Then the RHS needs 2 Ks, so put a 2 in front of KNO_3. Once you've done that, everything balances.

Page 20 — Titrations

1 Moles of NaOH = $0.500 \times \frac{14.6}{1000}$ = 0.00730 ***[1 mark]***
From the equation, 1 mole of NaOH neutralises 1 mole of CH_3COOH, so 0.00730 moles NaOH must neutralise 0.00730 moles of CH_3COOH ***[1 mark]***.
Concentration CH_3COOH = 0.00730 ÷ $\frac{25.4}{1000}$
= **0.287 mol dm^{-3}** ***[1 mark]***

2 M_r of $CaCO_3$ = 40.1 + 12.0 + (3 × 16.0) = 100.1 ***[1 mark]***
Moles of $CaCO_3$ = $\frac{0.75}{100.1}$ = 0.0075 ***[1 mark]***
From the equation, 1 mole $CaCO_3$ reacts with 1 mole H_2SO_4 so, 0.0075 moles of $CaCO_3$ must react with 0.0075 moles of H_2SO_4 ***[1 mark]***.
Volume needed = $\frac{0.0075}{0.25}$ = **0.030 dm^3** (or **30 cm^3**) ***[1 mark]***
If the question mentions concentration, you can bet your last clean pair of underwear that you'll need to use this formula: number of moles = conc. × volume (dm^3)

3 a) Titration 3 (42.90 cm^3) is anomalous. It is not concordant with / is significantly different to the other three results ***[1 mark]***.
b) Mean titre = $\frac{45.00 + 45.10 + 44.90}{3}$ = **45.00 cm^3** ***[1 mark]***
c) Moles of NaOH = $0.400 \times \frac{45.00}{1000}$ = 0.01800 ***[1 mark]***
From the equation, 1 mole of NaOH reacts with 1 mole of HNO_3, so 0.01800 moles NaOH must react with 0.0180 moles of HNO_3 ***[1 mark]***.
Moles of HNO_3 = 0.01800 ÷ $\frac{50.0}{1000}$= **0.3600 mol dm^{-3}** ***[1 mark]***

Page 23 — Formulas, Yield and Atom Economy

1 Start by working out how many moles of carbon and hydrogen there would be in 100 g of the hydrocarbon:
Number of moles of C = $\frac{92.3}{12.0}$ = 7.69 moles ***[1 mark]***
Number of moles of H = $\frac{7.70}{1.0}$ = 7.70 moles ***[1 mark]***
Divide both by 7.69: C: 7.69 ÷ 7.69 = 1. H: 7.70 ÷ 7.69 = 1.00.
So ratio C : H = 1 : 1
Empirical formula = CH ***[1 mark]***
Empirical mass = 12.0 + 1.0 = 13.0
Number of empirical units in molecule = $\frac{78.0}{13.0}$ = 6
So the molecular formula = 6 × CH = **C_6H_6** ***[1 mark]***

2 a) Moles PCl_3 = $\frac{\text{mass}}{\text{Mr}} = \frac{0.275}{137.5}$ = **0.00200 mol** ***[1 mark]***
From the formula, 1 mole of PCl_3 reacts with Cl_2 to form 1 mole of PCl_5, so 0.00200 moles of PCl_3 will react to form 0.00200 moles of PCl_5 ***[1 mark]***.
M_r of PCl_5 = 31.0 + (5 × 35.5) = 208.5
Theoretical yield = 0.00200 × 208.5 = **0.417 g** ***[1 mark]***
b) percentage yield = (0.198 ÷ 0.417) × 100 = **47.5%** ***[1 mark]***
c) 100% ***[1 mark]***. Since the equation shows that this reaction has only one product, its atom economy must be 100% ***[1 mark]***.

Answers

Unit 1: Section 3 — Bonding

Page 25 — Ionic Bonding

1 a) E.g.

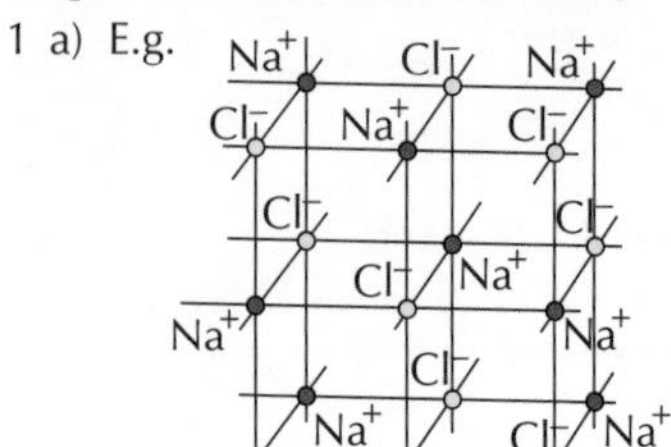

Your diagram should show the following:

- cubic structure with ions at corners ***[1 mark]***
- sodium ions and chloride ions labelled ***[1 mark]***
- alternating sodium ions and chloride ions ***[1 mark]***

b) giant ionic lattice ***[1 mark]***

c) You'd expect it to have a high melting point because the electrostatic attraction between ions ***[1 mark]*** is strong ***[1 mark]*** so a lot of energy is needed to overcome this attraction ***[1 mark]***.

2 $MgCO_3$ ***[1 mark]***

Magnesium is in group 2 of the periodic table, so a magnesium ion has a charge of 2+. A carbonate ion has a charge of 2–.

3 In a solid, ions are held in place by strong ionic bonds ***[1 mark]***. When the solid is heated to melting point, the ions gain enough energy to overcome these forces and move ***[1 mark]***, carrying charge (and so electricity) through the substance ***[1 mark]***.

Page 27 — Covalent Bonding

1 a) A shared pair of electrons between two atoms ***[1 mark]***

b) A single covalent bond only contains one pair of shared electrons, A double covalent bond contains two pairs of shared electrons ***[1 mark]***.

2 a) Dative covalent/coordinate bonding ***[1 mark]***

b) One atom donates a pair of electrons to the bond. / Both the electrons in the bond come from one atom. ***[1 mark]***.

3 a) Macromolecular/giant covalent ***[1 mark]***

b) Diamond

Graphite

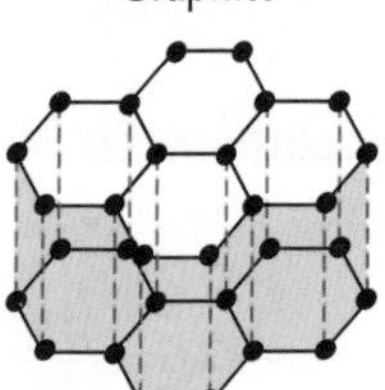

[1 mark for each correctly drawn diagram]

Diamond's a bit awkward to draw without it looking like a load of ballet dancing spiders — just make sure each carbon is connected to four others.

c) Diamond only has electrons in covalent bonds ***[1 mark]***, so is a poor electrical conductor ***[1 mark]***. Graphite has some delocalised electrons which can flow within the sheets ***[1 mark]***, making it an electrical conductor ***[1 mark]***.

Page 29 — Shapes of Molecules

1 a) NCl_3:

N, Cl, Cl, Cl, 107° ***[1 mark]***

shape: (trigonal) pyramidal ***[1 mark]***,
bond angle: 107° (accept between 105° and 109°) ***[1 mark]***

b) BCl_3:

B, Cl, Cl, Cl, 120° ***[1 mark]***

shape: trigonal planar ***[1 mark]***,
bond angle: 120° exactly ***[1 mark]***

c) BCl_3 has three bonding electron pairs around B which repel each other equally ***[1 mark]***. NCl_3 has three bonding electron pairs and one lone pair ***[1 mark]***. The lone pair repels the bonding pair more strongly than the bonding pairs repel each other ***[1 mark]***.

Page 32 — Polarisation and Intermolecular Forces

1 Decene has a higher boiling point. Decene is a larger molecule than octene, so it has more electrons/a larger surface area ***[1 mark]***. This means that the van der Waals forces between molecules of decene will be stronger ***[1 mark]***, so it will take more energy to overcome them ***[1 mark]***.

The more energy you need to overcome the intermolecular forces between the molecules, the higher the boiling point of the compound will be.

2 a) The C–Cl bond will be polar, because chlorine has a much higher electronegativity than carbon ***[1 mark]***.

b) The molecule CCl_4 is not polar. Each of the C–Cl bonds in the CCl_4 molecule is polar ***[1 mark]***, but the polar bonds are arranged symmetrically all around the molecule, so the charges cancel out ***[1 mark]***.

2 a) E.g.

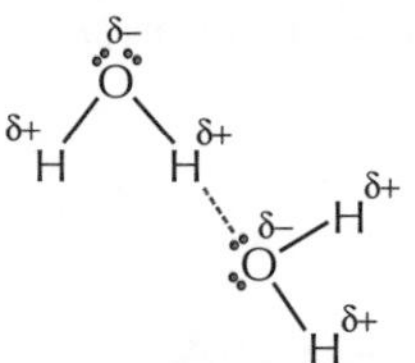

Your diagram should show the following:

- A hydrogen bond, shown as a dotted line, between an H atom on one molecule and an O atom on the other ***[1 mark]***.
- Two lone pairs on each oxygen atom ***[1 mark]***.
- Partial charges on all the atoms ***[1 mark]***.

b) Hydrogen bonding is present in water but not in any of the other group 6 hydrides ***[1 mark]***. Hydrogen bonds are stronger than other intermolecular forces ***[1 mark]***, so more energy is needed to break the intermolecular forces between water molecules than for other group 6 hydride molecules ***[1 mark]***.

Page 35 — Metallic Bonding and Properties of Materials

1 a) E.g.

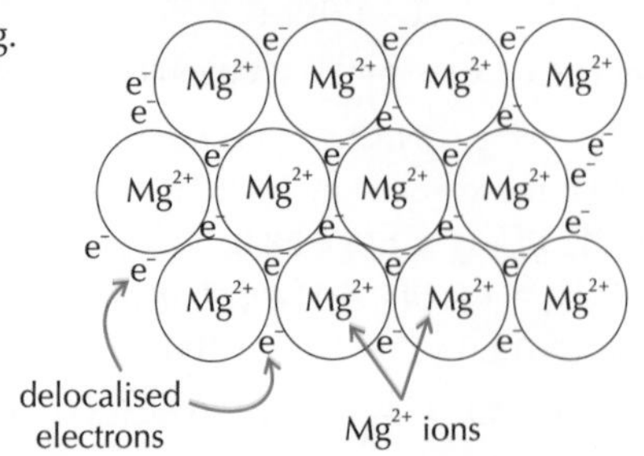

Your diagram should show the following:

- Closely packed Mg^{2+} ions ***[1 mark]***.
- Delocalised electrons ***[1 mark]***.

b) Metals contain delocalised electrons, which can move through the structure to carry a current ***[1 mark]***.

2 A = ionic ***[1 mark]***, B = simple molecular/covalent ***[1 mark]***, C = metallic ***[1 mark]***, D macromolecular/giant covalent ***[1 mark]***.

Answers

3 Iodine is a simple molecular substance ***[1 mark]***. To boil iodine, you only need to break the weak intermolecular forces holding the molecules together, which doesn't need much energy ***[1 mark]***. Graphite is a giant covalent substance ***[1 mark]***. To boil graphite, you need to break the strong covalent bonds between atoms, which needs a lot of energy ***[1 mark]***.

Unit 1: Section 4 — Energetics

Page 37 — Enthalpy Changes

1 a) Total energy required to break bonds = (4 × 435) + (2 × 498)
= 2736 kJ mol^{-1} ***[1 mark]***
Energy released when bonds form = (2 × 805) + (4 × 464)
= 3466 kJ mol^{-1} ***[1 mark]***
Enthalpy change = 2736 – 3466 = **–730 kJ mol^{-1}** ***[1 mark]***

b) The reaction is exothermic, because the enthalpy change is negative / more energy is given out than is taken in ***[1 mark]***.

2 a) $CH_3OH_{(l)} + 1\frac{1}{2}O_{2(g)} \rightarrow CO_{2(g)} + 2H_2O_{(l)}$ ***[1 mark]***

It's fine to use halves to balance equations. Make sure you've only got 1 mole of CH_3OH being burned, since the question is asking for the standard enthalpy of combustion.

b) $C_{(s)} + 2H_{2(g)} + \frac{1}{2}O_{2(g)} \rightarrow CH_3OH_{(l)}$ ***[1 mark]***

Again, make sure you're only forming 1 mole of CH_3OH, since the question is asking for the standard enthalpy of formation.

3 The standard enthalpy change of combustion is the enthalpy change when 1 mole of a substance is burned. This equation shows 2 moles of propane being burned ***[1 mark]***.

You really need to know the definitions of the standard enthalpy changes by heart. There's loads of teeny little details they could ask you about.

Page 39 — Calorimetry

1 ΔT = 25.5 – 19.0 = 6.50 °C = 6.50 K ***[1 mark]***
25.0 + 25.0 = 50.0 cm^3 of solution
which has a mass of 50.0 g ***[1 mark]***
$q = mc\Delta T$
= 50.0 × 4.18 × 6.50 = 1358.5 J ***[1 mark]***
1358.5 J ÷ 1000 = 1.3585 kJ
moles of HCl = concentration (mol dm^{-3}) × volume (dm^3)
1.00 × (25.0 ÷ 1000) = 0.0250 ***[1 mark]***
Enthalpy change = –1.3585 ÷ 0.0250
= **–54.3 kJ mol^{-1}** (3 s.f.) ***[1 mark]***

It's important that you remember to add a minus sign to the enthalpy change here (because the reaction's exothermic).

2 Heat produced by reaction = $mc\Delta T$
= 50.0 × 4.18 × 2.60 = 543.4 J ***[1 mark]***
543.4 J ÷ 1000 = 0.5434 kJ
moles of $CuSO_4$ = concentration (mol dm^{-3}) × volume (dm^3)
= 2.00 × (50.0 ÷ 1000) = 0.0100 mole ***[1 mark]***
From the equation, 1 mole of $CuSO_4$ reacts with 1 mole of Zn. So 0.0100 mole of $CuSO_4$ reacts with 0.0100 mole of Zn ***[1 mark]***.
Enthalpy change = –0.5434 ÷ 0.0100 = **–54.3 kJ** ***[1 mark]***

Once again, you can tell this reaction is exothermic, because it releases heat — so remember to add the minus sign to the enthalpy change.

Page 41 — Hess's Law

PQ3 $\Delta_r H^{\circ}$ = –965 + 890 = **–75 kJ mol^{-1}**

1 $\Delta_r H^{\circ}$ = sum of $\Delta_f H^{\circ}$(products) – sum of $\Delta_f H^{\circ}$(reactants) ***[1 mark]***
= [0 + (3 × –602)] – [–1676 + 0]
= **–130 kJ mol^{-1}** ***[1 mark]***

Don't forget the units — if you leave them out, you might well lose marks.

2 $\Delta_r H^{\circ} = \Delta_c H^{\circ}(C_2H_4) + \Delta_c H^{\circ}(H_2) - \Delta_c H^{\circ}(C_2H_6)$ ***[1 mark]***
= (–1400) + (–286) – (–1560)
= **–126 kJ mol^{-1}** ***[1 mark]***

Unit 1: Section 5 — Kinetics, Equilibria and Redox Reactions

Page 43 — Reaction Rates

1 The molecules don't always have enough energy to react ***[1 mark]***.

2

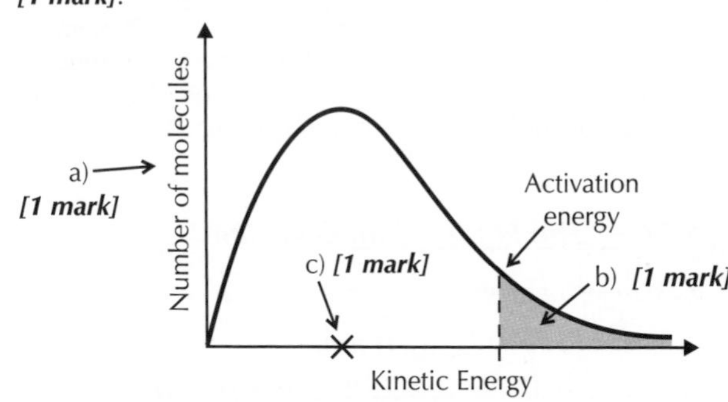

For part c), the important thing is to get your mark on the axis as close to the peak of the curve as possible.

Page 46 — More on Reaction Rates

1 a)

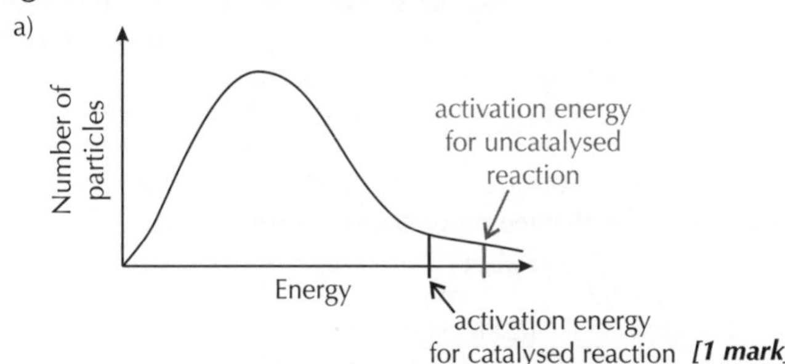

It doesn't matter exactly where you put the line showing the activation energy for the catalysed reaction here — as long as it's a bit lower than the one for the uncatalysed activation energy.

b) Manganese(IV) oxide lowers the activation energy of the reaction by providing an alternative reaction pathway ***[1 mark]***.

c) Raising the temperature will increase the rate of reaction ***[1 mark]***. When you raise the temperature you increase the kinetic energy of the molecules ***[1 mark]***. This means that more molecules will have at least the activation energy/enough energy to react ***[1 mark]***. Since the molecules are moving faster, it also means that collisions will occur more frequently ***[1 mark]***.

Page 49 — Reversible Reactions

1 a) If a reaction at equilibrium is subjected to a change in concentration, pressure or temperature, the position of equilibrium will move to counteract the change ***[1 mark]***.

You could well be asked to give definitions of things like this in an exam, so make sure you learn them — they're relatively easy marks.

b) i) There are the same number of molecules/moles on each side of the equation ***[1 mark]***, so the position of equilibrium will not move ***[1 mark]***.

ii) The reverse reaction is exothermic ***[1 mark]*** so the position of equilibrium will shift to the left to increase the temperature ***[1 mark]***.

iii) The position of equilibrium will shift to the right ***[1 mark]*** to increase the concentration of nitrogen monoxide ***[1 mark]***.

c) No effect ***[1 mark]***.

Catalysts don't affect the equilibrium position. They just help the reaction to get there sooner.

Answers

2 a) The forward reaction is exothermic *[1 mark]*. If you decrease the temperature, the position of equilibrium will move to the right/ the forward reaction will speed up in order to produce heat *[1 mark]*. This will increase the amount of ethanol/product that is made *[1 mark]*.

b) i) Increasing pressure increases the yield of ethanol *[1 mark]*. There are fewer moles of gas on the right hand side of the equation so equilibrium is shifted to the right *[1 mark]*.

ii) E.g. producing a high pressure is expensive. / The equipment needed to produce a high pressure is expensive. / The cost of producing the extra pressure is greater than the value of the extra yield *[1 mark]*.

Page 51 — The Equilibrium Constant

1 $K_c = \frac{[H_2][I_2]}{[HI]^2}$ ***[1 mark]***

At equilibrium, $[H_2] = [I_2]$ ***[1 mark]***

$[HI]^2 = \frac{[H_2][I_2]}{K_c} = \frac{0.770 \times 0.770}{0.0200} = 29.6$ ***[1 mark]***

$[HI] = \sqrt{29.6} = \mathbf{5.44\ mol\,dm^{-3}}$ ***[1 mark]***

2 a) i) mass ÷ M_r = 34.5 ÷ [14.0 + (2 × 16.0)] = 34.5 ÷ 46.0 = **0.750 mol** ***[1 mark]***

ii) moles of O_2 = mass ÷ M_r = 7.04 ÷ (2 × 16.0) = **0.220** ***[1 mark]***
moles of NO = 2 × moles of O_2 = 2 × 0.220 = **0.440** ***[1 mark]***
moles of NO_2 = 0.750 – 0.440 = **0.310** ***[1 mark]***

b) Concentration of O_2 = 0.220 ÷ 9.80 = 0.0224 $mol\,dm^{-3}$
Concentration of NO = 0.440 ÷ 9.80 = 0.0449 $mol\,dm^{-3}$
Concentration of NO_2 = 0.310 ÷ 9.80 = 0.0316 $mol\,dm^{-3}$
[1 mark for all three concentrations correct]

$K_c = \frac{[NO]^2[O_2]}{[NO_2]^2}$ ***[1 mark]***

$= \frac{0.0449^2 \times 0.0224}{0.0316^2} = 0.0452$

Units $= \frac{[mol\,dm^{-3}]^2[mol\,dm^{-3}]}{[mol\,dm^{-3}]^2} = mol\,dm^{-3}$

So K_c = $\mathbf{0.0452\ mol\,dm^{-3}}$
[1 mark for the correct value of K_c, 1 mark for units.]

Page 53 — Redox Reactions

1 a) Oxidation is the loss of electrons *[1 mark]*.

b) i) 0 ***[1 mark]***

ii) +1 ***[1 mark]***

c) Oxygen is being reduced *[1 mark]*.
$O_2 + 4e^- \rightarrow 2O^{2-}$ ***[1 mark]***

2 a) An oxidising agent accepts electrons and gets reduced *[1 mark]*.

b) $In \rightarrow In^{3+} + 3e^-$ ***[1 mark]***

c) $2In + 3Cl_2 \rightarrow 2InCl_3$ ***[2 marks — 1 mark for correct reactants and products, 1 mark for correct balancing.]***

Unit 1: Section 6 — Thermodynamics

Page 57 — Lattice Enthalpy and Born-Haber Cycles

1 a)

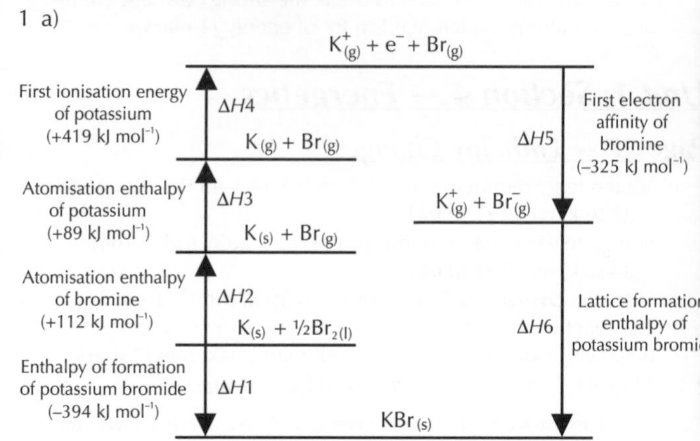

[1 mark for correct enthalpy changes. 1 mark for formulas/ state symbols. 1 mark for correct directions of arrows.]

b) Lattice formation enthalpy, $\Delta H6$
$= -\Delta H5 - \Delta H4 - \Delta H3 - \Delta H2 + \Delta H1$
= –(–325) – (+419) – (+89) – (+112) + (–394) ***[1 mark]***
= **–689 kJ mol⁻¹** ***[1 mark]***

2 a)

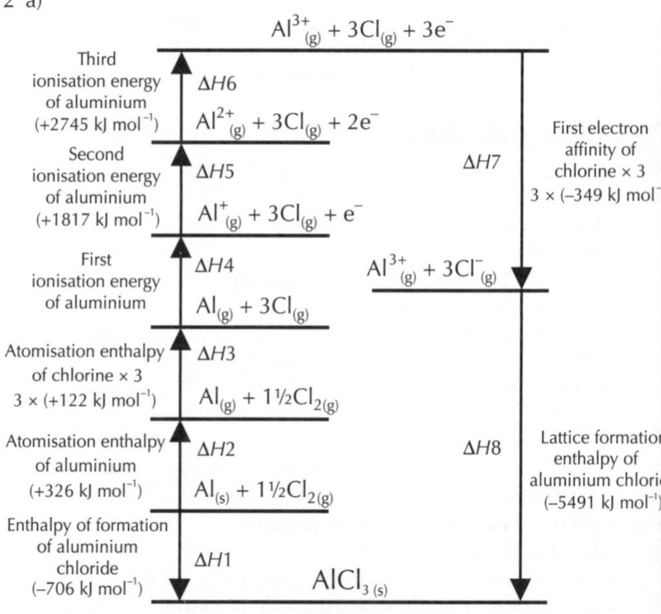

[1 mark for correct enthalpy changes and correctly multiplying all the enthalpies. 1 mark for formulas/state symbols. 1 mark for correct directions of arrows.]

There are often a couple of steps in a Born-Haber cycle that you can do in any order. For example, you could swap round $\Delta H2$ and $\Delta H3$ here and still be right.

b) First ionisation energy (aluminium), $\Delta H4$
$= -\Delta H3 - \Delta H2 + \Delta H1 - \Delta H8 - \Delta H7 - \Delta H6 - \Delta H5$
= –3(+122) – (+326) + (–706) – (–5491) –3(–349) – (+2745) – (+1817) ***[1 mark]***
= **+578 kJ mol⁻¹** ***[1 mark]***

Answers

Page 59 — Enthalpies of Solution

1 a)

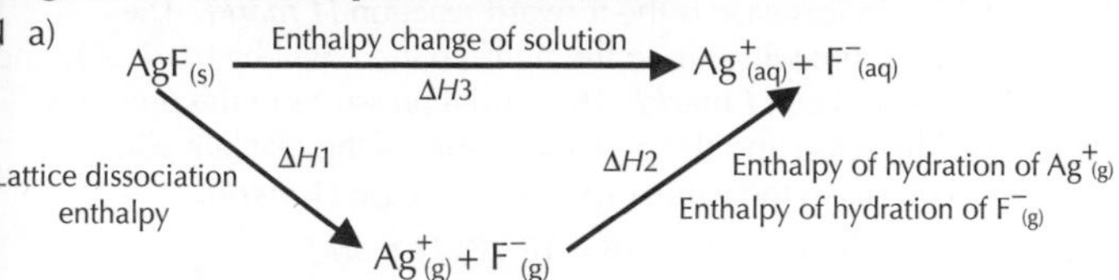

[1 mark for a complete correct cycle, 1 mark for correctly labelled arrows.]

b) $\Delta H3 = \Delta H1 + \Delta H2$
$= 960 + (-506) + (-464) =$ **-10 kJ mol^{-1}** ***[1 mark]***

2 Enthalpy change of solution ($SrF_{2(s)}$)
= lattice dissociation enthalpy ($SrF_{2(s)}$)
+ enthalpy of hydration ($Sr^{2+}_{(g)}$)
+ [2 × enthalpy of hydration ($F^-_{(g)}$)] ***[1 mark]***
$= 2492 + (-1480) + (2 \times -506) =$ **0 kJ mol^{-1}** ***[1 mark]***

Don't forget — you have to double the enthalpy of hydration for F^- because there are two in SrF_2.

3 By Hess's law:
Enthalpy change of solution ($MgCl_{2(s)}$)
= –lattice formation enthalpy ($MgCl_{2(s)}$)
+ enthalpy of hydration ($Mg^{2+}_{(g)}$)
+ [2 × enthalpy of hydration ($Cl^-_{(g)}$)] ***[1 mark]***

You've been given a negative lattice enthalpy value so it must be the lattice formation enthalpy.

So enthalpy of hydration ($Cl^-_{(g)}$)
= [enthalpy change of solution ($MgCl_{2(s)}$)
+ lattice formation enthalpy ($MgCl_{2(s)}$)
– enthalpy of hydration ($Mg^{2+}_{(g)}$)] ÷ 2
$= [(-122) + (-2526) - (-1920)] \div 2$ ***[1 mark]***
$= -728 \div 2 =$ **-364 kJ mol^{-1}** ***[1 mark]***

Page 61 — Entropy

1 You would expect the entropy change to be positive, because there are more moles of products than moles of reactant ***[1 mark]*** and the reactant is a solid while one of the products is a gas ***[1 mark]***.

2 $\Delta S = S_{products} - S_{reactants}$
$= 26.9 - (32.7 + (\frac{1}{2} \times 205))$ ***[1 mark]***
$= -108.3$ J K^{-1} mol^{-1} = **-108 J K^{-1} mol^{-1}** (3 s.f.) ***[1 mark]***

3 $\Delta S = S_{products} - S_{reactants}$
$= ((2 \times 69.9) + 205) - (2 \times 110)$ ***[1 mark]***
$= 124.8$ J K^{-1} mol^{-1} = **125 J K^{-1} mol^{-1}** (3 s.f.) ***[1 mark]***

Page 63 — Free-Energy Change

1 a) i) $\Delta G = \Delta H - (T \times \Delta S)$
$= 117\,000 - (500 \times 175)$
= **+29 500 J mol^{-1}** (= **+29.5 kJ mol^{-1}**) ***[1 mark]***

ii) $\Delta G = \Delta H - (T \times \Delta S)$
$= 117\,000 - (760 \times 175)$
= **$-16\,000$ J mol^{-1}** (= **-16 kJ mol^{-1}**) ***[1 mark]***

b) The reaction is feasible at 760 K ***[1 mark]*** because the free energy change at this temperature is negative ***[1 mark]***.

2 a) $\Delta S = \Delta S_{products} - \Delta S_{reactants}$
$= [214 + (2 \times 69.9)] - [186 + (2 \times 205)]$
$= -242.2$ J K^{-1} mol^{-1} ***[1 mark]***
$\Delta G = \Delta H - (T \times \Delta S)$
$= -730\,000 - (298 \times -242.2)$
= **$-658\,000$ J mol^{-1}** (= **-65.8 kJ mol^{-1}**) (3 s.f.) ***[1 mark]***

b) $T = \Delta H \div \Delta S$
$= -730\,000 \div -242.2$ ***[1 mark]***
= **3010 K** (3 s.f.) ***[1 mark]***

Unit 1: Section 7 — Rate Equations and K_p

Page 66 — Rate Equations

1 a) E.g.

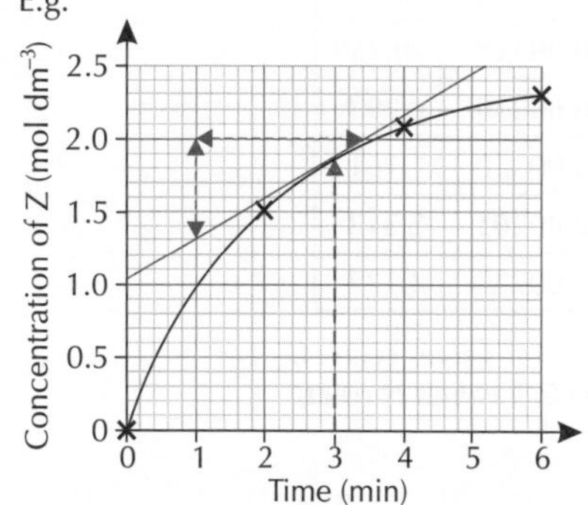

[1 mark for tangent drawn at 3 mins.]
rate of reaction = gradient of tangent at 3 mins
gradient = change in y ÷ change in x
e.g. $= (2.0 - 1.3) \div (3.4 - 1.0)$
= **0.29 mol dm^{-3} min^{-1}**
[1 mark for an answer correctly calculated from the tangent, 1 mark for units.]

Different people will draw ever-so-slightly different tangents and pick different points on their tangent to do the gradient calculation. So as long as you've drawn the tangent accurately, any answer that you've calculated from it correctly gets the mark.

b) rate = k[X][Y] ***[1 mark]***

2 a) rate = $k[NO]^2[H_2]$ ***[1 mark]***

b) $0.00267 = k \times 0.00400^2 \times 0.00200$
So $k = 0.00267 \div (0.00400^2 \times 0.00200) = 8.34 \times 10^4$
Units: mol dm^{-3} s^{-1}/[(mol dm^{-3})^2 × (mol dm^{-3})] = dm^6 mol^{-2} s^{-1}
k = **8.34×10^4 dm^6 mol^{-2} s^{-1}**
[1 mark for the correct value of k, 1 mark for units.]

Page 69 — Rate Experiments

1 a) 1st order ***[1 mark]***
[D] doubles between experiments 1 and 2 and the initial rate doubles (with [E] remaining constant).

b) 0 order ***[1 mark]***
[E] and [D] both halve between experiments 1 and 3 and the initial rate halves. Halving [D] alone would halve the rate, so changing [E] cannot affect the rate.

c) 1.30×3 = **3.90×10^{-3} mol dm^{-3} s^{-1}** ***[1 mark]***
[D] is tripled and [E] is halved between experiments 1 and 4. The rate is proportional to [D] (and [E] doesn't affect the rate), so the initial rate is tripled.

Page 71 — The Rate Determining Step

1 H^+ is acting as a catalyst ***[1 mark]***. You know this because it is not one of the reactants in the chemical equation, but it does affect the rate of reaction/appear in the rate equation ***[1 mark]***.

2 a) One molecule of H_2 and one molecule of ICl (or something derived from these molecules) ***[1 mark]***. If the molecule is in the rate equation, it must be in the rate determining step ***[1 mark]***. The orders of the reaction tell you how many molecules of each reactant are in the rate determining step ***[1 mark]***.

b) Incorrect ***[1 mark]***. H_2 and ICl are both in the rate equation, so they must both be in the rate determining step OR the order of the reaction with respect to ICl is 1, so there must be only one molecule of ICl in the rate determining step ***[1 mark]***.

Answers

Page 73 — The Arrhenius Equation

1 a)

T (K)	*k*	1/T (K^{-1})	ln *k*
305	0.181	0.00328	–1.709
313	0.468	**0.00319**	**–0.759**
323	1.34	**0.00310**	**0.293**
333	3.29	0.00300	1.191
344	10.1	**0.00291**	**2.313**
353	22.7	0.00283	3.122

[1 mark for all three 1/T values correct, 1 mark for all three ln k values correct]

b)

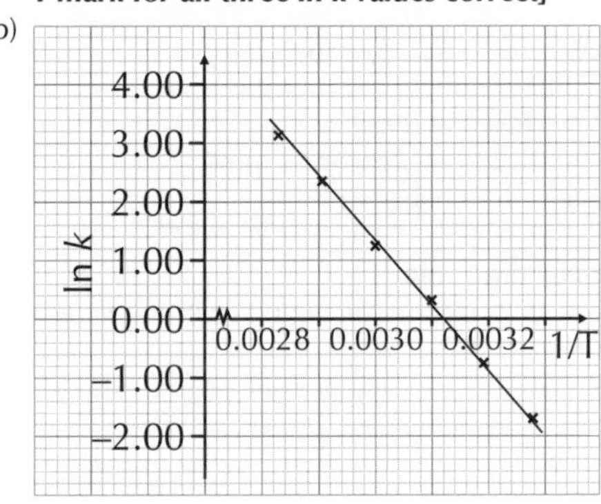

[1 mark for correct axes, 1 mark for correctly plotted points, 1 mark for line of best fit]

c) E.g. gradient = (2.313 – (–0.759)) ÷ (0.00291 – 0.00319)
= **–11 000** ***[1 mark]***

You can have the mark here for any gradient correctly calculated from two points on your graph.

$\frac{-E_a}{R} = -11\ 000$

E_a = 11 000 × 8.31 = **91 400 J mol⁻¹** OR **91.4 kJ mol⁻¹** ***[1 mark]***

Again, you can have the mark here for any value of E_a correctly calculated from your value for the gradient of the graph.

d) By substituting values into the expression $\ln k = \frac{-E_a}{RT} + \ln A$

E.g. 2.313 = (–11 000 × 0.00291) + ln A

ln A = 34.3

A = $e^{34.3}$ = **8 × 10¹⁴** ***[1 mark]***

You've guessed it — you get the mark here for a value of A correctly calculated using your value for the gradient from part c) and values for ln k and 1/T from a data point in the table or on the graph.

Page 75 — Gas Equilibria and K_p

PQ3 $K_p = \frac{(p_{NH_3})^2}{p_{N_2} \times (p_{H_2})^3}$

1 a) $K_p = \frac{p_{SO_2} \times p_{Cl_2}}{p_{SO_2Cl_2}}$ ***[1 mark]***

b) Cl_2 and SO_2 are produced in equal amounts so
$p_{Cl_2} = p_{SO_2}$ = 60.2 kPa ***[1 mark]***
Total pressure = $p_{SO_2Cl_2} + p_{Cl_2} + p_{SO_2}$ so
$p_{SO_2Cl_2}$ = 141 – 60.2 – 60.2 = **20.6 kPa** ***[1 mark]***

c) $K_p = \frac{60.2\,\text{kPa} \times 60.2\,\text{kPa}}{20.6\,\text{kPa}}$ = **176 kPa**

[2 marks — 1 mark for 176 correct, 1 mark for kPa correct]

The units are kPa because (kPa × kPa)/kPa = kPa.

d) Increasing the temperature favours the endothermic reaction, which in this case is the forward reaction ***[1 mark]***. The equilibrium will shift to the right, so there will be more SO_2 and Cl_2 produced ***[1 mark]***. The partial pressures of the products will increase and the partial pressure of the reactant will decrease, so the value of K_p will increase ***[1 mark]***.

2 a) p_{O_2} = ½ × p_{NO} = ½ × 36 = **18 kPa** ***[1 mark]***

b) p_{NO_2} = total pressure – p_{NO} – p_{O_2}
= 99 – 36 – 18 = **45 kPa** ***[1 mark]***

c) $K_p = \frac{(p_{NO_2})^2}{(p_{NO})^2 p_{O_2}}$

$= \frac{(45\,\text{kPa})^2}{(36\,\text{kPa})^2 \times (18\,\text{kPa})}$ = **0.087 kPa⁻¹**

[3 marks — 1 mark for K_p expression correct, 1 mark for 0.087 correct, 1 mark for kPa⁻¹ correct]

The units are kPa⁻¹ because kPa²/(kPa² × kPa) = kPa⁻¹.

Unit 1: Section 8 — Electrode Potentials and Cells

Page 77 — Electrode Potentials

1 Iron ***[1 mark]*** as it has a more negative electrode potential/it loses electrons more easily than lead ***[1 mark]***.

2 a) $Zn_{(s)} + 2Ag^+_{(aq)} \rightarrow Zn^{2+}_{(aq)} + 2Ag_{(s)}$ ***[1 mark]***

b) The silver half-cell. It has a more positive standard electrode potential/it's more easily reduced ***[1 mark]***.

Page 79 — The Electrochemical Series

1 a) $Zn_{(s)} + Ni^{2+}_{(aq)} \rightleftharpoons Zn^{2+}_{(aq)} + Ni_{(s)}$

$E^\ominus$ = (–0.25) – (–0.76) = **+0.51 V** ***[1 mark]***

This reaction takes places because Zn^{2+}/Zn has a more negative electrode potential so this half-reaction will go in the oxidisation direction. So Ni^{2+}/Ni will go in the reduction direction ***[1 mark]***.

b) $2MnO_4^-{}_{(aq)} + 16H^+_{(aq)} + 5Sn^{2+}_{(aq)} \rightleftharpoons 2Mn^{2+}_{(aq)} + 8H_2O_{(l)} + 5Sn^{4+}_{(aq)}$

$E^\ominus$ = (+1.51) – (+0.14) = **+1.37 V** ***[1 mark]***

This reaction takes places because MnO_4^-/Mn^{2+} has a more negative electrode potential so this half-reaction will go in the oxidisation direction. So Sn^{4+}/Sn^{2+} will go in the reduction direction ***[1 mark]***.

c) No reaction ***[1 mark]***. Both reactants are in their oxidised form ***[1 mark]***.

2 $KMnO_4$ ***[1 mark]*** because it has a more positive/less negative electrode potential ***[1 mark]***.

3 a) i) $Fe_{(s)} \mid Fe^{2+}_{(aq)} \,||\, O_{2(g)} \mid OH^-_{(aq)}$ ***[1 mark]***

ii) $E^\ominus_{cell} = E^\ominus_{reduced} - E^\ominus_{oxidised}$ =
0.40 – (–0.44) = **+0.84 V** ***[1 mark]***

You should come out with positive value when you calculate the EMF of a cell — check you've done the subtraction the right way around if you get a negative value.

b) The iron half-cell has a more negative electrode potential than the oxygen/water half-cell, so iron is oxidised ***[1 mark]***.

Answers

Page 81 — Batteries and Fuel Cells

1 a) i) and ii)

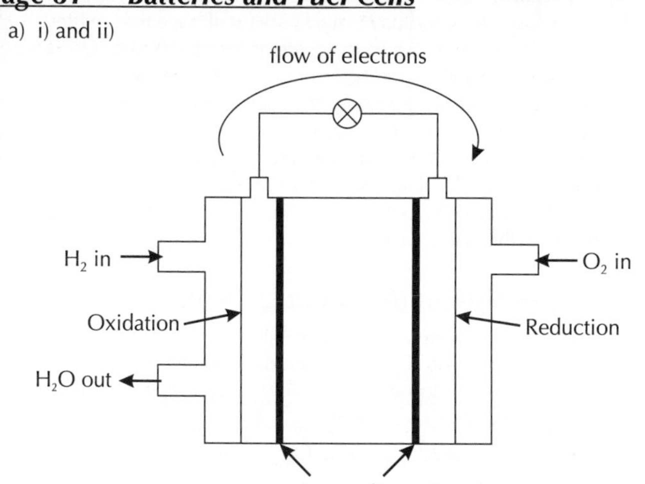

[1 mark for labelling the sites of reduction and oxidation correctly. 1 mark for drawing the arrow showing the direction of electron flow correctly.]

b) Negative electrode: $H_{2(g)} + 4OH^-_{(aq)} \rightarrow 4H_2O_{(l)} + 4e^-$ ***[1 mark]***
Positive electrode: $O_{2(g)} + 2H_2O_{(l)} + 4e^- \rightarrow 4OH^-_{(aq)}$ ***[1 mark]***

c) It only allows the OH^- across and not O_2 and H_2 gases ***[1 mark]***.

2 a) negative electrode: $Li_{(s)} \rightarrow Li^+_{(aq)} + e^-$ ***[1 mark]***
positive electrode: $Li^+_{(aq)} + CoO_{2(s)} + e^- \rightarrow Li^+[CoO_2]^-_{(s)}$ ***[1 mark]***

b) A current is supplied to force the electrons to flow in the opposite direction to when the cell is supplying electricity ***[1 mark]***.

c) They are reversed ***[1 mark]***.

Unit 1: Section 9 — Acids, Bases and pH

Page 83 — Acids, Bases and K_w

1 a) $HSO_4^- \rightarrow H^+ + SO_4^{2-}$ or $HSO_4^- + H_2O \rightarrow H_3O^+ + SO_4^{2-}$ ***[1 mark]***

b) $HSO_4^- + H^+ \rightarrow H_2SO_4$ or $HSO_4^- + H_2O \rightarrow H_2SO_4 + OH^-$ ***[1 mark]***

2 a) Weak acids dissociate (or ionise) a small amount to produce hydrogen ions (or protons) ***[1 mark]***.

b) $HCN \rightleftharpoons H^+ + CN^-$ ***[1 mark]***

3 Moles of NaOH = mass ÷ M_r
= 2.50 ÷ 40.0 = 0.0625 mol ***[1 mark]***
1 mol of NaOH gives 1 mol of OH^-.
So $[OH^-]$ = [NaOH] = **0.0625 mol dm^{-3}** ***[1 mark]***.

Page 85 — pH Calculations

1 a) HBr is a strong monoprotic acid,
so $[H^+]$ = [HBr] = 0.32 mol dm^{-3}.
pH = $-\log_{10}$ (0.32) = **0.49** ***[1 mark]***

b) HBr is a stronger acid than ethanoic acid, so will be more dissociated in solution. This means the concentration of hydrogen ions will be higher, so the pH will be lower ***[1 mark]***.

2 a) A monoprotic acid means that each molecule of acid will release one proton when it dissociates OR each mole of acid will produce one mole of protons when it dissociates ***[1 mark]***.

b) $[H^+] = 10^{-pH} = 10^{-0.55}$ = **0.28 mol dm^{-3}** ***[1 mark]***.
A strong acid will ionise fully in solution — so $[H^+]$ = [Acid].
That's why the question tells you that HNO_3 is a strong acid.

3 a) Moles of KOH = m ÷ M_r
= 11.22 ÷ 56.1 = 0.200 mol ***[1 mark]***
1 mol of KOH gives 1 mol of OH^-.
So $[OH^-]$ = [KOH] = **0.200 mol dm^{-3}** ***[1 mark]***.

b) $K_w = [H^+][OH^-]$
$[H^+] = (1.0 \times 10^{-14}) \div 0.200 = 5.0 \times 10^{-14}$ ***[1 mark]***
pH = $-\log_{10}(5.00 \times 10^{-14})$ = **13.3** ***[1 mark]***
If you got the answer to part a) wrong, you can have the mark for part b) if you used your wrong value in the right calculation.

Page 87 — More pH Calculations

1 a) $K_a = \frac{[H^+][A^-]}{[HA]}$ or $K_a = \frac{[H^+]^2}{[HA]}$ ***[1 mark]***

b) $K_a = \frac{[H^+]^2}{[HA]}$
[HA] is 0.280 because only a small amount of HA will dissociate
$[H^+] = \sqrt{(5.60\times10^{-4})\times(0.280)} = 0.0125\ \text{mol dm}^{-3}$ ***[1 mark]***
pH = $-\log_{10}[H^+] = -\log_{10}(0.0125)$ = **1.90** ***[1 mark]***

2 a) $[H^+] = 10^{-2.64} = 2.3 \times 10^{-3}$ mol dm^{-3} ***[1 mark]***
$K_a = \frac{[H^+]^2}{[HX]} = \frac{(2.3\times10^{-3})^2}{0.150}$
= **3.5 × 10^{-5} mol dm^{-3}** ***[1 mark]***

b) $pK_a = -\log_{10} K_a = -\log_{10}(3.5 \times 10^{-5})$ = **4.46** ***[1 mark]***

3 $K_a = 10^{-pK_a} = 10^{-4.2} = 6.3 \times 10^{-5}$ mol dm^{-3} ***[1 mark]***
$K_a = \frac{[H^+]^2}{[HA]}$ so $[H^+] = \sqrt{K_a \times [HA]}$
$= \sqrt{(6.3\times10^{-5})\times(1.6\times10^{-4})} = \sqrt{1.0\times10^{-8}}$
$= 1.0 \times 10^{-4}$ mol dm^{-3} ***[1 mark]***
pH = $-\log_{10}[H^+] = -\log_{10}(1.0 \times 10^{-4})$ = **4.0** ***[1 mark]***

Page 89 — pH Curves and Indicators

1 Nitric acid:

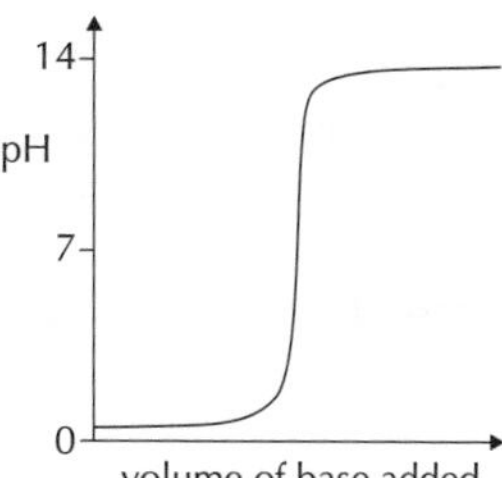

[1 mark]

Ethanoic acid:

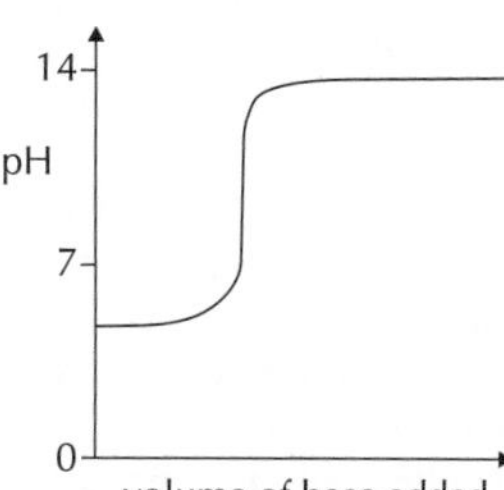

[1 mark]

2 Thymol blue ***[1 mark]***. It's a weak acid/strong base titration so the equivalence point/end point is above pH 8 ***[1 mark]***.

3 a)

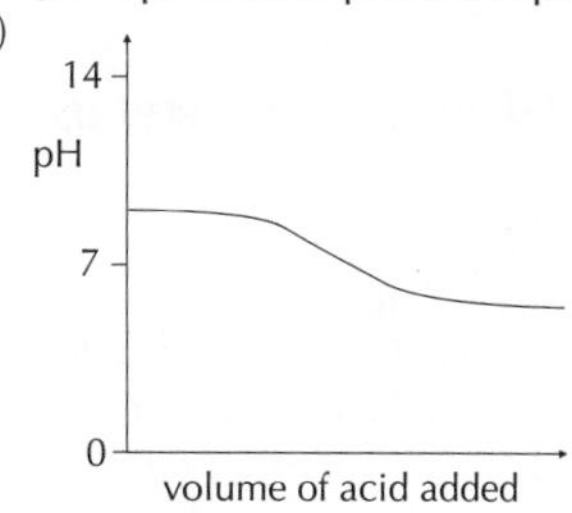

[1 mark]

Answers

b) In a weak acid/weak base titration, the change in pH is gradual, not sharp ***[1 mark]***. This makes it very difficult to determine the exact point the acid is neutralised, using an indicator ***[1 mark]***.

Page 91 — Titration Calculations

1 a) $HCl_{(aq)} + NaOH_{(aq)} \rightarrow NaCl_{(aq)} + H_2O_{(l)}$ ***[1 mark]***
b) i) $(25.60 + 25.65 + 25.55) \div 3 = \mathbf{25.60\ cm^3}$ ***[1 mark]***
ii) Moles NaOH = $(0.10 \times 25.60) \div 1000 = \mathbf{2.560 \times 10^{-3}}$ ***[1 mark]***
c) Conc. HCl = $((2.56 \times 10^{-3}) \times 1000) \div 25 = \mathbf{0.102\ mol\ dm^{-3}}$ ***[1 mark]***

2 a) $H_2SO_{4(aq)} + 2NaOH_{(aq)} \rightarrow Na_2SO_{4(aq)} + 2H_2O_{(l)}$ ***[1 mark]***
b) i) Moles NaOH = $(0.100 \times 35.6) \div 1000 = \mathbf{3.56 \times 10^{-3}}$ ***[1 mark]***
ii) Moles H_2SO_4 = Moles NaOH $\div 2 = \mathbf{1.78 \times 10^{-3}}$ ***[1 mark]***
iii) $[H_2SO_4] = ((1.78 \times 10^{-3}) \times 1000) \div 25.0$
$= \mathbf{0.0712\ mol\ dm^{-3}}$ ***[1 mark]***

Page 93 — Buffer Action

1 a) $K_a = ([H^+] \times [CH_3CH_2COO^-]) \div [CH_3CH_2COOH]$ ***[1 mark]***
So, $[H^+] = K_a \times ([CH_3CH_2COOH] \div [CH_3CH_2COO^-])$
$= (1.3 \times 10^{-5}) \times (0.40 \div 0.20)$
$= 2.6 \times 10^{-5}\ mol\ dm^{-3}$ ***[1 mark]***
$pH = -\log_{10}(2.6 \times 10^{-5}) = \mathbf{4.59}$ ***[1 mark]***
b) Adding H_2SO_4 increases the concentration of H^+ ***[1 mark]***. This will cause more propanoic acid in the buffer to dissociate / shift the $C_6H_5COOH \rightleftharpoons H^+ + C_6H_5COO^-$ equilibrium in the buffer to the left ***[1 mark]***.

2 a) $CH_3(CH_2)_2COOH \rightleftharpoons H^+ + CH_3(CH_2)_2COO^-$ ***[1 mark]***
b) $[CH_3(CH_2)_2COOH] = [CH_3(CH_2)_2COO^-]$ ***[1 mark]***
so $[CH_3(CH_2)_2COOH] \div [CH_3(CH_2)_2COO^-] = 1$
and $K_a = [H^+]$ ***[1 mark]***.
$pH = -\log_{10}(1.5 \times 10^{-5}) = \mathbf{4.8}$ ***[1 mark]***
If the concentrations of the weak acid and the salt of the weak acid are equal, they cancel from the K_a expression and the buffer pH = pK_a.

Unit 2: Section 1 — Periodicity

Page 95 — Periodicity

1 Magnesium ions have a 2+ charge whereas sodium ions only have a 1+ charge ***[1 mark]***. Magnesium also has more delocalised electrons than sodium ***[1 mark]***. So the metal-metal bonds are stronger in magnesium than in sodium and more energy is needed to break them ***[1 mark]***.

2 a) Si has a macromolecular (or giant molecular) structure ***[1 mark]*** consisting of very strong covalent bonds ***[1 mark]***.
b) Sulfur (S_8) molecules are larger than phosphorus (P_4) molecules ***[1 mark]***, which results in stronger van der Waals forces of attraction between molecules ***[1 mark]***.

3 The atomic radius decreases across the period from left to right ***[1 mark]***. The number of protons increases, so nuclear charge increases ***[1 mark]***. Electrons are pulled closer to the nucleus ***[1 mark]***.

Unit 2: Section 2 — Group 2 and Group 7 Elements

Page 97 — Group 2 — The Alkaline Earth Metals

1 a) Mg: $1s^2\ 2s^2\ 2p^6\ 3s^2$ ***[1 mark]***
Ca: $1s^2\ 2s^2\ 2p^6\ 3s^2\ 3p^6\ 4s^2$ ***[1 mark]***
b) First ionisation energy of Ca is lower ***[1 mark]*** because Ca has an extra electron shell ***[1 mark]***. This reduces the attraction between the nucleus and the outer electrons because it increases the shielding effect and because the outer electrons of Ca are further from the nucleus ***[1 mark]***.

2 a) Y ***[1 mark]***
b) Y has the largest radius ***[1 mark]*** so it will have the smallest ionisation energy/lose its outer electrons more easily ***[1 mark]***.

3 Barium ***[1 mark]*** is oxidised from 0 to +2 ***[1 mark]***.

4 The melting points decrease from calcium to strontium to barium. This is because the metallic ions get bigger ***[1 mark]*** but the number of delocalised electrons per atom doesn't change ***[1 mark]***. This means there's reduced attraction of the positive ions to the delocalised electrons, so it takes less energy to break the bonds ***[1 mark]***.

Page 99 — Uses of the Group 2 Elements

1 Add acidified barium chloride solution to both.
Zinc chloride would not change/no reaction.
Zinc sulfate solution would give a white precipitate ***[1 mark]***.
$BaCl_{2(aq)} + ZnSO_{4(aq)} \rightarrow BaSO_{4(s)} + ZnCl_{2(aq)}$
OR $Ba^{2+}_{(aq)} + SO_4{}^{2-}_{(aq)} \rightarrow BaSO_{4(s)}$ ***[1 mark]***

2 a) D barium ***[1 mark]***
b) B calcium ***[1 mark]***
c) A magnesium ***[1 mark]***

3 Patients can swallow a suspension of barium sulfate which will coat the tissues of the digestive system ***[1 mark]***. This will make them show up on the X-rays so any problems can be diagnosed ***[1 mark]***.

Page 101 — Group 7 — The Halogens

1 a) $I_2 + 2At^- \rightarrow 2I^- + At_2$ ***[1 mark]***
b) The astatide ***[1 mark]***

2 a) i) Boiling point increases down the group ***[1 mark]*** because the size and relative mass of the molecules increases ***[1 mark]***, so the van der Waals forces holding the molecules together get stronger ***[1 mark]***.
ii) Electronegativity decreases down the group ***[1 mark]*** because the atoms get larger ***[1 mark]***, and the shielding effect of the inner electrons increases ***[1 mark]***.
b) Fluorine ***[1 mark]***

3 a) $Cl_2 + H_2O \rightarrow 2H^+ + Cl^- + ClO^-$ ***[1 mark]***
b) Chlorine (or the chlorate(I) ions) kill bacteria ***[1 mark]***.
Too much chlorine would be dangerous as it's toxic ***[1 mark]***.

Page 103 — Halide Ions

1 First, separately dissolve each solid in water ***[1 mark]***.
Add dilute nitric acid each solution. Then add a few drops of aqueous $AgNO_3$ ***[1 mark]***. With sodium chloride, the silver nitrate gives white precipitate ***[1 mark]*** which dissolves in dilute ammonia solution ***[1 mark]***. With sodium bromide, the silver nitrate gives cream precipitate ***[1 mark]*** which is only soluble in concentrated ammonia solution ***[1 mark]***.

2 a) C $NaHSO_{4(aq)}$, $At_{2(s)}$, $H_2S_{(g)}$, $2H_2O_{(l)}$ ***[1 mark]***
b) AgAt would not dissolve ***[1 mark]***. E.g. AgI is insoluble in concentrated ammonia solution and the solubility of halides in ammonia solution decreases down the group ***[1 mark]***.
Question 2 is the kind of question that could throw you if you're not really clued up on the facts. If you really know page 102, then in part a) you'll go, "Ah - ha! Reactions of halides with H_2SO_4 — reducing power increases down the group..." If not, you basically won't have a clue. The moral is, it really is just about learning all the facts. Boring, but true.

Answers

'age 105 — Tests for Ions

D Iodide *[1 mark]*

$BaCO_3$ ***[2 marks — 1 mark for correct cation, 1 mark for correct anion]***

Add a few drops of sodium hydroxide to a sample of the solution Warm the mixture and test any gas that is given off using damp red litmus paper ***[1 mark]***. If the solution contained ammonium ions, the litmus paper will turn blue ***[1 mark]***. Take another sample of the solution and add dilute hydrochloric acid followed by barium chloride solution ***[1 mark]***. If the solution contained sulfate ions, a white precipitate will form ***[1 mark]***.

'Jnit 2: Section 3 — Period 3 Elements

'age 107 — Period 3 Elements and Oxides

a) Compound X is SO_3 ***[1 mark]***.
$SO_3 + H_2O \rightarrow H_2SO_4$ ***[1 mark]***

b) i) Compound Y is Na_2O ***[1 mark]***.
$Na_2O + H_2O \rightarrow 2\ NaOH$ ***[1 mark]***

ii) Na_2O/compound Y has a giant lattice structure with strong ionic bonds ***[1 mark]*** that take a lot of energy to break ***[1 mark]***.

'Jnit 2: Section 4 — Transition Metals

'age 109 — Transition Metals — The Basics

a) $1s^2\ 2s^2\ 2p^6\ 3s^2\ 3p^6\ 3d^{10}$ or $[Ar]3d^{10}$ ***[1 mark]***

b) No ***[1 mark]***. Cu^+ ions have a full 3d subshell ***[1 mark]***.

c) copper(II) sulfate ($CuSO_{4(aq)}$) ***[1 mark]***

'age 111 — Complex Ions

a) i) A species (atom, ion or molecule) that donates a lone pair of electrons to form a co-ordinate bond with a metal atom or ion ***[1 mark]***, such as NH_3 in $[Ag(NH_3)_2]^+$ ***[1 mark]***.

ii) A covalent bond in which both electrons come from the same species ***[1 mark]***. In the Ag–NH_3 bond, both electrons come from nitrogen ***[1 mark]***.

iii) The number of co-ordinate bonds formed with the central metal atom or ion ***[1 mark]***. In $[Ag(NH_3)_2]^+$, the co-ordination number is 2 because two NH_3 ligands are bonded to Ag^+ ***[1 mark]***.

b) Linear ***[1 mark]***

a) Co-ordination number: 6 ***[1 mark]***
Shape: octahedral ***[1 mark]***

b) Co-ordination number: 4 ***[1 mark]***
Shape: tetrahedral ***[1 mark]***
Formula: $[CuCl_4]^{2-}$ ***[1 mark]***

c) 109.5° ***[1 mark]***

d) Cl^- ligands are larger than water ligands ***[1 mark]***, so only 4 Cl^- ligands can fit around the Cu^{2+} ion ***[1 mark]***.

'age 113 — More on Complex Ions

a) A bidentate ligand is an atom, ion or molecule that can form two co-ordinate bonds with a transition metal ion ***[1 mark]***.

b) i)

[1 mark]

[1 mark]

ii) Optical isomerism ***[1 mark]***.

2

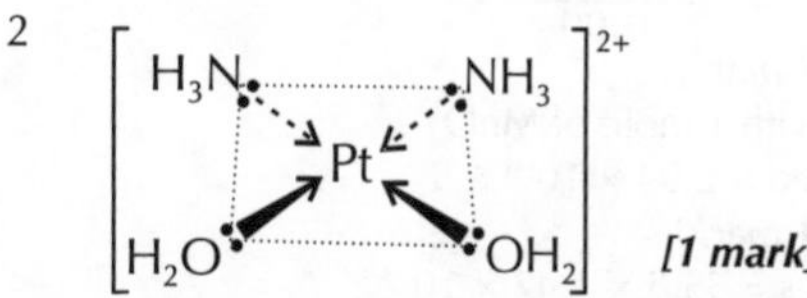

[1 mark]

Page 115 — Formation of Coloured Ions

1 a) Energy is absorbed from visible light when electrons move from the ground state to a higher energy level ***[1 mark]***.

b) Change in oxidation state ***[1 mark]***, ligand ***[1 mark]*** or co-ordination number ***[1 mark]***.

2 Prepare a range of dilutions of known concentrations ***[1 mark]***. Measure the absorbance of the solutions ***[1 mark]***. Plot a graph of concentration versus absorbance ***[1 mark]***.

3 a) ΔE is the energy absorbed when an electron moves from the ground state to a higher energy level/excited state ***[1 mark]***.
OR ΔE is the difference between the ground state energy and the energy of an excited electron ***[1 mark]***.

b) i) $[Ar]\ 3d^{10}$ ***[1 mark]***

ii) $[Ar]\ 3d^9$ ***[1 mark]***

c) Cu^{2+} because it has an incomplete d-subshell ***[1 mark]***.

Page 117 — Substitution Reactions

1 a) $[Fe(H_2O)_6]^{3+} + EDTA^{4-} \rightarrow [FeEDTA]^- + 6H_2O$ ***[1 mark]***

b) The formation of $[FeEDTA]^-$ results in an increase in entropy, because the number of particles increases from two to seven ***[1 mark]***.

2 a) $[Co(H_2O)_6]^{2+} + 6NH_3 \rightarrow [Co(NH_3)_6]^{2+} + 6H_2O$ ***[1 mark]***
You can fit the same number of H_2O and NH_3 ligands around a Co^{2+} ion because they are a similar size and both uncharged.

b) $[Co(H_2O)_6]^{2+} + 3NH_2CH_2CH_2NH_2 \rightarrow [Co(NH_2CH_2CH_2NH_2)_3]^{2+} + 6H_2O$ ***[1 mark]***

c) $[Co(H_2O)_6]^{2+} + 4Cl^- \rightarrow [CoCl_4]^{2-} + 6H_2O$ ***[1 mark]***

Answers

Page 119 — Variable Oxidation States

1 a) i) $[Ag(NH_3)_2]^+$ ***[1 mark]***
ii) Silver is in the +1 oxidation state ***[1 mark]***.
b) Aldehydes are oxidised by Tollens' reagent in the following redox reaction:
$RCHO_{(aq)} + 2[Ag(NH_3)_2]^+_{(aq)} + 3OH^-_{(aq)} \rightarrow$
$RCOO^-_{(aq)} + 2Ag_{(s)} + 4NH_{3(aq)} + 2H_2O_{(l)}$ ***[1 mark]***.
This causes a silver mirror to form on the inside of the test tube ***[1 mark]***. Ketones aren't oxidised by Tollens' reagent, so no silver mirror is formed ***[1 mark]***.
2 a) Zinc metal ***[1 mark]*** and an acidic solution ***[1 mark]***.
b) The blue solution is due to the presence of $VO^{2+}_{(aq)}$ ions ***[1 mark]***, and the green solution is due to the presence of $V^{3+}_{(aq)}$ ions ***[1 mark]***.
c) $2VO_2^+{}_{(aq)} + Zn_{(s)} + 4H^+_{(aq)} \rightarrow 2VO^{2+}_{(aq)} + Zn^{2+}_{(aq)} + 2H_2O_{(l)}$
[1 mark]
$2VO^{2+}_{(aq)} + Zn_{(s)} + 4H^+_{(aq)} \rightarrow 2V^{3+}_{(aq)} + Zn^{2+}_{(aq)} + 2H_2O_{(l)}$
[1 mark]
$2V^{3+}_{(aq)} + Zn_{(s)} \rightarrow 2V^{2+}_{(aq)} + Zn^{2+}_{(aq)}$ ***[1 mark]***
3 B ***[1 mark]***

Page 121 — Titrations with Transition Metals

1 a) Moles of MnO_4^- added = $\frac{0.0100 \times 29.4}{1000}$
= **2.94×10^{-4} moles** ***[1 mark]***
b) 5 moles of Fe^{2+} react with 1 mole of MnO_4^-.
So moles of Fe^{2+} reacted = $2.94 \times 10^{-4} \times 5$
= **1.47×10^{-3} moles** ***[1 mark]***
c) Mass of Fe = A_r × moles = $55.8 \times 1.47 \times 10^{-3}$
= **0.0820 g** ***[1 mark]***
d) % Fe = $\frac{0.0820}{0.100} \times 100$
= **82.0%** ***[1 mark]***
2 Moles of MnO_4^- added = $\frac{0.0200 \times 18.30}{1000}$
= 3.66×10^{-4} moles ***[1 mark]***.
5 moles of $C_2O_4^{2-}$ react with 2 moles of MnO_4^-
so moles of $C_2O_4^{2-}$ added = $\frac{(3.66 \times 10^{-4}) \times 5}{2}$
= 9.15×10^{-4} moles ***[1 mark]***.
M_r of $Na_2C_2O_4$ = (23.0 × 2) + (12.0 × 2) + (16.0 × 4)
= 134.0 ***[1 mark]***.
Mass of $Na_2C_2O_4$ = M_r × moles
= $134.0 \times 9.15 \times 10^{-4}$
= **0.123 g** ***[1 mark]***.
3 Add dilute sulfuric acid to the iron(II) sulfate solution ***[1 mark]***. Titrate with potassium manganate(VII) ***[1 mark]***. The remaining Fe^{2+} will be oxidised to Fe^{3+} ***[1 mark]***. Calculate the number of moles of potassium manganate(VII) that react and use this to calculate the number of moles of iron(II), and hence the concentration ***[1 mark]***.

Page 123 — Catalysts

1 a) Vanadium(V) is reduced to vanadium(IV) and oxidises SO_2 to SO_3: $V_2O_5 + SO_2 \rightarrow V_2O_4 + SO_3$ ***[1 mark]***
Vanadium(IV) is oxidised to vanadium(V) by oxygen gas:
$V_2O_4 + \frac{1}{2}O_2 \rightarrow V_2O_5$ ***[1 mark]***
b) i) Impurities are adsorbed onto the surface of the catalyst ***[1 mark]***. For example, sulfur in hydrogen poisons the iron catalyst in the Haber Process ***[1 mark]***.
ii) E.g. Reduced efficiency ***[1 mark]***, increased cost ***[1 mark]***.
2 The overall equation for the reaction is:
$2MnO_4^-{}_{(aq)} + 16H^+_{(aq)} + 5C_2O_4^{2-}{}_{(aq)} \rightarrow 2Mn^{2+}_{(aq)} + 8H_2O_{(l)} + 10CO_{2(g)}$ ***[1 mark]***.
This is slow to begin with, because the MnO_4^- and $C_2O_4^{2-}$ ions are both negatively charged, so repel each other and don't collide very frequently ***[1 mark]***. The Mn^{2+} product, however, is able to catalyse the reaction. It reduces MnO_4^- to Mn^{3+}:
$MnO_4^-{}_{(aq)} + 4Mn^{2+}_{(aq)} + 8H^+_{(aq)} \rightarrow 5Mn^{3+}_{(aq)} + 4H_2O_{(l)}$ ***[1 mark]***.
The Mn^{3+} ions are reduced back to Mn^{2+} by reaction with $C_2O_4^{2-}$:
$2Mn^{3+}_{(aq)} + C_2O_4^{2-}{}_{(aq)} \rightarrow 2Mn^{2+}_{(aq)} + 2CO_{2(g)}$ ***[1 mark]***.
This means the reaction is an autocatalysis reaction. As more Mn^{2+} is produced, there is more catalyst available and so the reaction rate will increase ***[1 mark]***.

Page 125 — Metal-Aqua Ions

1 Fe^{3+} has a higher charge density than Fe^{2+} ***[1 mark]***. This means Fe^{3+} polarises water molecules more, weakening the O–H bond more and making it more likely that H^+ ions are released into the solution ***[1 mark]***.
2 a) $[Fe(H_2O)_6]^{2+}_{(aq)} + CO_3^{2-}{}_{(aq)} \rightarrow FeCO_{3(s)} + 6H_2O_{(l)}$ ***[1 mark]***
b) The CO_3^{2-} ions remove H_3O^+ from the solution:
$CO_3^{2-}{}_{(aq)} + 2H_3O^+_{(aq)} \rightarrow CO_{2(g)} + 2H_2O_{(l)}$ ***[1 mark]***.
This causes the following two equilibrium reactions to shift to the right:
$[Fe(OH)(H_2O)_5]^{2+}_{(aq)} + H_2O_{(l)} \rightleftharpoons [Fe(OH)_2(H_2O)_4]^+_{(aq)} + H_3O^+_{(aq)}$
$[Fe(OH)_2(H_2O)_4]^+_{(aq)} + H_2O_{(l)} \rightleftharpoons Fe(OH)_3(H_2O)_{3(s)} + H_3O^+_{(aq)}$
[1 mark].

Page 127 — More on Metal-Aqua Ions

1 A blue precipitate ***[1 mark]*** of copper(II) hydroxide forms in the blue solution of copper sulfate:
$[Cu(H_2O)_6]^{2+}_{(aq)} + 2H_2O_{(l)} \rightleftharpoons Cu(OH)_2(H_2O)_{4(s)} + 2H_3O^+_{(aq)}$
[1 mark].
On addition of excess ammonia, the precipitate dissolves to give a deep blue solution ***[1 mark]***:
$Cu(OH)_2(H_2O)_{4(s)} + 4NH_{3(aq)} \rightleftharpoons [Cu(NH_3)_4(H_2O)_2]^{2+}_{(aq)} + 2H_2O_{(l)} + 2OH^-_{(aq)}$ ***[1 mark]***
2 a) i) $[Al(H_2O)_6]^{3+}$ ***[1 mark]***
ii) $Al(OH)_3(H_2O)_3$ ***[1 mark]***
iii) $[Al(OH)_4(H_2O)_2]^-$ ***[1 mark]***
b) $Al(OH)_3(H_2O)_{3(s)} + OH^-_{(aq)} \rightleftharpoons [Al(OH)_4(H_2O)_2]^-_{(aq)} + H_2O_{(l)}$
[1 mark]
3 a) i) Formation of a brown precipitate ***[1 mark]***.
ii) Formation of a green precipitate ***[1 mark]***.
b) $[Fe(H_2O)_6]^{2+}_{(aq)} + CO_3^{2-}{}_{(aq)} \rightarrow FeCO_{3(s)} + 6H_2O_{(aq)}$ ***[1 mark]***
c) i) Formation of a brown precipitate ***[1 mark]***.
ii) Fe^{2+} has been oxidised to Fe^{3+} by the oxygen in the air ***[1 mark]***.

Answers

Unit 3: Section 1 — Introduction to Organic Chemistry

Page 130 — Basic Stuff

1 a) 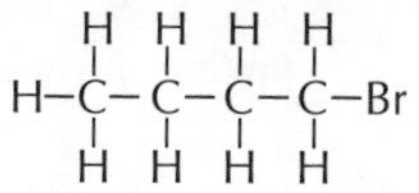
```
    H  H  H  H
    |  |  |  |
 H–C–C–C–C–Br
    |  |  |  |
    H  H  H  H
```
[1 mark]

b) Halogenoalkanes ***[1 mark]***

c) but-1-ene ***[1 mark]***

2 a) A group of compounds that have the same functional group ***[1 mark].***

b) i) C_5H_{12} ***[1 mark]***
ii) Pentane ***[1 mark]***

3 a) 1,2-dichloroethane ***[1 mark]***

b) 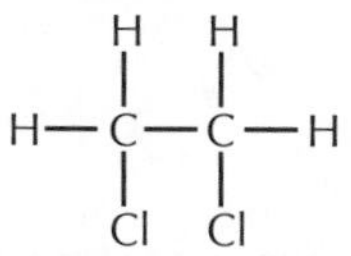
```
    H   H
    |   |
 H—C—C—H
    |   |
    Cl  Cl
```
[1 mark

c) Cl, Cl (skeletal formula) ***[1 mark]***

Page 133 — Isomerism

1 a) 1-chlorobutane, 2-chlorobutane, 1-chloro-2-methylpropane, 2-chloro-2-methylpropane ***[1 mark for each correct isomer]***

b) 1-chloro-2-methylpropane and 2-chloro-2-methylpropane OR 1-chlorobutane and 2-chlorobutane ***[1 mark]***

c) 1-chlorobutane and 1-chloro-2-methylpropane OR 2-chlorobutane and 2-chloro-2-methylpropane ***[1 mark]***

2 a) 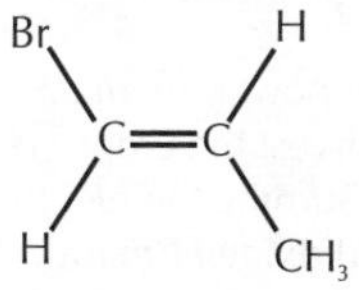
```
 Br      H
   \    /
    C=C
   /    \
 H       CH3
```
E-1-bromopropene ***[1 mark]***

b) 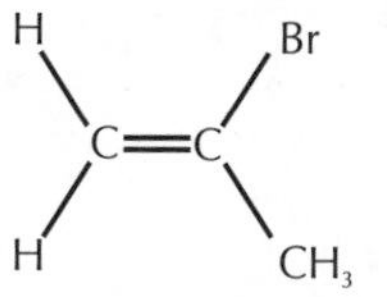
```
 H       Br
   \    /
    C=C
   /    \
 H       CH3
```
[1 mark]

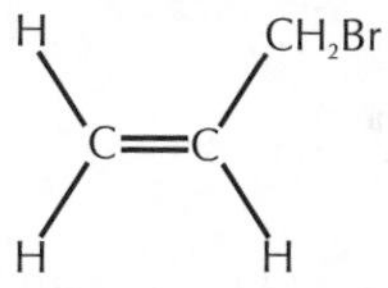

```
 H       CH2Br
   \    /
    C=C
   /    \
 H       H
```
[1 mark]

If one of the C=C carbons has the same two groups attached, then the alkene won't have E/Z isomers.

3 a) 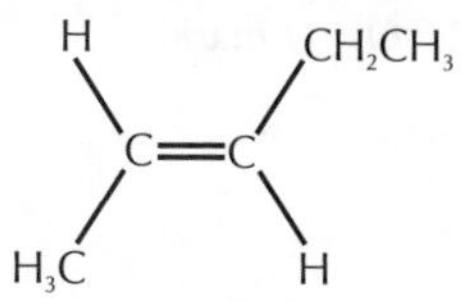
```
 H       CH2CH3
   \    /
    C=C
   /    \
 H3C     H
```
E-pent-2-ene ***[1 mark]***

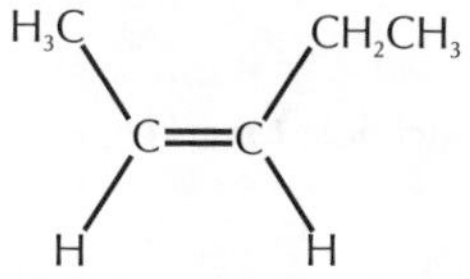

```
 H3C     CH2CH3
   \    /
    C=C
   /    \
 H       H
```
Z-pent-2-ene ***[1 mark]***

b) E/Z isomers occur because atoms can't rotate about C=C double bonds ***[1 mark]***. Alkenes contain C=C double bonds and alkanes don't, so alkenes can form E/Z isomers and alkanes can't ***[1 mark]***.

Unit 3: Section 2 — Alkanes and Halogenoalkanes

Page 135 — Alkanes and Petroleum

1 a) As a mixture, crude oil is not very useful, but the different alkanes it's made up of are useful ***[1 mark]***.

b) Boiling point ***[1 mark]***.

c) i) C_8H_{18} ***[1 mark]***
ii) Near the top. This is because the molecules in petrol have a relatively low boiling point ***[1 mark]*** and the fractionating column is cooler at the top than the bottom ***[1 mark]***.

2 a) There's greater demand for the smaller fractions ***[1 mark]***.

b) $C_{12}H_{26} \rightarrow C_6H_{14} + C_4H_8 + C_2H_4$ ***[1 mark]***.

Page 137 — Alkanes as Fuels

1 a) $C_7H_{16} + 11O_2 \rightarrow 7CO_2 + 8H_2O$ ***[1 mark]***

b) i) carbon monoxide ***[1 mark]***
ii) By fitting a catalytic converter ***[1 mark]***.

2 a) The high pressures and temperatures in the engine of a car ***[1 mark]*** cause nitrogen and oxygen from the air to react together ***[1 mark]***.

b) Powdered calcium carbonate is mixed with water to make an alkaline slurry ***[1 mark]***. When the flue gases mix with the alkaline slurry, the acidic sulfur dioxide gas reacts with the calcium carbonate to form a salt/calcium sulfate ***[1 mark]***.

Page 139 — Chloroalkanes and CFCs

1 $Cl\bullet + O_3 \rightarrow O_2 + ClO\bullet$ ***[1 mark]***
$ClO\bullet + O_3 \rightarrow 2O_2 + Cl\bullet$ ***[1 mark]***

2 a) Free radical substitution ***[1 mark]***

b) Initiation: $Cl_2 \xrightarrow{UV} 2Cl\bullet$ ***[1 mark]***
Propagation: $C_2H_6 + Cl\bullet \rightarrow C_2H_5\bullet + HCl$ ***[1 mark]***
$C_2H_5\bullet + Cl_2 \rightarrow C_2H_5Cl + Cl\bullet$ ***[1 mark]***
Termination: $C_2H_5\bullet + Cl\bullet \rightarrow C_2H_5Cl$ /
$C_2H_5\bullet + C_2H_5\bullet \rightarrow C_4H_{10}$ ***[1 mark]***

You can't have a termination reaction where two chlorine radicals react together to form Cl_2 here, because the question asked for one that forms an organic compound.

Page 142 — Halogenoalkanes

1 a) Reaction 1:
Reagent — NaOH/KOH ***[1 mark]***
Solvent — Water/aqueous solution ***[1 mark]***
Reaction 2:
Reagent — Ammonia/NH_3 ***[1 mark]***
Solvent — Ethanol/alcohol ***[1 mark]***
Reaction 3:
Reagent — NaOH/KOH ***[1 mark]***
Solvent — Ethanol/alcohol ***[1 mark]***

b) The reaction would be faster ***[1 mark]*** because the C–I bond is weaker than C–Br/C–I bond enthalpy is lower ***[1 mark]***.

2 B ***[1 mark]***

Answers

Unit 3: Section 3 — Alkenes and Alcohols

Page 145 — Alkenes

1 a) Shake the alkene with bromine water ***[1 mark]***. The solution goes from orange to colourless if a double bond is present ***[1 mark]***.

b) Electrophilic addition ***[1 mark]***

c) i)

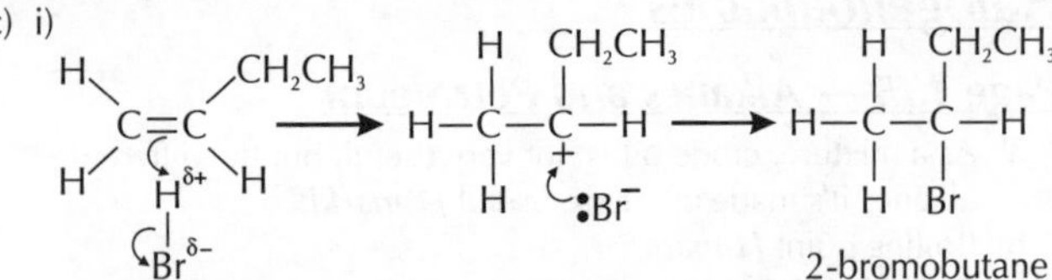

[1 mark for each correct curly arrow and 1 mark for the product being correctly named.]

Check that your curly arrows are exactly right, or you'll lose marks. They have to go from exactly where the electrons are from, to where they're going to.

ii) The secondary carbocation OR the carbocation with the most attached alkyl groups ***[1 mark]*** is the most stable intermediate and so is the most likely to form ***[1 mark]***.

2 A ***[1 mark]***

Page 147 — Addition Polymers

1 a)
```
H       H
 \     /
  C = C
 /     \
H       CH3
```
[1 mark]

b)
```
   H  H
   |  |
 — C— C —
   |  |
   H  CH3
```
[1 mark]

2 a)
```
   H  H
   |  |
 — C— C —
   |  |
   H  Cl
```
[1 mark]

b) When you add a plasticiser, PVC becomes more flexible ***[1 mark]***. This is because the plasticisers get between the polymer chains and push them apart ***[1 mark]***. This reduces the strength of the intermolecular forces between the chains, so they can slide over each other more easily ***[1 mark]***.

c) Any two from: e.g. electrical cable insulation / flooring tiles / clothing ***[1 mark for each]***.

Page 149 — Alcohols

1 a) i) Primary ***[1 mark]***

ii) Tertiary ***[1 mark]***

iii) Secondary ***[1 mark]***

iv) Primary ***[1 mark]***

b) i)
```
  H  H  H
  |  |  |      H
H—C——C——C=C/
  |  |      \
  H  H       H
```
[1 mark]

ii)
```
   H
   |
 H—C—H
 H |
 | |    H
H—C—C=C/
 |      \
 H       H
```
[1 mark]

iii)
```
  H  H  H
  |  |  |      H
H—C——C——C=C/
  |  |      \
  H  H       H
```
or
```
  H     H  H
  |     |  |
H—C——C=C——C—H
  |  |     |
  H  H     H
```
[1 mark]

iv)
```
     H   H
      \ /
       C
  H    ||   H
  |    ||   |
H—C————C————C—H
  |         |
  H         H
```
[1 mark]

2 Each liquid in a mixture has a different boiling point ***[1 mark]***. Collecting only the liquid (fraction) that boils at a particular temperature will separate it from the mixture ***[1 mark]***.

Page 151 — Ethanol Production

1 a) $C_6H_{12}O_{6(aq)} \rightarrow 2C_2H_5OH_{(aq)} + 2CO_{2(g)}$ ***[1 mark]***

b) In the presence of yeast ***[1 mark]***, at a temperature of between 30 and 40 °C ***[1 mark]***, anaerobic conditions / air/oxygen excluded ***[1 mark]***.

c) Fractional distillation ***[1 mark]***

2 a) E.g. as they grow, the plants used to produce bioethanol take in the same amount of carbon dioxide as burning the fuel you produce from them gives out. ***[1 mark]***.

b) E.g. fossil fuels will need to be burned to power the machinery used to make fertilisers for the crops / harvest the crops / refine and transport the bioethanol ***[1 mark]***. Burning the fuel to power this machinery produces carbon dioxide ***[1 mark]***.

Page 153 — Oxidation of Alcohols

1 a) i) Acidified potassium dichromate(VI) ***[1 mark]***

ii)
```
      H    H
   H   \   |
    \   C—C=O
     C /
   /  \
  H  H  H
```
[1 mark]

b) i) Warm with Fehling's/Benedict's solution: turns from blue to brick-red ***[1 mark].***
OR warm with Tollens' reagent: a silver mirror is produced ***[1 mark]***.

ii) Propanoic acid ***[1 mark]***

iii) $CH_3CH_2CH_2OH + [O] \rightarrow CH_3CH_2CHO + H_2O$ ***[1 mark]***
$CH_3CH_2CHO + [O] \rightarrow CH_3CH_2COOH$ ***[1 mark]***

iv) Distillation ***[1 mark]***

c) i)
```
    |
    |
   /|\
  / | \
    |
    OH
```
[1 mark]

ii) 2-methylpropan-2-ol is a tertiary alcohol ***[1 mark]***.

Answers

Unit 3: Section 4 — Organic Analysis

Page 156 — Tests for Functional Groups

1 D *[1 mark]*
Propanone is a ketone and so nothing happens when it is warmed with Fehling's solution.

2 B *[1 mark]*
Cyclohexene is an alkene so it decolourises bromine water.

3 E.g. put a sample of the solution that you want to test in a test tube and add some sodium carbonate ***[1 mark]***. If the solution is a carboxylic acid, the mixture will fizz ***[1 mark]***. If you collect the gas produced and bubble it through limewater, the limewater should turn cloudy ***[1 mark]***.

4 E.g. add excess alcohol to acidified potassium dichromate solution ***[1 mark]*** in a round bottomed flask. Set up the flask as part of distillation apparatus, gently heat it and collect the product ***[1 mark]***. Place some Fehling's solution/Benedict's solution/Tollens' reagent in a test tube and add a few drops of the product ***[1 mark]***. Put the test tube in a hot water bath to warm it for a few minutes ***[1 mark]***. If the blue solution gives a brick red precipitate/if a silver mirror forms on the inside of the tube, the alcohol was a primary alcohol. If there is no change, the alcohol was a secondary alcohol ***[1 mark for a correct observation that matches the reagent used]***.

Page 159— Analytical Techniques

1 a) C ***[1 mark]***
The relative molecular mass of the compound is equal to the m/z value of the molecular ion, so calculate the precise M_r of each possible molecular formula:
A: (3 × 12.0000) + (6 × 1.0078) + (2 × 15.9990) = 74.0448
B: (4 × 12.0000) + (10 × 1.0078) + 15.9990 = 74.077
C: (3 × 12.0000) + (10 × 1.0078) + (2 × 14.0064) = 74.0908
D: (2 × 12.0000) + (6 × 1.0078) + (2 × 14.0064) + 15.9990 = 74.0586

b) The four options given in part a) all have the same M_r to the nearest whole number, so their molecular ions would all have the same m/z value on a low resolution mass spectrum ***[1 mark]***.

c) 64.0364 – [(2 × 12.0000) + (5 × 1.0078) + 15.9990] = 64.0364 – 45.038 = **18.9984** ***[1 mark]***
The m/z value of the molecular ion is equal to the molecular mass of the compound, so to find the precise atomic mass of F you can just subtract the precise atomic masses of the other elements:

2 a) A is caused by an O–H bond in an alcohol ***[1 mark]***.
B is caused by a C–O bond ***[1 mark]***.

b) E.g. the molecule must be an alcohol ***[1 mark]***. An O–H group has a mass of 17, so the rest of the molecule must be a hydrocarbon chain with a mass of 57 ***[1 mark]***. So the molecule is probably C_4H_9OH ***[1 mark]***.
You can tell that the molecule is an alcohol because the spectrum has peaks caused by an O–H bond in an alcohol and a C–O bond.

Unit 3: Section 5 — Isomerism and Carbonyl Compounds

Page 161 — Optical Isomerism

1 a) The property of having stereoisomers, which are molecules with the same molecular formula and with their atoms arranged in the same way ***[1 mark]***, but with a different orientation of the bonds in space ***[1 mark]***.

b) i)

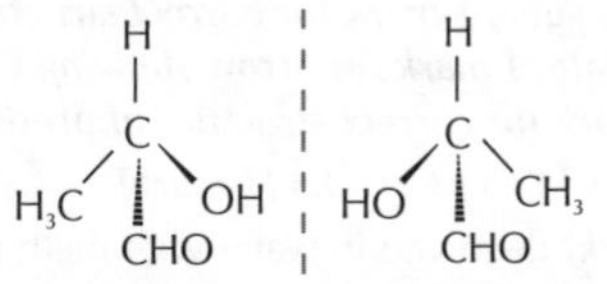

[1 mark for each correctly drawn structure]
Your isomers don't have to be orientated in the same way as in the diagram above, just as long as the molecules are mirror images of each other.

ii) Shine (monochromatic) plane-polarised light through a solution of the molecule ***[1 mark]***. The enantiomers will rotate the light in opposite directions ***[1 mark]***.

2 a)

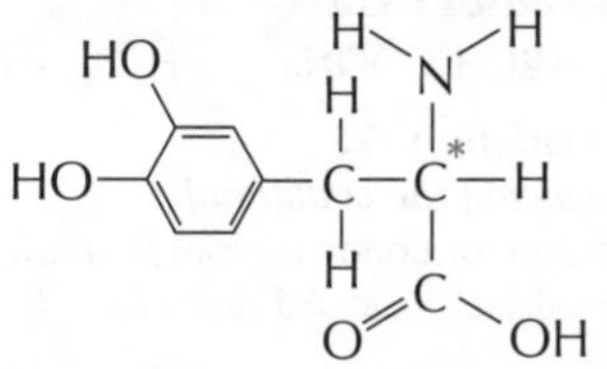

[1 mark for chiral carbon clearly marked]

b) A mixture of equal quantities of each enantiomer of an optically active compound ***[1 mark]***.

3 a) E.g. Pentanal has a planar C=O bond ***[1 mark]***. Depending on which direction the CN^- ion attacks the δ+ carbon from, two different enantiomers can be formed ***[1 mark]***. The CN^- ion is equally likely to attack from either direction ***[1 mark]***, so equal amounts of each enantiomer are formed, giving an optically inactive racemic mixture ***[1 mark]***.

b) The reaction of pentan-3-one would not produce an optically active product ***[1 mark]*** as it is a symmetrical ketone/the product does not contain a chiral centre ***[1 mark]***.

Page 163 — Aldehydes and Ketones

1 a) Propanal ***[1 mark]***

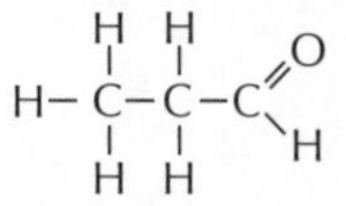

[1 mark]

Propanone ***[1 mark]***

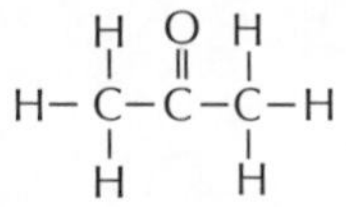

[1 mark]

Answers

b) i) Nucleophilic addition ***[1 mark]***.

ii)

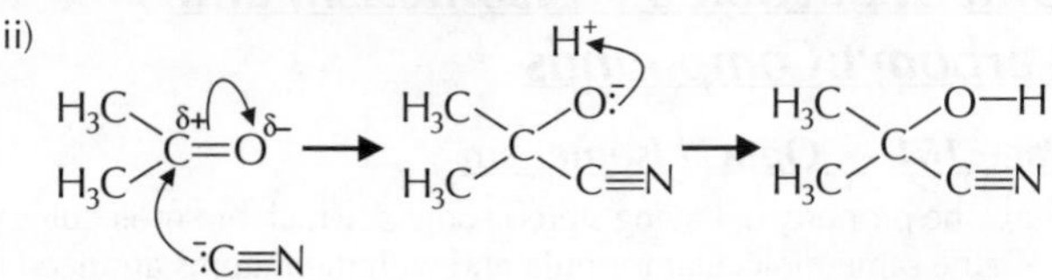

[1 mark for arrow showing attack of $^-$:CN on δ+ carbon, 1 mark for arrow showing electrons moving from the C=O double bond to the oxygen, 1 mark for correct structure of charged intermediate, 1 mark for arrow showing O:$^-$ attacking an H$^+$, 1 mark for correct structure of product.]

c) $CH_3CH_2CHO + 2[H] \rightarrow CH_3CH_2CH_2OH$ ***[1 mark]***
[H] = e.g. $NaBH_4$ ***[1 mark]*** dissolved in water with methanol ***[1 mark]***.

2 a) Butanal ***[1 mark]*** and butanone (or butan-2-one) ***[1 mark]***

b) EITHER: Tollens' reagent ***[1 mark]*** gives a silver mirror with butanal ***[1 mark]*** but no reaction with butanone ***[1 mark]***
OR Fehling's solution ***[1 mark]*** gives a brick-red precipitate with butanal ***[1 mark]*** but no reaction with butanone ***[1 mark]***.

Page 165 — Carboxylic Acids and Esters

1 a) $2CH_3COOH_{(aq)} + Na_2CO_{3(s)} \rightarrow 2CH_3COONa_{(aq)} + H_2O_{(l)} + CO_{2(g)}$
[1 mark for CH_3COONa, 1 mark for CO_2, and 1 mark for correctly balancing the equation.]

b) methanol ***[1 mark]***, esterification (or condensation) ***[1 mark]***.
Substance X is a carboxylic acid (ethanoic acid) and substance Y is an ester (methyl ethanoate).

2 a)

```
     H   H   H    O
     |   |   |   //
 H — C — C — C — C      H   H
     |   |   |    \     |   |
     H   H   H     O —  C — C — H
                        |   |
                        H   H
```

[1 mark]

b) Flavouring / perfume / plasticiser ***[1 mark]***

3 a) $CH_3COOH + CH_3CH(CH_3)CH_2CH_2OH \rightleftharpoons CH_3COOCH_2CH_2CH(CH_3)CH_3 + H_2O$ ***[1 mark]***
Heat OR warm OR reflux ***[1 mark]***
and (concentrated sulfuric) acid catalyst ***[1 mark]***.

b) ethanoic acid ***[1 mark]***

Page 167 — More on Esters

1 a) 2-methylpropyl ethanoate ***[1 mark]***

b) Ethanoic acid ***[1 mark]***

```
    H    O
    |   //
H — C — C
    |    \
    H     OH
```

[1 mark]

2-methylpropan-1-ol ***[1 mark]***

```
     H  H   H
     |  |   |
HO — C — C — C — H
     |  |   |
     H  CH3 H
```

[1 mark]

This is acid hydrolysis ***[1 mark]***

c) With sodium hydroxide, sodium ethanoate/ethanoate ions are produced, but in the reaction in part b), ethanoic acid is produced ***[1 mark]***.

2 a)

```
    H
    |
H — C — OH
    |
H — C — OH
    |
H — C — OH
    |
    H
```

[1 mark]

b) $CH_3(CH_2)_7CH{=}CH(CH_2)_7COONa + H^+ \rightarrow CH_3(CH_2)_7CH{=}CH(CH_2)_7COOH + Na^+$ ***[1 mark]***

c) Shake with bromine water ***[1 mark]***. The bromine water will turn from orange to colourless with oleic acid, but not with stearic acid, $CH_3(CH_2)_{16}COOH$ ***[1 mark]***.
You might never have heard of these two fatty acids, but you are told in the question that one of them has a double bond — and you already know how to test for unsaturation...

Page 169 — Acyl Chlorides

1 a) Ethanoyl chloride:
$CH_3COCl + CH_3OH \rightarrow CH_3COOCH_3 + HCl$ ***[1 mark]***
Ethanoic anhydride:
$(CH_3CO)_2O + CH_3OH \rightarrow CH_3COOCH_3 + CH_3COOH$ ***[1 mark]***
methyl ethanoate ***[1 mark]***

b) Vigorous reaction OR (HCl) gas/fumes produced ***[1 mark]***

c) Irreversible reaction OR faster reaction ***[1 mark]***

2 a) $CH_3COCl + CH_3CH_2NH_2 \rightarrow CH_3CONHCH_2CH_3 + HCl$ ***[1 mark]***
N-ethylethanamide ***[1 mark]***

b)

```
     Oδ–               O:–               O                 O
  δ+ ||                |                 ||                ||
H3C—C—Clδ– →   H3C—C—Cl   →   H3C—C        →   H3C—C
     |                 |                 |                 |
 H—N̈—H            H—N⁺—H          H—N⁺—H            H—N     + HCl
     |                 |           Cl:–  |                 |
   CH2CH3            CH2CH3            CH2CH3            CH2CH3
```

[1 mark for each curly arrow on first diagram, 1 mark for curly arrows on second diagram, 1 mark for correct curly arrows on third diagram, 1 mark for correct structures and charges.]

Page 171 — Purifying Organic Compounds

1 B ***[1 mark]***

2 a) The scientist used the minimum possible amount of hot solvent to make sure that the solution would be saturated ***[1 mark]***.

b) Filter the hot solution through a heated funnel to remove any insoluble impurities ***[1 mark]***. Leave the solution to cool down slowly until crystals of the product have formed ***[1 mark]***. Filter the mixture under reduced pressure ***[1 mark]***. Wash the crystals with ice-cold solvent ***[1 mark]***. Leave the crystals to dry ***[1 mark]***.

c) The melting point range of the impure product will be lower and broader than that of the pure product ***[1 mark]***.

Answers

Unit 3: Section 6 — Aromatic Compounds and Amines

Page 174 — Aromatic Compounds

1 a) Conditions: non-aqueous solvent (e.g. dry ether), reflux ***[1 mark]***

b) The acyl chloride molecule isn't polarised enough/isn't a strong enough electrophile to attack the benzene ***[1 mark]***. The halogen carrier makes the acyl chloride electrophile stronger ***[1 mark]***.

c) $H_3C-\overset{+}{C}=O$

[1 mark]

2 a) A: nitrobenzene ***[1 mark]***

B + C: concentrated nitric acid and concentrated sulfuric acid ***[1 mark]***

D: warm, not more than 55 °C ***[1 mark]***

When you're asked to name a compound, give the name, not the formula.

b) $HNO_3 + H_2SO_4 \rightarrow H_2NO_3^+ + HSO_4^-$ ***[1 mark]***

$H_2NO_3^+ \rightarrow NO_2^+ + H_2O$ ***[1 mark]***

c)

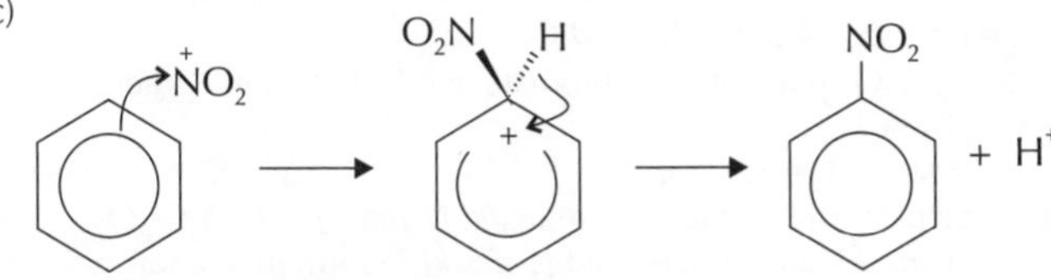

[2 marks — 1 mark for each correct step.]

Page 177 — Amines and Amides

1 a) It can accept protons/H⁺ ions, or it can donate a lone pair of electrons ***[1 mark]***.

b) Methylamine is stronger, as the methyl group/CH_3 pushes electrons onto/increases electron density on the nitrogen, making the lone pair more available ***[1 mark]***. Phenylamine is weaker, as the nitrogen lone pair is less available — nitrogen's electron density is decreased as it's partially delocalised around the benzene ring ***[1 mark]***.

2 a) You get a mixture of primary, secondary and tertiary amines, and quaternary ammonium salts ***[1 mark]***.

b) i) $LiAlH_4$ and a non-aqueous solvent (e.g. dry ether), followed by dilute acid ***[1 mark]***.

ii) It's too expensive ***[1 mark]***.

iii) Hydrogen gas ***[1 mark]***, metal catalyst such as platinum or nickel and high temperature and pressure ***[1 mark]***.

3 $2CH_3CH_2NH_2 + CH_3CH_2Br \rightarrow (CH_3CH_2)_2NH + CH_3CH_2NH_3Br$

[1 mark]

Mechanism:

$H_3CH_2C-\ddot{N}H_2 + CH_3CH_2Br \rightarrow H_3CH_2C-\overset{+}{N}H_2-CH_2CH_3$

[1 mark]

Then:

$H_3CH_2C-\overset{+}{N}H_2-CH_2CH_3 + H_2\ddot{N}-CH_2CH_3 \rightarrow H_3CH_2C-NH-CH_2CH_3 + CH_3CH_2\overset{+}{N}H_3Br^-$

[1 mark]

Unit 3: Section 7 — Polymers

Page 179 — Condensation Polymers

1 a) A polyamide ***[1 mark]***.

b) A dicarboxylic acid and a diamine ***[1 mark]***.

c) Hydrolysis ***[1 mark]***.

2 $H_2N-(CH_2)_6-NH_2$

[1 mark]

$HOOC-(CH_2)_4-COOH$

[1 mark]

3 a) $-\overset{O}{\overset{\|}{C}}-(CH_2)_4-\overset{O}{\overset{\|}{C}}-O-(CH_2)_6-O-$

or

$-O-(CH_2)_6-O-\overset{O}{\overset{\|}{C}}-(CH_2)_4-\overset{O}{\overset{\|}{C}}-$ ***[1 mark]***

b) For each link formed, one small molecule (water) is eliminated ***[1 mark]***.

Page 181 — Disposing of Polymers

1 a) E.g. heat energy produced can be used to generate electricity / saves on space in landfill ***[1 mark]***

b) Burning plastics that contain chlorine produces toxic gases/ HCl gas ***[1 mark]***. Waste gases from combustion can be passed through scrubbers which neutralise the gases/HCl by allowing them/it to react with a base ***[1 mark]***.

2 Any sensible advantage e.g. it is cheap and easy / it doesn't require waste plastics to be separated or sorted ***[1 mark]***.
Any sensible disadvantage e.g. it requires large areas of land / decomposing waste may release methane/greenhouse gases / leaks from landfill sites can contaminate water supplies ***[1 mark]***.

3 a) Polymer A ***[1 mark]***. Polymer A is more reactive than polymer B because it has polar bonds in its chain, so it can be attacked by nucleophiles/hydrolysed ***[1 mark]***.

b) hydrolysis ***[1 mark]***.

Unit 3: Section 8 — Amino Acids, Proteins and DNA

Page 183 — Amino Acids

1 a) $CH_3-CH(CH_3)-C^*H(NH_2)-COOH$ (i.e. $H_3C-CH(CH_3)$ attached to $H_2N-\overset{*}{C}H-COOH$)

[2 marks — 1 mark for the correct structure and 1 mark for labelling the chiral carbon]

b) 2-amino-3-methylbutanoic acid ***[1 mark]***

2 a) $H_3C-CH(CH_3)-CH_2-CH(NH_2)-COOH$

[1 mark]

Answers

b)

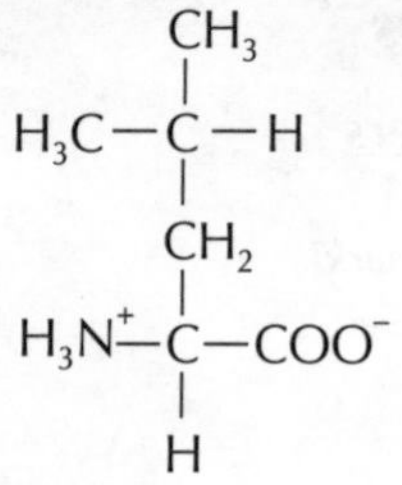

```
        CH3
        |
  H3C—C—H
        |
        CH2
        |
  H3N+—C—COO-
        |
        H
```

[1 mark]

c)

```
        CH3
        |
  H3C—C—H
        |
        CH2
        |
  H2N—C—COO-
        |
        H
```

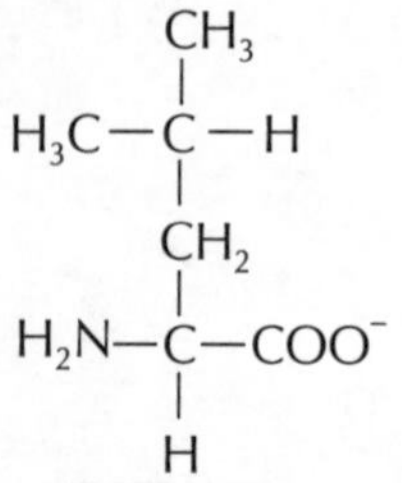

[1 mark]

Page 186 — Proteins and Enzymes

1

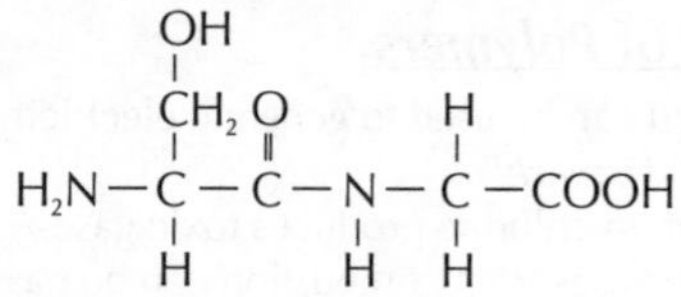

```
        OH
        |
        CH2 O       H
        |   ||      |
  H2N — C — C — N — C — COOH
        |       |   |
        H       H   H
```

[1 mark]

```
                    OH
                    |
        H   O       CH2
        |   ||      |
  H2N — C — C — N — C — COOH
        |       |   |
        H       H   H
```

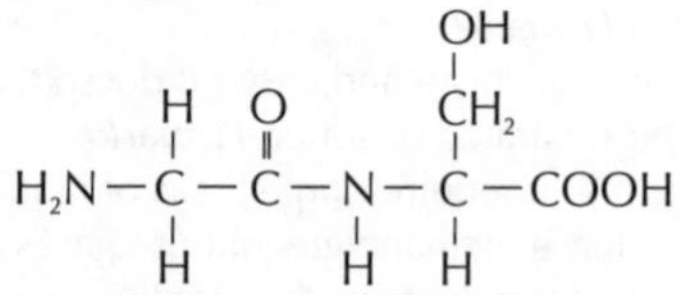

[1 mark]

The amino acids can join together in either order — serine first then glycine, or glycine first then serine. That's why there are two possible dipeptides that can be made here.

2 a) inhibitors *[1 mark]*

b) i) A drug molecule may have a very similar shape to the enzyme's substrate and fit into its active site *[1 mark]*. This blocks the active site, preventing the substrate from entering *[1 mark]*.

ii) The active site of an enzyme is stereospecific *[1 mark]* so only one of the enantiomers will be able to fit into the active site of the enzyme *[1 mark]*.

Page 189 — DNA

1 a)

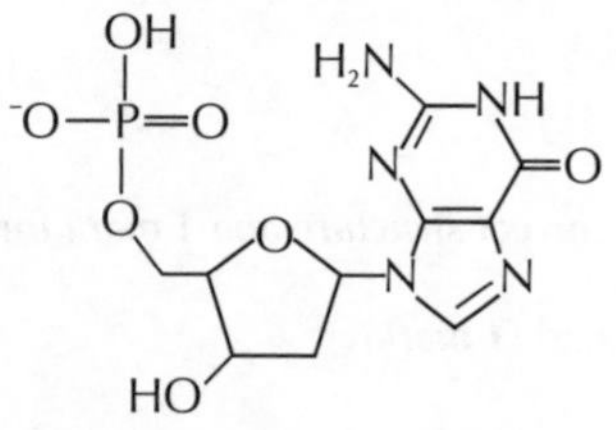

OH, $^{-}$O—P=O, O, HO, H_2N, NH, N, O, N, N

[1 mark for phosphate group correctly attached to 2-deoxyribose. 1 mark for guanine correctly attached to 2-deoxyribose.]

b) E.g.

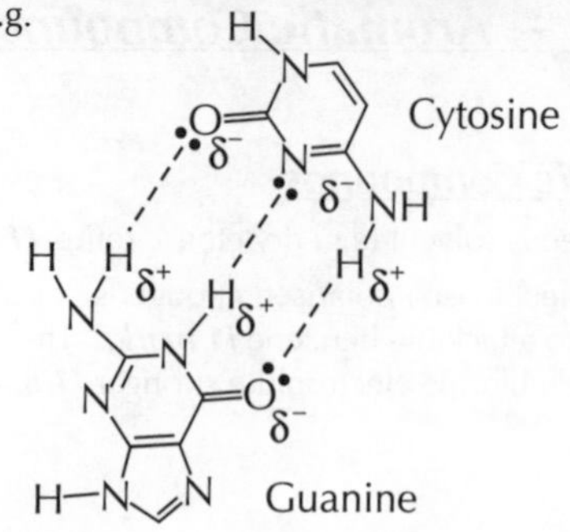

H, N, O, δ^-, N, δ^-, NH, Cytosine, H, H, δ^+, H, δ^+, H, δ^+, N, N, N, O, δ^-, H—N, N, Guanine

[3 marks — 1 mark for each hydrogen bond correctly shown.]

The bonds between the bases are hydrogen bonds *[1 mark]*.

Unit 3: Section 9 — Further Synthesis and Analysis

Page 191 — Organic Synthesis

1 a) Heat under reflux *[1 mark]*

b) $K_2Cr_2O_7$/potassium dichromate and H_2SO_4/sulfuric acid *[1 mark]*.
Heat and reflux *[1 mark]*.

2 Step 1: The methanol is refluxed *[1 mark]* with $K_2Cr_2O_7$ *[1 mark]* and sulfuric acid *[1 mark]* to form methanoic acid *[1 mark]*.
Step 2: The methanoic acid is reacted under reflux *[1 mark]* with ethanol *[1 mark]* using an acid catalyst *[1 mark]*.

3 Step 1: React propane with bromine *[1 mark]* in the presence of UV light *[1 mark]*. Bromine is toxic and corrosive *[1 mark]* so great care should be taken. Bromopropane is formed *[1 mark]*.
Step 2: Bromopropane is then refluxed *[1 mark]* with sodium hydroxide solution *[1 mark]*, again a corrosive substance so take care *[1 mark]*, to form propanol *[1 mark]*.

Page 194 — NMR Spectroscopy

1 a)

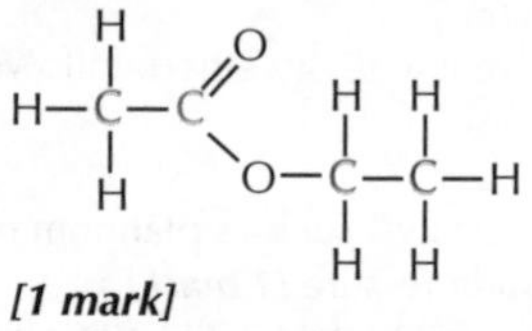

```
    H    O
    |   //      H   H
  H—C—C         |   |
    |   \   O — C — C — H
    H           |   |
                H   H
```

[1 mark]

4 peaks *[1 mark]*

This molecule has no symmetry — each carbon is joined to different groups and in a unique environment. So its ^{13}C NMR spectrum has four peaks.

b)

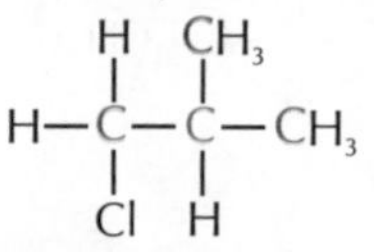

```
    H   CH3
    |   |
  H—C — C—CH3
    |   |
    Cl  H
```

[1 mark]

3 peaks *[1 mark]*

One peak is for the red carbon (joined to $H_2ClCH(CH_3)_2$), another is for the blue carbon (joined to $H(CH_3)_2CH_2Cl$), and the third is for both green carbons, which are in the same environment (both joined to $H_3C(CH_3)HCH_2Cl$).

Answers

c)

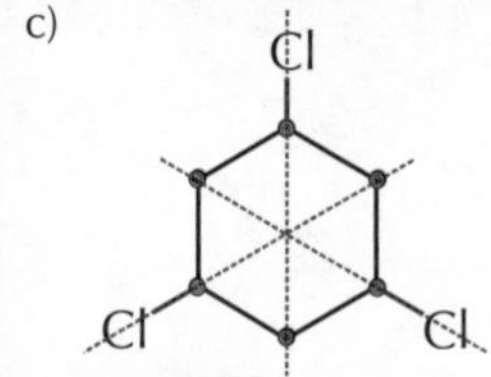

[1 mark]

2 peaks ***[1 mark]***

This molecule has three lines of symmetry — the three (blue) carbons with Cl atoms attached to them are all in the same environment, and the three (red) carbons that don't are all in the same environment. So its ^{13}C NMR spectrum has two peaks.

2 a)

```
    H  OH H
    |  |  |
H—C—C—C—H
    |  |  |
    H  H  H
```
[1 mark]

```
    H  H  H
    |  |  |
H—C—C—C—OH
    |  |  |
    H  H  H
```
[1 mark]

b)

```
    H  OH H
    |  |  |
H—C—C—C—H
    |  |  |
    H  H  H
```
OR propan-2-ol ***[1 mark]***

There are only two peaks on the spectrum. So there must be exactly 2 different carbon environments in the isomer ***[1 mark]***.

c) Tetramethylsilane / TMS / $Si(CH_3)_4$ ***[1 mark]***

Page 197 — ^{1}H NMR

1 a) A CH_2 group adjacent to a halogen / R–CH_2–X ***[1 mark]***.
You gotta read the question carefully — it tells you it's a halogenoalkane. So the group at 3.6 ppm can't have oxygen in it. It can't be halogen-CH_3 either, as this has 3 hydrogens in it.

b) A CH_3 group / R–CH_3 ***[1 mark]***.

c) CH_2 added to CH_3 gives a mass of 29, so the halogen must be chlorine with a mass of 35.5 ***[1 mark]***. So a likely structure is CH_3CH_2Cl ***[1 mark]***.

d) The quartet at 3.6 ppm is caused by 3 protons on the adjacent carbon. The n + 1 rule tells you that 3 protons give 3 + 1 = 4 peaks ***[1 mark]***. Similarly the triplet at 1.0 ppm is due to 2 adjacent protons giving 2 + 1 = 3 peaks ***[1 mark]***.

2 2 ***[1 mark]***

This molecule is symmetrical, so the hydrogens at opposite ends of the molecule are in the same environment:

```
    H  H  O  H  H
    |  |  ||  |  |
H—C—C—C—C—C—H
    |  |     |  |
    H  H     H  H
```

3 a) 3:2:3 ***[1 mark]***

b) A: singlet ***[1 mark]***, B: quartet ***[1 mark]***, C: triplet ***[1 mark]***.

Page 199 — Chromatography

1 a) To prevent contamination from any substances on their hands ***[1 mark]***.

b) To prevent the solvent evaporating away ***[1 mark]***.

c) Place the plate in a fume cupboard and leave it to dry ***[1 mark]***.

2 a) X ***[1 mark]***

b) 1 and 2 ***[1 mark]*** since the spots present in mixture Y are at the same height (and so would have the same R_f values) as 1 and 2 ***[1 mark]***.

c) R_f = spot distance ÷ solvent distance
= 5.6 ÷ 8 = **0.7** ***[1 mark]***

There are no units as it's a ratio.

Page 201 — More on Chromatography

1 a) The mixture is injected into a stream of carrier gas, which takes it through a tube over the stationary phase ***[1 mark]***. The components of the mixture dissolve in the stationary phase ***[1 mark]***, evaporate into the mobile phase ***[1 mark]***, and redissolve, gradually travelling along the tube to the detector ***[1 mark]***.

b) The substances separate because they have different solubilities in the mobile phase and retention to the stationary phase ***[1 mark]***, so they take different amounts of time to move through the tube ***[1 mark]***.

c) The areas under the peaks will be proportional to the relative amount of each substance in the mixture / the area under the benzene peak will be three times greater than the area under the ethanol peak ***[1 mark]***.

2 a) Gas Chromatography ***[1 mark]***

b) Different substances have different retention times ***[1 mark]***. The retention time of substances in the sample is compared against that for ethanol ***[1 mark]***.

c) It is unreactive/does not react with the sample ***[1 mark]***.

3 B ***[1 mark]***

Index

A

B

C

Index

Index

Index

K

L

M

N

Index

Index

The Periodic Table

relative atomic mass →
atomic (proton) number →

1	2											3	4	5	6	7	0
							1.0 H Hydrogen 1										4.0 He helium 2
6.9 Li lithium 3	9.0 Be beryllium 4											10.8 B boron 5	12.0 C carbon 6	14.0 N nitrogen 7	16.0 O oxygen 8	19.0 F fluorine 9	20.2 Ne neon 10
23.0 Na sodium 11	24.3 Mg magnesium 12											27.0 Al aluminium 13	28.1 Si silicon 14	31.0 P phosphorus 15	32.1 S sulfur 16	35.5 Cl chlorine 17	39.9 Ar argon 18
39.1 K potassium 19	40.1 Ca calcium 20	45.0 Sc scandium 21	47.9 Ti titanium 22	50.9 V vanadium 23	52.0 Cr chromium 24	54.9 Mn manganese 25	55.8 Fe iron 26	58.9 Co cobalt 27	58.7 Ni nickel 28	63.5 Cu copper 29	65.4 Zn zinc 30	69.7 Ga gallium 31	72.6 Ge germanium 32	74.9 As arsenic 33	79.0 Se selenium 34	79.9 Br bromine 35	83.8 Kr krypton 36
85.5 Rb rubidium 37	87.6 Sr strontium 38	88.9 Y yttrium 39	91.2 Zr zirconium 40	92.9 Nb niobium 41	96.0 Mo molybdenum 42	[98] Tc technetium 43	101.1 Ru ruthenium 44	102.9 Rh rhodium 45	106.4 Pd palladium 46	107.9 Ag silver 47	112.4 Cd cadmium 48	114.8 In Indium 49	118.7 Sn tin 50	121.8 Sb antimony 51	127.6 Te tellurium 52	126.9 I Iodine 53	131.3 Xe xenon 54
132.9 Cs caesium 55	137.3 Ba barium 56	138.9 La* lanthanum 57	178.5 Hf hafnium 72	180.9 Ta tantalum 73	183.8 W tungsten 74	186.2 Re rhenium 75	190.2 Os osmium 76	192.2 Ir iridium 77	195.1 Pt platinum 78	197.0 Au gold 79	200.6 Hg mercury 80	204.4 Tl thallium 81	207.2 Pb lead 82	209.0 Bi bismuth 83	[209] Po polonium 84	[210] At astatine 85	[222] Rn radon 86
[223] Fr francium 87	[226] Ra radium 88	[227] Ac† actinium 89	[267] Rf rutherfordium 104	[268] Db dubnium 105	[271] Sg seaborgium 106	[272] Bh bohrium 107	[270] Hs hassium 108	[276] Mt meitnerium 109	[281] Ds darmstadtium 110	[280] Rg roentgenium 111	Elements with atomic numbers 112–116 have been reported but not fully authenticated						

* 58 – 71 Lanthanides	140.1 Ce cerium 58	140.9 Pr praseodymium 59	144.2 Nd neodymium 60	[145] Pm promethium 61	150.4 Sm samarium 62	152.0 Eu europium 63	157.3 Gd gadolinium 64	158.9 Tb terbium 65	162.5 Dy dysprosium 66	164.9 Ho holmium 67	167.3 Er erbium 68	168.9 Tm thulium 69	173.1 Yb ytterbium 70	175.0 Lu lutetium 71
† 90 – 103 Actinides	232.0 Th thorium 90	231.0 Pa protactinium 91	238.0 U uranium 92	[237] Np neptunium 93	[244] Pu plutonium 94	[243] Am americium 95	[247] Cm curium 96	[247] Bk berkelium 97	[251] Cf californium 98	[252] Es einsteinium 99	[257] Fm fermium 100	[258] Md mendelevium 101	[259] No nobelium 102	[262] Lr lawrencium 103